Professional Practice in Earth Sciences

Series Editor

James W. LaMoreaux, Tuscaloosa, USA

Books in Springer's Professional Practice in Earth Sciences Series present state-of-the-art guidelines to be applied in multiple disciplines of the earth system sciences. The series portfolio contains practical training guidebooks and supporting material for academic courses, laboratory manuals, work procedures and protocols for environmental sciences and engineering. Items published in the series are directed at researchers, students, and anyone interested in the practical application of science. Books in the series cover the applied components of selected fields in the earth sciences and enable practitioners to better plan, optimize and interpret their results. The series is subdivided into the different fields of applied earth system sciences: Laboratory Manuals and work procedures, Environmental methods and protocols and training guidebooks.

Peiyue Li · Vetrimurugan Elumalai

Groundwater Quality Under Agricultural Activities—Cases from China and South Africa

Springer

Peiyue Li
School of Water and Environment
Chang'an University
Xi'an, China

Vetrimurugan Elumalai
Department of Hydrology
University of Zululand
Richards Bay, South Africa

ISSN 2364-0073 ISSN 2364-0081 (electronic)
Professional Practice in Earth Sciences
ISBN 978-3-031-98992-6 ISBN 978-3-031-98993-3 (eBook)
https://doi.org/10.1007/978-3-031-98993-3

This Springer imprint is published by the registered company Springer Nature Switzerland AG
The registered company address is: Gewerbestrasse 11, 6330 Cham, Switzerland

Foreword

It is with great pleasure that I write this Foreword for the book *Groundwater Quality Under Agricultural Activities—Cases from China and South Africa*, coauthored by Prof. Peiyue Li and Prof. Vetrimurugan Elumalai. This remarkable work is a testament to the power of international collaboration and the shared commitment to addressing pressing global challenges in water resource management.

The partnership between Chang'an University, China, and the University of Zululand, South Africa, exemplified by the collaboration between Prof. Li and Prof. Elumalai, is a shining example of how academic institutions can come together to advance knowledge and develop solutions for sustainable development. This book, supported by the National Natural Science Foundation of China (NSFC) and the National Research Foundation of South Africa (NRF), is the third in a series of collaborative works by these distinguished scholars. Their previous collaborations have already made significant contributions to the fields of hydrology, environmental science, and water resource management, and this latest work continues to build on that strong foundation.

The significance of this book cannot be overstated. Groundwater is a critical resource for agriculture, drinking water, and ecosystem health, yet it is increasingly threatened by pollution, over-extraction, and land use changes and climate change. By focusing on the selected areas in China and selected regions in South Africa, this book provides a comprehensive and comparative analysis of groundwater quality under agricultural activities. It not only highlights the hydrogeochemical characteristics and contamination mechanisms but also addresses the health risks associated with groundwater pollution and offers practical strategies for sustainable management.

What makes this book particularly valuable is its blend of rigorous scientific research and real-world applications. The case studies presented here are grounded in extensive fieldwork and advanced modeling techniques, making the findings both scientifically robust and practically relevant. The authors have succeeded in creating a resource that will benefit a wide range of stakeholders, including researchers, policymakers, environmental scientists, and public health officials.

I would like to extend my heartfelt congratulations to Prof. Li and Prof. Elumalai for their outstanding work on this project. Their dedication to advancing knowledge and their ability to bridge the gap between theory and practice are truly commendable. I also wish to acknowledge the contributions of their respective institutions, which have provided the support and infrastructure necessary for this successful collaboration.

As we look to the future, I am optimistic that this book will serve as a catalyst for further research and collaboration between our universities and beyond. The challenges we face in protecting and managing groundwater resources are complex and multifaceted, but through continued partnership and innovation, we can develop solutions that ensure the sustainability of this vital resource for generations to come.

It is my sincere hope that this book will inspire more collaborative efforts between scholars, institutions, and nations, and that it will contribute to a deeper understanding of groundwater systems and their sustainable management. I wish Prof. Li and Prof. Elumalai continued success in their research endeavors and look forward to seeing the impact of this important work on both academic and practical fronts.

Prof. Xoliswa Mtose
Vice Chancellor
University of Zululand
Richards Bay, South Africa

Preface

Groundwater is an indispensable resource, silently sustaining life beneath our feet. It nourishes ecosystems, secures drinking water for billions, and serves as the backbone of agriculture in many regions, particularly where surface water is scarce or unreliable. As the global demand for food continues to rise, agricultural intensification has surged, bringing with it a growing concern: the degradation of groundwater quality. Fertilizers, pesticides, irrigation practices, and even treated wastewater used in agriculture are emerging as significant sources of groundwater pollution. This degradation not only jeopardizes the safety of water supplies, but also compromises the health of soils, ecosystems, and communities who depend on them.

Nowhere are these challenges more acutely felt than in developing regions where agriculture drives local economies, and water governance systems are often under-resourced. China and South Africa exemplify this dynamic: both countries are agriculturally intensive, highly dependent on groundwater, and increasingly vulnerable to the twin pressures of population growth and climate variability. Yet, both are also sites of innovation and insight, offering important case studies of how groundwater quality is shaped by agricultural practices and how scientific assessments can guide sustainable management.

This book, *Groundwater Quality Under Agricultural Activities—Cases from China and South Africa*, was conceived to provide a focused, data-rich, and scientifically grounded exploration of these dynamics. It combines field-based hydrogeological research, advanced analytical techniques, and grounded case studies to unravel the interplay between agriculture and groundwater pollution. Our aim is not only to document challenges, but to inspire new research, support evidence-based policy, and ultimately, help safeguard this precious hidden resource. This book is the outcome of the international collaboration project sponsored by the National Natural Science Foundation of China (NSFC) and the National Research Foundation (NRF) of South Africa.

This book is divided into 10 chapters. The opening chapter (Chap. 1) lays the conceptual foundation for the entire book. It underscores the centrality of groundwater in global water systems and food security, while drawing attention to the paradox of its simultaneous vulnerability to agricultural intensification. The chapter comprehensively defines groundwater, its classification systems, and its unique role in the hydrological cycle, especially within the context of arid and semi-arid regions like China's Loess Plateau and South Africa's inland farming zones. But this chapter is more than just an introduction; it's a call to action. By elaborating on contamination pathways, from nitrate and pesticide infiltration to salinization and emerging contaminants such as antibiotics and hormones, it paints a picture of the silent transformations happening beneath agricultural fields.

Chapters 2–5 take readers deep into the Weining Plain, a historically irrigated region in northwest China with over two millennia of agricultural legacy. Combining hydrogeological, chemical, and isotopic data, Chap. 2 offers a rigorous examination of the groundwater system's structure, flow patterns, recharge dynamics, and geochemical evolution. What makes this chapter especially valuable is its capacity to bridge historical and contemporary perspectives. It shows how long-term irrigation, industrial developments like papermaking, and afforestation projects have collectively altered groundwater flow and chemistry. For example, the chapter presents striking data on the infiltration of paper wastewater used in forest irrigation, with clear implications for groundwater salinity and chemical oxygen demand (COD). It also lays out detailed spatial analyses of water types, identifying dominant hydrochemical facies and revealing how human activities have reshaped natural geochemical gradients. The Weining Plain thus emerges not just as a site of study but also as a microcosm of global agricultural water challenges.

Building directly on Chap. 2, Chap. 3 zooms in on the actual state of groundwater quality and its public health implications. Utilizing a comprehensive dataset of 223 samples and employing the entropy-weighted TOPSIS method, a robust multicriteria decision analysis tool, the chapter evaluates the degree of pollution and ranks the groundwater samples by suitability for drinking. One of the chapter's strengths is its clarity in demonstrating how science can translate into actionable insight. It reveals that more than half of the samples exceed national thresholds for sulfate, total dissolved solids, and hardness, while others show elevated levels of nitrates, ammonia, fluoride, and manganese. These findings matter for local families relying on groundwater as their sole drinking source. Through detailed mapping and classification of water quality zones, the chapter offers an evidence-based framework that could inform groundwater management not only in Weining but in similar agro-ecological zones worldwide.

In Chap. 4, the health risk was assessed. The chapter begins by identifying key pollutants, fluoride, nitrates, nitrites, and manganese, with some exceedance rates surpassing 50%. These contaminants are linked to severe health issues, from skeletal fluorosis to methemoglobinemia, particularly affecting vulnerable populations like children. This chapter paints a spatial picture of risk, highlighting high-risk zones in Zhongwei City and Zhongning County. It underscores the urgent need for targeted interventions, such as stricter controls on fertilizer use and industrial discharge, to safeguard drinking water. This chapter sets a new benchmark for groundwater health risk assessment, combining probabilistic models with spatial analysis to pinpoint risk hotspots. It serves as both a "battle map" for policymakers and a wake-up call for public awareness, embodying the transition from data to action, a profound reminder that groundwater management must be rooted in science and prioritized in practice.

Chapter 5 investigates the profound impact of land use and land cover (LULC) changes on groundwater quality in the Weining Plain. The study employs advanced geostatistical methods, including the Curved Streamline Searchlight Model (CS-SLM), to analyze spatial pollutant distribution and quantify LULC changes from 2000 to 2020. Key pollutants such as nitrate (NO_3^-), fluoride (F^-), and heavy metals (e.g., Mn) are linked to agricultural intensification, urbanization, and industrial activities. This chapter transcends traditional land-use studies by integrating hydrogeological flow dynamics into pollution assessment, offering a nuanced understanding of how human activities reshape groundwater systems. The CS-SLM's innovative approach addresses a critical gap in previous models by accounting for curved groundwater flow paths, providing more accurate pollution source attribution. The findings underscore the dual role of LULC changes—both as drivers of contamination (e.g., fertilizer leaching) and as potential solutions (e.g., afforestation for pollution dilution). By bridging GIS technology, field data, and statistical rigor, Chap. 5 serves as a blueprint for sustainable land-use planning in arid regions, emphasizing the need to balance agricultural productivity with groundwater protection.

Chapters 6–9 are case studies in coastal aquifers in South Africa. Chapter 6 focuses on three catchments, Maputaland (upper), uMhlathuze (middle), and Tugela (lower), each revealing distinct hydrogeochemical signatures. Through meticulous sampling and advanced analytical techniques, the chapter uncovers the dominance of Na-Cl water types, a telltale sign of seawater intrusion (SWI). The Seawater Mixing Index (SMI) paints a stark picture: while 64.84% of samples remain freshwater, over 35% show varying degrees of saline pollution, with the lower catchment bearing the brunt. Ionic ratios and Gibbs plots further decode the story, highlighting the roles of evaporation, rock weathering, and anthropogenic inputs in reshaping groundwater chemistry. It is a revelation of how coastal aquifers, under pressure from agriculture and industry, are transforming into reservoirs of salinity and contamination, with far-reaching consequences for food security and public health.

Groundwater quality in coastal aquifers is a critical yet often overlooked aspect of sustainable water management, particularly in agriculturally dominant regions like KwaZulu-Natal, South Africa. Chapter 7 delves into the geochemical characteristics and suitability of groundwater in the Maputoland, uMhlathuze, and Tugela aquifers for domestic, agricultural, and industrial uses. Through a comprehensive analysis of physicochemical parameters and hydrochemical indices—such as the Water Quality Index (WQI), Sodium Adsorption Ratio (SAR), and Corrosivity Ratio—the study reveals stark spatial variations in groundwater quality. Elevated salinity and contamination hotspots, particularly in central and southern aquifers, underscore the threats posed by seawater intrusion, agricultural runoff, and industrial activities. The findings highlight the urgent need for regular monitoring and targeted mitigation strategies to safeguard water resources. Beyond its scientific rigor, this chapter serves as a wake-up call, emphasizing how groundwater degradation intertwines with food security, public health, and infrastructure resilience. By mapping contamination zones and assessing irrigation suitability, the study provides actionable insights for policymakers and stakeholders, making it a pivotal read for anyone invested in the sustainable future of coastal regions facing growing anthropogenic pressures.

The silent threat of heavy metal contamination in groundwater is a growing global crisis, with profound implications for human health, particularly in agriculturally intensive regions. Chapter 8 investigates the health risks posed by metals such as cadmium (Cd), chromium (Cr), and lead (Pb) in the coastal aquifers of KwaZulu-Natal, South Africa, blending advanced methodologies like Positive Matrix Factorization (PMF) and Monte Carlo simulations to pinpoint contamination sources and quantify risks. The findings are alarming: children face significantly higher non-carcinogenic and carcinogenic risks, with Cd and Cr emerging as primary culprits, largely linked to agricultural runoff and industrial activities. Spatial mapping reveals contamination hotspots in southern and central aquifers, underscoring the urgent need for targeted interventions. Beyond the data, this chapter serves as a stark reminder of the invisible dangers lurking in our water supply, connecting environmental degradation to public health vulnerabilities. It calls for immediate action to safeguard vulnerable populations and align water management practices with sustainable development goals, making it a crucial read for researchers, policymakers, and communities alike.

Like the Weining Plain in China, the rapid transformation of land use and land cover (LULC) in KwaZulu-Natal's coastal region is also reshaping both the landscape and the quality of its vital groundwater resources. Chapter 9 unravels the dramatic shifts from 2001 to 2023, revealing a 5% surge in agricultural land and a fourfold increase in urban areas, while bushland dwindled by 12%. These changes are not just numbers, but they directly threaten groundwater, with elevated salinity, chloride, and lead levels linked to fertilizers, urbanization, and commercial forestry. Remote sensing and statistical analyses pinpoint contamination hotspots, underscoring the urgent need for sustainable land management. Beyond data, this chapter sounds an alarm: unchecked LULC changes jeopardize water security, public health, and ecosystem resilience, calling for immediate policy action to balance development with environmental preservation in this fragile coastal zone.

As the concluding chapter of this book, Chap. 10 synthesizes the critical groundwater challenges posed by agricultural activities in China's Weining Plain and South Africa's KwaZulu-Natal coastal aquifers, highlighting their shared struggles with nitrate pollution, salinity, and heavy metal contamination. Beyond diagnosing these issues, the chapter charts a path forward, advocating for precision agriculture, managed aquifer recharge, and community-driven governance to mitigate risks. It underscores the urgency of integrating technology, policy, and local engagement to safeguard water resources. By bridging scientific insights with actionable strategies, this chapter not only encapsulates the book's core themes but also serves as a call to action—emphasizing that sustainable groundwater management is not just an environmental imperative but a cornerstone of public health and food security in vulnerable regions.

As the global population continues to grow, the pressure on groundwater resources will only intensify. This book serves as a timely reminder of the urgent need to balance agricultural productivity with environmental sustainability. By bridging scientific research and practical applications, it aims to inspire further exploration and collaboration in the field of groundwater quality management. We hope that readers will find this work not only informative but also a catalyst for positive change in their own regions.

The journey to safeguarding our groundwater is far from over, and the insights shared here are but a stepping stone toward a more sustainable future. Let this book be a guide and a challenge—to researchers, policymakers, and communities alike, to innovate, collaborate, and act decisively in protecting this vital resource for generations to come.

Xi'an, China Prof. Peiyue Li
Durban, South Africa Prof. Vetrimurugan Elumalai

Acknowledgments The completion of this book, *Groundwater Quality Under Agricultural Activities—Cases from China and South Africa*, would not have been possible without the support and encouragement of many individuals and organizations. We would like to take this opportunity to express our deepest gratitude to those who have contributed to this journey.

First and foremost, we owe a heartfelt thank you to our families, especially our wives, who have tirelessly supported us and taken care of our sons. Their understanding and sacrifices have allowed us to dedicate more time to our research and writing. This book is as much a product of their love and patience as it is of our efforts.

Several esteemed colleagues have generously contributed their time and expertise by reviewing some chapters of this book which partially originates from the Ph.D. dissertation of Prof. Peiyue Li. We extend our sincere thanks to Prof. Peicheng Li, Prof. Jingsheng Li, Prof. Qiang Huang, Prof. Hui Qian, Prof. Yaoguo Wu, Prof. Wei Wang, and Prof. Hongbo Zhang for their insightful feedback and constructive suggestions. Their input has greatly enhanced the quality of this work. We would also like to pay special tribute to Prof. Peicheng Li, who passed away in February 2024. Professor Li was not only a mentor and colleague but also a dear friend who provided unwavering support to our research over the years. His wisdom, encouragement, and dedication to the field of hydrology and hydrogeology have left an indelible mark on our work. We deeply miss him and are profoundly grateful for his contributions to our academic journey.

Sincere thanks are due to Ashok Arumairaj and Annett Buettner, our editorial contacts at Springer, for their unwavering support and guidance throughout the publication process. Their prompt responses and insightful advice have been invaluable in addressing our queries and ensuring the smooth progression of this project. We are also deeply grateful to our students, Zhiwen Zheng, Jiajia Kong and Mengyu Gong, for their invaluable assistance in preparing the manuscripts. Dr. Song He, Dr. Xiaodong He, Dr. Fengmei Su, Dr. Fei Xu, Dr. Yuanhang Wang, Dr. Lingxi Li, Dr. Yueyue Meng, Dr. Linjie Xie, Dr. Dan Wang, Dr. Jing Ning, and Dr. Xiaofei Ren participated in field investigation. Their dedication has been instrumental in bringing this project to fruition.

This book is a testament to the strong collaboration between Chang'an University, China, and the University of Zululand, South Africa. We are deeply appreciative of the support provided by the governments of China and South Africa, particularly the National Natural Science Foundation of China (NSFC) and the National Research Foundation (NRF) of South Africa. Their funding and encouragement have enabled us to sustain a fruitful partnership over the past 9 years, resulting in impactful research outcomes.

The signing of the Memorandum of Understanding (MOU) between Chang'an University and the University of Zululand on September 13, 2024, marked a significant milestone in our collaboration. We are grateful to both institutions for fostering an environment that supports international research and academic exchange.

Finally, this work has also been made possible through the generous support of several research grants, including those from the National Natural Science Foundation of China (Grants 42472316, 42072286, 41761144059, and 41502234), the National Key Research and Development Program of China (2023YFC3706901), the State Top Foreign Experts Recruitment Program (H20240343, H20240353, G2023171005L, G2023171004L, G2022171007L, G2021171005L), the Qinchuangyuan "Scientist + Engineer" Team Development Program of the Shaanxi Provincial Department of Science and Technology (2022KXJ-005), the National Ten Thousand Talents Program (W03070125), and the Fok Ying Tong Education Foundation (161098). These grants have provided the necessary resources to conduct our collaborative research and produce this book.

We hope this book will serve as a valuable resource for researchers, policymakers, and practitioners working on groundwater quality and sustainable water resource management. It is our sincere wish that this work will inspire further collaboration and innovation in addressing the challenges of groundwater sustainability in the face of agricultural activities, land use changes, and climate change.

Xi'an, China Prof. Peiyue Li
Richards Bay, South Africa Prof. Vetrimurugan Elumalai

Competing Interests The authors have no competing interests to declare that are relevant to the content of this manuscript.

Contents

About the Authors

Prof. Peiyue Li is a Full Professor in the field of Hydrogeology and Environmental Sciences at Chang'an University, China. He serves as the Vice Dean of School of Water and Environment at this university. He holds a Ph.D. in Groundwater Science and Engineering (2014), and has a wide experience in groundwater quality assessment, hydrogeochemistry, and groundwater modeling. He has edited and co-edited 11 textbooks and academic monographs, and published over 210 articles in refereed journals on topics that range from groundwater quality assessment and groundwater hydrochemistry to groundwater pumping tests and in situ tracer tests. Peiyue Li has been selected for the National Ten Thousand Talents Program as a Young Top Talent and recognized as a Young Changjiang Scholar by the Ministry of Education of China. He has received numerous national and international honors. He at present serves as Associate Editor for *Exposure and Health, Mine Water and the Environment, Archives of Environmental Contamination and Toxicology, Environmental Monitoring and Assessment, Human and Ecological Risk Assessment, Applied Water Science, Water Practice and Technology, and Discover Water*. He is also book series editors for *Springer Hydrogeology, Cave and Karst Systems of the World, Advances in Karst Science*. He has been awarded one of the most highly cited researchers by Clarivate, Elsevier, and other organizations.

Prof. Vetrimurugan Elumalai is an established Hydrogeologist expertise in contaminant hydrology and groundwater modeling. He is excelling as an established researcher and mentor and Head of the Department of Hydrology, University of Zululand, South Africa. He is South African National Chair of Hydrology (SARChI-Tier 1). His project experience covers groundwater resource investigation and development, groundwater characterization, contaminant hydrogeology and urban development, surface water groundwater interactions, remote sensing, and GIS application in Water Resource Management. He has published about 80 research papers in reputed journals and edited two books. He is registered Scientist with SACNASP. He is a Life Member of International Association of Hydrological Science, International Association of Hydrogeologists, International Association of Hydrological Sciences and Geological Survey of South Africa.

Chapter 1
Groundwater Quality Under Agricultural Activities: An Introduction

Groundwater is a vital resource for agricultural irrigation, drinking water supply, and ecosystem sustainability, playing a critical role in supporting global food production and environmental health. However, its quality is increasingly threatened by intensive agricultural activities, which are essential for meeting the food demands of a growing population. Agricultural practices such as the application of fertilizers, pesticides, and irrigation water often lead to the contamination of groundwater resources through the leaching of pollutants like nitrates, phosphates, and agrochemicals into aquifers. These contaminants not only degrade water quality but also pose significant risks to human health, ecosystems, and long-term agricultural productivity. Understanding the complex interplay between agriculture and groundwater quality is, therefore, crucial for developing sustainable water management strategies that balance food security with environmental protection. This chapter provides an overview of these critical issues and introduces key concepts such as groundwater contamination pathways, and major agricultural pollutants. By outlining the objectives and scope of the book, this chapter also introduces the main structure of the book, providing readers with the necessary background and context to engage with the detailed discussions in subsequent chapters.

1.1 Importance of Groundwater

1.1.1 Definition of Groundwater and Its Role in the Hydrological Cycle

Groundwater is explicitly defined as the water that occupies the voids, pores, and fractures within the saturated zones of soil and rock formations below the Earth's surface (Fitts 2023). It is located beneath the water table, where all available spaces are

P. Li and V. Elumalai, *Groundwater Quality Under Agricultural Activities—Cases from China and South Africa*, Professional Practice in Earth Sciences,
https://doi.org/10.1007/978-3-031-98993-3_1

completely filled with water. It is about 100 times larger than the volume of fresh water available in rivers and lakes, indicating the significance of groundwater as a resource for economic activities, including agriculture (Madramootoo 2012). Groundwater is primarily recharged by the infiltration of precipitation, surface water (such as rivers and lakes), and, in some cases, artificial sources (e.g., irrigation or managed aquifer recharge). It moves slowly through permeable layers known as aquifers, which can store and transmit significant quantities of water. Groundwater plays a critical role in the hydrological cycle by sustaining baseflow in rivers, supporting ecosystems, and serving as a vital source of freshwater for human consumption, agriculture, and industry. Its quality and availability are influenced by geological conditions, climate, and human activities (Chao et al. 2020; Ghasemizade and Schirmer 2013; Xie et al. 2024).

Groundwater can be classified based on different criteria such as aquifer conditions, geological media, aquifer depth, hydrochemical characteristics. By aquifer conditions, groundwater can be categorized as unconfined, confined, or perched groundwater (Haque 2014). Unconfined groundwater resides in aquifers with a free water table, directly recharged by precipitation or surface water infiltration, and is highly responsive to seasonal changes. In contrast, confined groundwater is trapped between impermeable layers (aquitards or aquicludes), often under pressure, creating artesian conditions where water levels can rise up above the ground surface. Perched groundwater forms locally in the vadose zone due to small impermeable layers, such as clay lenses, and is often temporary. Unconfined and confined groundwaters are typical resources for various purposes. By geological media, groundwater usually exists in porous media, fractured media or karst systems. Thus, it can be classified into porous groundwater, fracture groundwater or karst groundwater (Hahn and Fuchs 2009). Porous groundwater flows through granular materials like sand or gravel, characterized by uniform, laminar flow and moderate contamination risk. Fracture groundwater moves through cracks in hard rocks such as granite or basalt, with irregular flow patterns and high vulnerability to contamination due to rapid transport along fractures. Karst groundwater occurs in soluble rocks like limestone, where dissolution creates conduits and caves, resulting in turbulent, fast-moving flow and very high contamination risk due to minimal natural filtration. By aquifer depth, groundwater is usually classified as shallow or deep groundwater. However, the classification of groundwater as shallow or deep is inherently subjective and context-dependent. It is influenced by local geology, hydrological conditions, and practical considerations rather than a fixed depth threshold. By water quality, groundwater ranges from fresh (TDS < 1000 mg/L, suitable for drinking and irrigation), to brackish (TDS in the range of 1000–10,000 mg/L, requiring treatment for use), to saline (TDS $> 10{,}000$ mg/L, often found in deep aquifers or coastal regions affected by saltwater intrusion) (Krishna Kumar et al. 2017).

Groundwater is a vital component of the hydrological cycle, fulfilling critical roles in storage, regulation, ecological support, water supply, and water quality purification (Karamouz et al. 2020). Its ability to store water over extended periods makes it an essential reservoir, particularly during droughts, when it sustains rivers, lakes, and ecosystems, ensuring water resource stability. For instance, in arid regions like

Chinese Loess Plateau, groundwater is indispensable for agricultural irrigation and domestic water supply, highlighting its importance in water-scarce environments (He and Wu 2019; Li et al. 2019, 2024a; Ning et al. 2025; Wu et al. 2020). Groundwater also plays a key regulatory role in hydrological systems. By slowly moving through aquifers, it replenishes river flows during dry seasons and absorbs excess water during wet seasons, mitigating flood risks (Chakraborty 2021). A notable example is the Kissimmee River Basin in Florida, USA, where groundwater absorption during the rainy season reduces flood pressure, while its contribution to baseflow during dry periods maintains the health of the river ecosystem and supports aquatic habitats (Boning 2016). Ecologically, groundwater is irreplaceable. It sustains wetlands, springs, and other ecosystems by providing a reliable water source for plant growth and animal habitats. The Sanjiang Plain wetlands in China, for example, depend on groundwater recharge to support migratory bird populations and preserve biodiversity (Ye et al. 2022; Qu et al. 2022). Similarly, in the arid regions of China, groundwater-fed oases are lifelines for both wildlife and human communities (Zhang et al. 2021). From a human perspective, groundwater is a cornerstone of drinking water supplies and agricultural irrigation. For instance, the groundwater irrigation systems in the Yinchuan Plain of China support large-scale agricultural production, ensuring food security for the region (Li et al. 2018a; Su et al. 2020). In areas with limited surface water, such as Saudi Arabia in the Middle East, groundwater is almost the sole water source, meeting domestic and agricultural needs (Kim and van der Beek 2018). Additionally, groundwater contributes to water quality improvement through natural filtration. As it percolates through soil and rock layers, it removes pollutants, enhancing water quality (Ray et al. 2003).

1.1.2 Overview of Groundwater Significance in Agriculture

Groundwater is a cornerstone of global agriculture, providing a reliable and often indispensable source of water for irrigation, livestock, and agro-processing. Its significance extends beyond mere quantity, encompassing aspects of quality, accessibility, and sustainability. In regions where surface water is scarce or seasonal, groundwater serves as a critical buffer to ensure agricultural productivity and food security. It is estimated that groundwater accounts for about 43% of the total water used for irrigation over the world (FAO 2020; Madramootoo 2012; Siebert et al. 2010). However, this rate in arid and semi-arid regions such as the Northwest China, Middle East and parts of India, can be up to over 50% of the irrigation water (Shah 2009).

The quantity of groundwater available for agriculture is a key determinant of its significance. Unlike surface water, groundwater is less susceptible to seasonal fluctuations, making it a dependable resource during droughts or dry seasons. However, over-extraction of groundwater for irrigation has led to declining water tables in many agricultural hotspots (Siebert et al. 2010). This over-reliance on groundwater highlights the need for sustainable management practices to balance agricultural demands with long-term resource availability.

Groundwater quality is equally critical for agricultural use. The chemical composition of groundwater, including its salinity, nutrient content, and presence of contaminants, directly impacts crop yields and soil health (Li and Elumalai 2025). For instance, high salinity levels in groundwater can lead to soil degradation and reduced crop productivity. In addition, the infiltration of irrigation water not only leaches soluble salts from the unsaturated zone, altering the concentrations of various hydrochemical ions in groundwater, but also introduces fertilizers into aquifers, directly impacting groundwater quality (Li and Elumalai 2025). Conversely, groundwater with optimal levels of dissolved nutrients, such as nitrates, can enhance crop growth, though excessive concentrations may pose environmental and health risks.

Beyond quantity and quality, groundwater accessibility and cost-effectiveness make it a preferred choice for farmers. In many developing regions, groundwater is accessed through shallow wells or tube wells, providing a decentralized and flexible water supply. However, the energy costs associated with pumping groundwater, particularly from deeper aquifers, can be a significant financial burden for smallholder farmers (Hagos and Mamo 2014).

1.1.3 Groundwater Quality Challenges Posed by Agricultural Activities

While groundwater is vital for agriculture, agricultural activities are among the leading contributors to groundwater quality degradation. The intensive use of fertilizers, pesticides, and irrigation practices has introduced a range of contaminants into groundwater systems, threatening both human health and ecosystem integrity (Hildebrandt et al. 2008).

One of the most pervasive issues is nutrient pollution, particularly from nitrates and phosphates. Excessive application of nitrogen-based fertilizers leads to nitrate leaching into groundwater, often exceeding safe drinking water standards (Alam et al. 2024). High nitrate levels in groundwater have been linked to methemoglobinemia (blue baby syndrome) and other health issues, as well as eutrophication in surface water bodies. For example, in many basins in China where agricultural activities are intense and extensive serious groundwater nitrate pollution has been widely reported (Alam et al. 2025; Wang et al. 2023, 2024; Zhou et al. 2024).

Pesticide contamination is another major concern in agriculture (Pérez-Indoval and Romero-López 2024; Khezami et al. 2024). Many pesticides, including herbicides, insecticides, and fungicides, are highly mobile and persistent in the environment. When applied to crops, these chemicals can infiltrate the soil and reach groundwater, posing risks to human health and aquatic ecosystems. Atrazine, a widely used herbicide, has been detected in groundwater across the United States, raising concerns about its long-term environmental and health impacts (Gonzalez and Smiley 2025).

Irrigation practices also contribute to groundwater quality degradation through salinization. In arid and semi-arid regions, the use of groundwater for irrigation can lead to the accumulation of salts in the root zone of the soils as well as in aquifers. Over time, this reduces the suitability of groundwater for both agricultural and domestic use. For instance, in parts of Pakistan and India, secondary salinization caused by intensive irrigation has rendered large areas of farmland unproductive (Qureshi et al. 2008; Singh 2024). The Weinig plain of China, which is the one of the study areas for this book, has also witnessed serious soil salinization (Wu et al. 2014; Li et al. 2016a). In this plain, long term irrigation elevated the groundwater level, inducing the accumulation of salts in the root zone due to intense evaporation.

In addition to direct pollution from agricultural activities, agricultural activities can exacerbate natural changes of groundwater quality (Li and Elumalai 2025). This is defined as indirect impacts of human activities on groundwater quality in this book. In regions with naturally occurring arsenic or fluoride in soils, irrigation can mobilize these contaminants, increasing their concentrations in groundwater. This phenomenon has been observed in the Bengal Basin, where irrigation-driven groundwater extraction has intensified arsenic contamination, affecting millions of people (Sarkar et al. 2022). Zhang et al. (2020) further demonstrated that increased irrigation can lead to a rise in groundwater levels, which alters the redox conditions of the aquifer. This shift facilitates the reductive dissolution of iron oxide minerals, enhancing the interaction between groundwater and arsenic-rich sediments.

1.2 Groundwater Pollution from Agriculture

The widespread use of fertilizers, pesticides, and irrigation practices have introduced a range of contaminants into groundwater systems, threatening both human health and ecosystem integrity. This section explores the key contaminants originating from agricultural activities and the pathways through which they enter groundwater, providing a foundation for understanding the scale and complexity of this environmental challenge.

1.2.1 Key Contaminants from Agriculture

Agricultural activities can introduce a wide range of pollutants into groundwater. The primary contaminants from agriculture usually include nutrients, pesticides, soluble salts, and emerging contaminants.

(1) **Nitrates**

Nitrates are one of the most widespread agricultural contaminants in groundwater, primarily originating from the excessive use of nitrogen-based fertilizers. High nitrate concentrations in drinking water can cause methemoglobinemia (blue baby syndrome) and other health issues. Globally, nitrate levels in groundwater in intensively farmed regions often exceed the World Health Organization (WHO) guideline of 50 mg/L. For example, in China's North China Plain, nitrate concentrations in groundwater have been reported to reach up to 300 mg/L due to over-fertilization and intensive irrigation (Zhang et al. 1996). Similarly, in South Africa, nitrate levels in groundwater have exceeded 150 mg/L in agricultural areas such as the Vyeboom region in Limpopo Province, primarily due to the use of nitrogen fertilizers and animal manure (Mutileni et al. 2023).

(2) **Pesticides**

In modern agriculture, the primary chemicals utilized as pesticides include organophosphates, organochlorines, pyrethroids, neonicotinoids, and carbamates. However, since the 1960s, organochlorines have been prohibited in developed nations due to their environmental and health risks (Arias-Estévez et al. 2008). Globally, pesticides play a crucial role in agricultural systems, serving to control pests and enhance crop yields. These chemicals are highly mobile and persistent, often leaching into groundwater through the soil profile. For example, chlorpyrifos and endosulfan, strongly adsorbed and slow-degrading pesticides, have been reported in nearly all surface and groundwater samples in the Robit Bata watershed in Ethiopian, with maximum concentrations in groundwater of 4.33 μg/L for chlorpyrifos and 1.71 μg/ for endosulfan (Sishu et al. 2022). However, the African continent was found to be the lowest user of pesticides as they contributed only about 2%–4% in the global market (Malla et al. 2021). This indicates that pesticide pollution in other continents is more serious.

(3) **Soluble Salts**

Irrigation can introduce more soluble salts into groundwater, causing groundwater salinization. Groundwater salinization is more significant in arid and semi-arid regions, where intensive irrigation leads to the accumulation of salts in aquifers. For example, groundwater salinization has been reported in the Guanzhong Basin in China, with shallow groundwater salinity reaching up to more than 3000 mg/L due to excessive irrigation and poor drainage, rendering the water unsuitable for drinking or irrigation (Mu et al. 2024). Similarly, in Luxor in Upper Egypt, a significant increase in groundwater salinity from an average of 1280.6 ± 997.2 mg/L in 1997 to 1828.6 ± 1124.7 mg/L in 2017 has been reported (Said et al. 2022).

(4) **Emerging Contaminants**

Emerging contaminants, such as antibiotics and hormones, are increasingly detected in groundwater due to their use in livestock farming and aquaculture. In China, significant antibiotic contamination in groundwater in Xiong'an New Area was reported with the highest concentrations found in shallow groundwater at depths of < 50 m, averaging 79.22 ± 56.46 ng/L (Fu et al. 2022). Surface water infiltration and soil leaching are likely sources of antibiotic pollution in groundwater in this area. In South Africa, hormones such as estradiol have been detected in groundwater near dairy farms, with concentrations reaching 0.05 μg/L (Barnes et al. 2008). Groundwater antibiotic contamination has also been reported in European countries such as in Romania, with an average concentration of 1671.17 ± 671.23 ng/L (Szekeres et al. 2018), followed by Spain (693.21 ± 1387.62 ng/L; Lopez-Serna et al. 2013), Serbia (12.96 ± 22.57 ng/L; Radovic et al. 2015), France (9.00 ± 4.5 ng/L; Charuaud et al. 2019), and the United Kingdom (5.70 ± 3.76 ng/L; Stuart et al. 2012), though some of them are not purely caused by agriculture. These contaminants can pose risks to human health and aquatic ecosystems, even at low concentrations. However, the emerging contaminants are at present not well researched, and more efforts are needed to understand the pathways and processes of the pollution in groundwater.

1.2.2 Pathways of Contamination

Agricultural contaminants enter groundwater primarily through two pathways. The first is direct infiltration, where pollutants dissolved in irrigation water percolate into the aquifer, leading to groundwater contamination. This pathway is most commonly associated with the use of untreated wastewater or reclaimed water for irrigation. Reclaimed water refers to treated wastewater, but both untreated and treated wastewater contain pollutants, with untreated wastewater often having significantly higher concentrations. When such water is used for irrigation, the pollutants infiltrate into the aquifer, inevitably causing groundwater contamination. For example, the paper mill in the Weining Plain used papermaking wastewater to irrigate fast-growing forests, resulting in severe groundwater contamination in the surrounding area (Fig. 1.1). The papermaking wastewater contained high levels of organic matter and other contaminants, causing very serious pollution in groundwater (Li et al. 2016b, 2018b). This type of contamination can often be mitigated through regular monitoring and control of irrigation water quality.

The second pathway is leaching, which is the primary mechanism of groundwater contamination from agricultural activities. Leaching refers to the process by which rainfall or surface water percolates through the unsaturated zone, interacting with the soil and dissolving soluble substances (Li et al. 2024b). These dissolved substances are then transported into the groundwater, altering its chemical composition. In agricultural practices, a significant portion of applied fertilizers and pesticides does not directly enter groundwater but accumulates on the surface or within the unsaturated zone (Singh and Craswell 2021). Irrigation or rainfall can transport nutrients such as nitrogen and phosphorus from these fertilizers into the groundwater, potentially leading to excessive nutrient levels and contamination. The widespread issue of nitrate pollution in groundwater globally is essentially a result of leaching. In addition to carrying fertilizers and pesticides into the aquifer, irrigation water can also interact with the unsaturated zone, dissolving and transporting soluble salts and mineral components into the groundwater, further degrading its quality. For instance, the unsaturated zone often contains high concentrations of soluble salts, particularly in regions with severe soil salinization. These salts can infiltrate into the aquifer with irrigation water, leading to elevated levels of sulfates, chlorides, and other contaminants in groundwater.

Fig. 1.1 Groundwater collected for analysis showing contamination by paper wastewater irrigation in the fast-growing forest of the Weining Plain, China

It is worth noting that agricultural pollutants undergo continuous changes both before and after entering groundwater. Irrigation, in particular, often causes fluctuations in groundwater levels. These fluctuations not only repeatedly leach soluble salts from the fluctuation zone but also alter the biogeochemical environment of the unsaturated zone and the groundwater fluctuation zone, influencing various biogeochemical processes. These biogeochemical processes then can have significant impacts on groundwater quality. However, there is currently a lack of in-depth research on the mechanisms of agricultural pollutant migration and transformation under fluctuating groundwater levels. Specifically, how groundwater level fluctuations in agricultural areas affect biogeochemical processes in the fluctuation zone and, consequently, groundwater quality remains a topic requiring further investigation. Therefore, to fully understand the evolution of groundwater environments in large-scale irrigation zones in arid and semi-arid regions, it is essential to precisely characterize the impact of agricultural irrigation on groundwater dynamics, and advance research on how groundwater flow, hydrochemical field, and biogeochemical processes interact with each other.

1.3 Objectives and Scope of the Book

This book is the outcome of an international collaboration research project between China and South Africa. It aims to provide a comprehensive understanding of the impacts of agricultural activities on groundwater quality, with a specific focus on case studies from China and South Africa. By integrating scientific research, field investigations, and advanced modeling techniques, the book seeks to address critical issues related to groundwater pollution, hydrogeochemical processes, and sustainable water resource management. The following subsections outline the purpose of the book, the key questions it addresses, and an overview of its structure.

1.3.1 Purpose of the Book

The primary purpose of this book is to explore the complex relationship between agricultural activities and groundwater quality, emphasizing the hydrogeochemical characteristics, pollution mechanisms, and health risks associated with groundwater contamination. By presenting case studies from China and South Africa, the book highlights the regional and global significance of groundwater quality management in the context of intensive agricultural practices and land use changes. Additionally, the book aims to provide practical strategies and tools for policymakers, researchers, and water resource managers to protect and sustainably manage groundwater resources. Through a combination of theoretical insights and empirical evidence, this book may serve as a valuable resource for understanding and mitigating the impacts of human activities on groundwater systems.

1.3.2 Key Questions Addressed in This Book

This book addresses several critical questions related to groundwater quality under agricultural activities, including:

- What are the primary hydrogeochemical processes influencing groundwater quality in agricultural regions?
- How do land use and land cover changes contribute to groundwater quality?
- What are the main contaminants of concern in groundwater systems in agricultural regions, and what are their sources?
- How can groundwater quality be effectively assessed and monitored in regions with intensive agricultural activities?
- What are the health risks associated with groundwater contamination, and how can they be mitigated?
- What strategies and technologies are available for protecting and managing groundwater quality in the face of agricultural pressures and land use changes?
- How can advanced groundwater modeling techniques improve our understanding of groundwater systems and inform decision-making?

1.3.3 Overview of the Book's Structure

The book is organized into thematic chapters that explore groundwater quality under agricultural activities, with a focus on case studies from China and South Africa. The structure is designed to provide a comparative perspective on groundwater quality issues, hydrogeochemical processes, and management strategies in these two regions. The book has 10 chapters.

This chapter is the introductory chapter. It introduces the background, significance, and objectives of the book, setting the stage for the subsequent discussions. From Chapters 2, 3, 4 and 5 the book focuses on the groundwater quality under agricultural activities in China with the Weining Plain as a study area. Specifically, Chapter 2 examines the fundamental hydrogeochemical processes governing groundwater quality in the Weining Plain, including the natural and anthropogenic factors influencing water chemistry. Chapter 3 provides an assessment of groundwater quality in the Weining Plain, identifying key contaminants, their sources, and the primary challenges related to agricultural activities. Chapter 4 evaluates the health risks associated with groundwater contamination in the Weining Plain, focusing on exposure pathways, vulnerable populations, and mitigation measures. Chapter 5 explores the impacts of agricultural land use and land cover changes on groundwater quality, highlighting the interplay between human activities and hydrogeological systems.

From Chapters 6, 7, 8 and 9 the book shifts its focus to South Africa, presenting case studies and analyses of groundwater quality under agricultural activities in the coastal aquifers of KwaZulu-Natal of South Africa. Chapter 6 provides a comprehensive hydrogeochemical assessment of groundwater in selected catchments of the KwaZulu-Natal coastal region through field sampling, laboratory analysis, and statistical methods. Chapter 7 assesses the geochemical characteristics and quality of groundwater in the selected coastal aquifers of South Africa to determine its suitability for domestic, agricultural, and industrial uses. Chapter 8 investigates the health risks associated with heavy metal contamination in the coastal aquifers of KwaZulu-Natal, South Africa. Chapter 9 investigates the impact of land use and land cover changes on groundwater quality in KwaZulu-Natal coastal region of South Africa by employing Landsat imagery and GIS techniques. Chapter 10 is the concluding chapter in which the groundwater risks from agricultural activities in China's Weining Plain and South Africa's KwaZulu-Natal coastal aquifers were summarized and some mitigation strategies were proposed to protect groundwater resources in the agricultural regions.

References

Alam SMK, Li P, Fida M (2024) Groundwater nitrate pollution due to excessive use of N-fertilizers in rural areas of Bangladesh: Pollution status, health risk, source contribution, and future impacts. Exposure and Health 16(1):159–182. https://doi.org/10.1007/s12403-023-00545-0

Alam SMK, Li P, Rahman M, Fida M, Elumalai V (2025) Key factors affecting groundwater nitrate levels in the Yinchuan Region, Northwest China: Research using the eXtreme Gradient Boosting (XGBoost) model with the SHapley Additive exPlanations (SHAP) method. Environ Pollut 364:125336. https://doi.org/10.1016/j.envpol.2024.125336

Arias-Estévez M, López-Periago E, Martínez-Carballo E, Simal-Gándara J, Mejuto JC, García-Río L (2008) The mobility and degradation of pesticides in soils and the pollution of groundwater resources. Agric Ecosyst Environ 123(4):247–260. https://doi.org/10.1016/j.agee.2007.07.011

Barnes KK, Kolpin DW, Furlong ET (2008) A national reconnaissance of pharmaceuticals and other organic wastewater contaminants in the United States—I) Groundwater. Sci Total Environ 402(2–3):192–200. https://doi.org/10.1016/j.scitotenv.2008.04.028

Boning C (2016) Florida's rivers, 2nd edn. Pineapple Press, Florida

Chakraborty SK (2021) Riverine ecology volume 1: eco-functionality of the physical environment of rivers. Springer, Cham. https://doi.org/10.1007/978-3-030-53897-2

Chao N, Chen G, Li J, Xiang L, Wang Z, Tian K (2020) Groundwater storage change in the Jinsha River Basin from GRACE, hydrologic models, and in situ data. Groundwater 58(5):735–748. https://doi.org/10.1111/gwat.12966

Charuaud L, Jarde E, Jaffrezic A, Liotaud M, Goyat Q, Mercier F et al (2019) Veterinary pharmaceutical residues in water resources and tap water in an intensive husbandry area in France. Sci Total Environ 664:605–615. https://doi.org/10.1016/j.scitotenv.2019.01.303

FAO (2020) The State of Food and Agriculture 2020: Overcoming water challenges in agriculture. Available at: https://www.fao.org. Accessed March 18, 2025

Fitts CR (2023) Groundwater science, 3rd edn. Academic Press, London. https://doi.org/10.1016/C2016-0-00985-3

Fu C, Xu B, Chen H et al (2022) Occurrence and distribution of antibiotics in groundwater, surface water, and sediment in Xiong'an New Area, China, and their relationship with antibiotic resistance genes. Sci Total Environ 807:151011. https://doi.org/10.1016/j.scitotenv.2021.151011

Ghasemizade M, Schirmer M (2013) Subsurface flow contribution in the hydrological cycle: lessons learned and challenges ahead—a review. Environ Earth Sci 69:707–718. https://doi.org/10.1007/s12665-013-2329-8

Gonzalez JM, Smiley PC (2025) Temporal trends of Atrazine, Simazine, and Metolachlor in a U.S. Midwest Agricultural Watershed: A 10-year study. Environ Process 12:12. https://doi.org/10.1007/s40710-025-00753-7

Hagos F, Mamo K (2014) Financial viability of groundwater irrigation and its impact on livelihoods of smallholder farmers: The case of eastern Ethiopia. Water Res Econ 7:55–65. https://doi.org/10.1016/j.wre.2014.08.001

Hahn HJ, Fuchs A (2009) Distribution patterns of groundwater communities across aquifer types in south-western Germany. Freshw Biol 54(4):848–860. https://doi.org/10.1111/j.1365-2427.2008.02132.x

Haque MI (2014) Mechanics of Groundwater in Porous Media. CRC Press, Boca Raton. https://doi.org/10.1201/b17200

He S, Wu J (2019) Relationships of groundwater quality and associated health risks with land use/land cover patterns: a case study in a loess area, northwest China. Hum Ecol Risk Assess 25(1–2):354–373. https://doi.org/10.1080/10807039.2019.1570463

Hildebrandt A, Guillamón M, Lacorte S, Tauler R, Barceló D (2008) Impact of pesticides used in agriculture and vineyards to surface and groundwater quality (North Spain). Water Res 42(13):3315–3326. https://doi.org/10.1016/j.watres.2008.04.009

Karamouz M, Ahmad Ai, Akhbari M (2020) Groundwater hydrology: Engineering, planning, and management, 2nd edn. CRC Press, Boca Raton. https://doi.org/10.1201/9780429265693

Khezami F, Gómez-Navarro O, Barbieri MV, Khiari N, Chkirbene A, Chiron S, Khadhar S, Pérez S (2024) Occurrence of contaminants of emerging concern and pesticides and relative risk assessment in Tunisian groundwater. Sci Total Environ 906:167319. https://doi.org/10.1016/j.scitotenv.2023.167319

Kim A, van der Beek H (2018) A holistic assessment of the water-for-agriculture dilemma in the Kingdom of Saudi Arabia. Center for International and Regional Studies, Georgetown University in Qatar. Available at https://repository.digital.georgetown.edu/handle/10822/1050168, Accessed 18 Mar 2025

Krishna Kumar S, Hari Babu S, Eswar Rao P et al (2017) Evaluation of water quality and hydrogeochemistry of surface and groundwater, Tiruvallur District, Tamil Nadu, India. Appl Water Sci 7:2533–2544. https://doi.org/10.1007/s13201-016-0447-7

Li P, Wu J, Qian H, Zhang Y, Yang N, Jing L, Yu P (2016b) Hydrogeochemical characterization of groundwater in and around a wastewater irrigated forest in the southeastern edge of the Tengger Desert, Northwest China. Exposure and Health 8:331–348. https://doi.org/10.1007/s12403-016-0193-y

Li P, Elumalai V (2025) Hydrogeology and the global significance of groundwater. In: P. Li et al. (eds) Sustainable groundwater and environment: Challenges and solutions. Springer Nature, Switzerland, pp. 1–20. https://doi.org/10.1007/978-3-031-82194-3_1

Li P, He S, He X, Tian R (2018b) Seasonal hydrochemical characterization and groundwater quality delineation based on matter element extension analysis in a paper wastewater irrigation area, northwest China. Exposure and Health 10(4):241–258. https://doi.org/10.1007/s12403-017-0258-6

Li P, He X, Guo W (2019) Spatial groundwater quality and potential health risks due to nitrate ingestion through drinking water: a case study in Yan'an City on the Loess Plateau of northwest China. Hum Ecol Risk Assess 25(1–2):11–31. https://doi.org/10.1080/10807039.2018.1553612

Li P, Kou X, Wang Y, Niu L (2024a) Building a more sustainable Chinese Loess Plateau. Journal of Earth Science 35(1):283–287. https://doi.org/10.1007/s12583-024-1970-3

Li P, Qian H, Wu J (2018a) Conjunctive use of groundwater and surface water to reduce soil salinization in the Yinchuan Plain, North-West China. Int J Water Resour Dev 34(3):337–353. https://doi.org/10.1080/07900627.2018.1443059

Li P, Wu J, He X (2024b) Groundwater and environment. Geological Publishing House, Beijing (in Chinese)

Li P, Wu J, Qian H (2016a) Regulation of secondary soil salinization in semi-arid regions: A simulation research in the Nanshantaizi area along the Silk Road, northwest China. Environ Earth Sci 75(8):698. https://doi.org/10.1007/s12665-016-5381-3

Lopez-Serna R, Jurado A, Vazquez-Sune E, Carrera J, Petrovic M, Barcelo D (2013) Occurrence of 95 pharmaceuticals and transformation products in urban groundwaters underlying the metropolis of Barcelona, Spain. Environ Pollut 174:305–315. https://doi.org/10.1016/j.envpol.2012.11.022

Madramootoo CA (2012) Sustainable groundwater use in agriculture. 61:26–33. https://doi.org/10.1002/ird.1658

Malla MA, Gupta S, Dubey A, Kumar A, Yadav S (2021) Chapter 7—contamination of groundwater resources by pesticides. In: Ahamad A, Siddiqui SI, Singh P (eds) Contamination of water: Health risk assessment and treatment strategies. Academic Press, London, pp 99–107. https://doi.org/10.1016/B978-0-12-824058-8.00023-2

Mu D, Li P, De Baets B, Li D, Li Z, He S (2024) A multi-perspective exploration of the salinization mechanisms of groundwater in the Guanzhong Basin, China. Science of the Total Environment 957:177421. https://doi.org/10.1016/j.scitotenv.2024.177421

Mutileni N, Mudau M, Edokpayi JN (2023) Water quality, geochemistry and human health risk of groundwater in the Vyeboom region, Limpopo province. South Africa. Sci Rep 13:19071. https://doi.org/10.1038/s41598-023-46386-4

Ning J, Li P, He X, Elumalai V, Fida M, Xu D, Zhang P (2025) Occurrence and enrichment mechanisms of groundwater hexavalent chromium in typical loess area of China. Environ Pollut 367:125654. https://doi.org/10.1016/j.envpol.2025.125654

Pérez-Indoval R, Romero-López R (2024) Assessment of pesticide contamination in groundwater bodies in the Jucar River Basin (Spain) and its spatial distribution. Groundw Sustain Dev 25:101118. https://doi.org/10.1016/j.gsd.2024.101118

Qu Y, Zheng Y, Gong P, Shi J, Li L, Wang S, Luo C, Zhang H, Xu L (2022) Estimation of wetland biodiversity based on the hydrological patterns and connectivity and its potential application in change detection and monitoring: A case study of the Sanjiang Plain, China. Science of the Total Environment 805:150291. https://doi.org/10.1016/j.scitotenv.2021.150291

Qureshi AS, McCornick PG, Qadir M, Aslam Z (2008) Managing salinity and waterlogging in the Indus Basin of Pakistan. Agric Water Manag 95(1):1–10. https://doi.org/10.1016/j.agwat.2007.09.014

Radovic T, Grujic S, Petkovic A, Dimkic M, Lausevic M (2015) Determination of pharmaceuticals and pesticides in river sediments and corresponding surface and ground water in the Danube River and tributaries in Serbia. Environ Monit Assess 187:4092. https://doi.org/10.1007/s10661-014-4092-z

Ray C, Melin G, Linsky RB (2003) Riverbank filtration: Improving source-water quality. Kluwer Academic Publishers, Dordrecht

Said I, Salman SA, Elnazer AA (2022) Salinization of groundwater during 20 years of agricultural irrigation, Luxor, Eqypt. Environ Geochem Health 44:3821–3835. https://doi.org/10.1007/s10653-021-01135-2

Sarkar A, Paul B, Darbha GK (2022) The groundwater arsenic contamination in the Bengal Basin-A review in brief. Chemosphere 299:134369. https://doi.org/10.1016/j.chemosphere.2022.134369

Shah T (2009) Taming the Anarchy: Groundwater Governance in South Asia. Routledge, p 324

Siebert S, Burke J, Faures JM, Frenken K, Hoogeveen J, Döll P, Portmann FT (2010) Groundwater use for irrigation-a global inventory. Hydrol Earth Syst Sci 14:1863–1880. https://doi.org/10.5194/hess-14-1863-2010

Singh A (2024) Effective management of water resources problems in irrigated agriculture through simulation modeling. Water Resour Manage 38:2869–2887. https://doi.org/10.1007/s11269-024-03796-x

Singh B, Craswell E (2021) Fertilizers and nitrate pollution of surface and ground water: an increasingly pervasive global problem. SN Appl Sci 3:518. https://doi.org/10.1007/s42452-021-04521-8

Sishu FK, Tilahun SA, Schmitter P, Assefa G, Steenhuis TS (2022) Pesticide contamination of surface and groundwater in an Ethiopian highlands' watershed. Water 14(21):3446. https://doi.org/10.3390/w14213446

Stuart M, Lapworth D, Crane E, Hart A (2012) Review of risk from potential emerging contaminants in UK groundwater. Sci Total Environ 416:1–21. https://doi.org/10.1016/j.scitotenv.2011.11.072

Su Z, Wu J, He X, Elumalai V (2020) Temporal changes of groundwater quality within the groundwater depression cone and prediction of confined groundwater salinity using Grey Markov model in Yinchuan area of northwest China. Exposure and Health 12(3):447–468. https://doi.org/10.1007/s12403-020-00355-8

Szekeres E, Chiriac CM, Baricz A, Szoke-Nagy T, Lung I, Soran ML et al (2018) Investigating antibiotics, antibiotic resistance genes, and microbial contaminants in groundwater in relation to the proximity of urban areas. Environ Pollut 236:734–744. https://doi.org/10.1016/j.envpol.2018.01.107

Wang D, Li P, Yang N, Yang C, Zhou Y, Li J (2023) Distribution, sources and main controlling factors of nitrate in a typical intensive agricultural region, northwestern China: Vertical profile perspectives. Environ Res 237:116911. https://doi.org/10.1016/j.envres.2023.116911

Wang D, Wu J, Li P, Li L, Yang J, Zhang P, He S, Kou X, Wang Y (2024) Seasonal nitrate variations, risks, and sources in groundwater under different land use types in a thousand-year-cultivated region, northwestern China. Environ Res 251:118699. https://doi.org/10.1016/j.envres.2024.118699

Wu J, Li P, Qian H, Fang Y (2014) Assessment of soil salinization based on a low-cost method and its influencing factors in a semi-arid agricultural area, northwest China. Environmental Earth Sciences 71(8):3465–3475. https://doi.org/10.1007/s12665-013-2736-x

Wu J, Li P, Wang D, Ren X, Wei M (2020) Statistical and multivariate statistical techniques to trace the sources and affecting factors of groundwater pollution in a rapidly growing city on the

Chinese Loess Plateau. Hum Ecol Risk Assess 26(6):1603–1621. https://doi.org/10.1080/10807039.2019.1594156

Xie J, Liu X, Jasechko S et al (2024) Majority of global river flow sustained by groundwater. Nat Geosci 17:770–777. https://doi.org/10.1038/s41561-024-01483-5

Ye S, Pei L, He L, Xie L, Zhao G, Yuan H, Ding X, Pei S, Yang S, Li X et al (2022) Wetlands in China: Evolution, carbon sequestrations and services, threats, and preservation/restoration. Water 14(7):1152. https://doi.org/10.3390/w14071152

Zhang H, Yu J, Wang P, Wang T, Li Y (2021) Groundwater-fed oasis in arid Northwest China: Insights into hydrological and hydrochemical processes. J Hydrol 597:126154. https://doi.org/10.1016/j.jhydrol.2021.126154

Zhang WL, Tian ZX, Zhang N, Li XQ (1996) Nitrate pollution of groundwater in northern China. Agr Ecosyst Environ 59(3):223–231. https://doi.org/10.1016/0167-8809(96)01052-3

Zhang W, Yu K, Li Y et al (2020) Research progress of groundwater environment in Hetao Plain. Environ Chem 39(2):489–499 (in Chinese)

Zhou Q, Zhang J, Kg X, Wei J, Yao Y (2024) Groundwater nitrate contamination in China: Spatial distribution, temporal trend, and driver analysis. Environ Res 262:119932. https://doi.org/10.1016/j.envres.2024.119932

Chapter 2
Hydrogeochemical Characteristics of Groundwater in the Weining Plain

Groundwater chemistry is a crucial component of the groundwater environment. Understanding the basic hydrogeochemical characteristics and their influencing factors is the basis of conducting groundwater quality research. The impact of agricultural activities on the groundwater quality is largely reflected through the changes in hydrochemical compositions. Therefore, studying the formation, distribution, and evolution of groundwater chemistry under the influence of agricultural activities has significant theoretical implications for a profound understanding of the mechanisms by which agricultural activities impact the groundwater environment. This chapter, taking the Weining Plain, which has over 2000 years of irrigation history, as an example, employs hydrogeochemical and isotopic techniques to investigate the formation, distribution, and evolution of groundwater chemistry under the influence of agricultural activities.

2.1 Overview of the Weining Plain

2.1.1 Geographical Location

The Weining Plain is located in the western part of Ningxia, in the upper and middle reaches of the Yellow River, and is a typical Yellow River irrigation area. It lies between 105°00′00″–106°00′00″ east longitude and 37°20′00″–37°50′00″ north latitude (Gong et al. 2025). The plain extends from west to east and then turns to northeast, with a general topography that is higher in the west and lower in the east, steeper in the south and gentler in the north. The plain is 2.5–14 km wide from north to south, with an elevation of 1130–1240 m above the mean sea level. Administratively, it belongs to Zhongwei City and Zhongning County. The location of the study

P. Li and V. Elumalai, *Groundwater Quality Under Agricultural Activities—Cases from China and South Africa*, Professional Practice in Earth Sciences,
https://doi.org/10.1007/978-3-031-98993-3_2

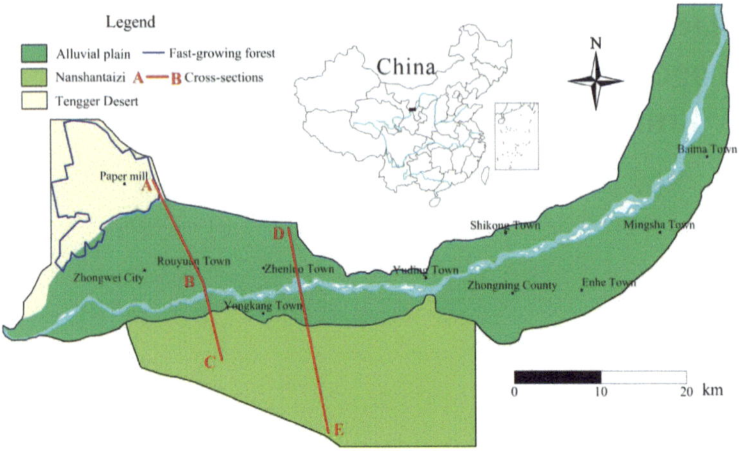

Fig. 2.1 Location of the study area

area is shown in Fig. 2.1. The alluvial plain stretching along the Yellow River is the main region for residence and agriculture.

The Nanshantaizi, located to the southern part of the Yellow River alluvial plain, is a typical irrigation area by lifting the Yellow River water, with an annual volume of approximately 65 million m^3. The substantial volume of water used for irrigation in the Nanshantaizi has caused a rise in the groundwater level in the Yellow River alluvial plain, leading to secondary soil salinization (Wu et al. 2014). Additionally, to the northwestern part of the Yellow River alluvial plain is the Tengger Desert. Since 2003, the Zhongye Meili Paper Group has been afforesting this area, establishing a fast-growing forest base covering more than 200,000 mu (approximately 13,333 ha). Before the establishment of the fast-growing forest, the paper mill's wastewater was directly discharged into the Yellow River after a simple treatment, significantly impacting the river's water quality and exerting considerable pressure on its aquatic environment. After the fast-growing forest base was established, the forest was used to produce raw materials for papermaking, while the wastewater from paper production was mixed with Yellow River water in a 1:1 ratio and used for irrigation. This measure not only provided a reasonable outlet for the paper mill's wastewater, avoiding direct pollution of the Yellow River, but also reduced the need for imported papermaking materials. Furthermore, it transformed a part of the desert into an oasis, greatly improving the local climate and ecological environment. However, the annual irrigation with large volumes of paper mill wastewater has had a noticeable impact on the local groundwater level and quality (Li et al. 2016a, 2018). The COD (chemical oxygen demand) content in the groundwater of the fast-growing forest base far exceeds the national groundwater standard. In addition, the groundwater in the fast-growing forest base flows from north to south. However, the southern boundary of the fast-growing forest is less than 4 km from the protection zone of Zhongwei City's water source. If harmful substances from the paper mill wastewater flow into the protection zone, they will inevitably affect the urban water supply of Zhongwei City. Therefore, considering the impacts of irrigation in both the Nanshantaizi and the paper mill forest, these two areas are also introduced in this chapter to give readers additional information into the overall environment in the Weining Plain.

2.1.2 *Climate and Hydrology*

The Weining Plain is located in inland of China, which is far from the ocean, and lies on the western edge of China's monsoon region. In winter, it is controlled by the Mongolian high-pressure system, making it a key passage for cold air moving southward. In summer, it is at the end of the southeast monsoon, resulting in a typical continental climate. The region is characterized by long and cold winters; short but hot summers, rapid temperature rising in springs, and cool autumns. This area is arid with limited rainfall, but the sunshine here is abundant and the evaporation is intense. The recorded maximum annual precipitation is 308 mm (in 1978), while the minimum annual precipitation is 83 mm (in 1982). The maximum daily precipitation is 68.3 mm, and the maximum frozen soil depth is 83 cm. According to data from the Zhongwei Meteorological Station from 1959 to 2009, the average annual temperature in Zhongwei City is 8.9 °C, and the average annual precipitation is 189.2 mm, with most rainfall concentrated between June and September, accounting for 73.6% of the annual total precipitation (He et al. 2022; Xu and Li 2024). The average annual evaporation is 1828.6 mm, with the majority occurring between April and August, accounting for 64.5% of the annual total (Fig. 2.2). Influenced by global climate change, the study area has experienced a rising temperature and a declining evaporation, but no significant change was observed in precipitation (Fig. 2.3).

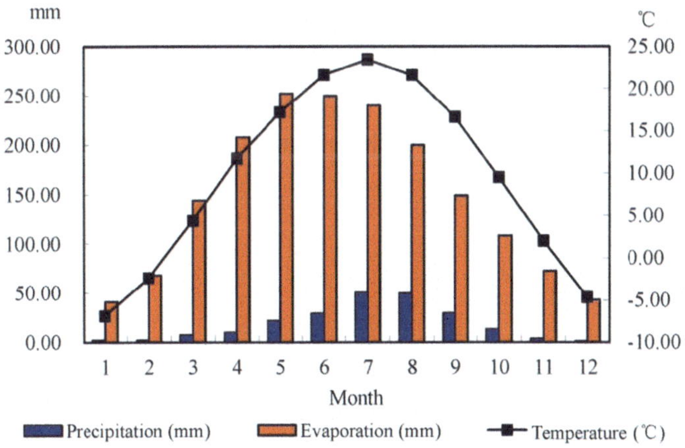

Fig. 2.2 Monthly climatic elements in Zhongwei city (Li 2014)

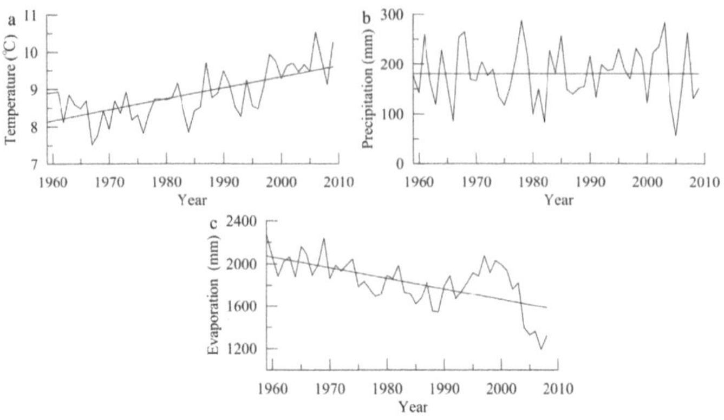

Fig. 2.3 Trend of climatic elements in the study area (Li 2014)

The main rivers in the study area are the Yellow River and its tributary, the Qingshui River. The Yellow River flows from west to east through the study area, covering approximately 106 km within the region, which accounts for 26.7% of its total length of 397 km in Ningxia. It is located about 2 km from Zhongwei City. The riverbed is generally 600 m wide, with an average water depth of 2–4 m and a maximum depth of 8.6 m. The riverbed gradient is 0.7%. The highest recorded flood river water level is 1296.7 m (in 1904), while the lowest water level is 1286.24 m. The maximum flow rate is 5820 m^3/s (in 1945), and the minimum flow rate is 140 m^3/s, with a normal flow rate of 1800 m^3/s. The flow velocity ranges from 0.34 to 2.16 m^3/s, and the sediment concentration ranges from 53.11 to 60.10 kg/m^3. The average annual flow rate of the Yellow River is 1039.8 m^3/s. This area is an important tier for hydropower development in the upper reaches of the Yellow River. The Shapotou Water Control Project, with a total investment of 1.356 billion Chinese yuan, has been completed and is designed to irrigate 1.07 million mu (approximately 71,333 ha) of land and supply 606 million kWh of electricity to the grid annually (Li 2014). Benefiting from the Yellow River, the study area is crisscrossed by numerous irrigation canals and drainage ditches, including major canals such as the Beigan Canal, Meili Canal, and Taiping Canal, as well as drainage ditches like the First, Second, Third, and Fourth Drainage Ditches.

2.1.3 Topography

The alluvial plain is distributed on both sides of the Yellow River, with the Xiangshan and Weining North Mountains acting as natural boundaries to the south and north, respectively. To the west lies the Tengger Desert, one of China's four major deserts, and to the east is the Niushou Mountain. The plain is higher in the west and lower in the east, and is higher in the north and south and lower in the middle. The elevation within the study area ranges from 1150 to 1460 m. The entire area is divided into three major geomorphological units: the Yellow River alluvial plain (the Weining Plain), hilly desert (Tengger Desert), and alluvial-proluvial high terraces (Nanshantaizi). The hilly desert is mainly distributed to the western part of the Weining Plain, and consists of fine sand accumulated by the northwest monsoon. The topography generally slopes from north to south, with landforms including crescent-shaped dunes, dune chains, and grassy dunes. In recent years, afforestation efforts by the Meili Paper Group have transformed this barren area into a green oasis with economic value. The Yellow River alluvial plain is flat, with a gentle slope in the south and a steeper slope in the north, and is primarily used for residential and agricultural purposes. To the south of the plain lies the Nanshantaizi, a high terrace formed by Quaternary alluvial and proluvial processes, with an elevation generally ranging from 1300 to 2600 m. This area is relatively flat and serves as a major irrigation zone by lifting Yellow River water in Ningxia, as well as an important base for watermelons.

2.1.4 Geology and Hydrogeology

(1) **Aquifer Structures**

According to borehole data, the Yellow River alluvial plain in the Weining Plain is a single unconfined aquifer. In the western part, the Quaternary stratum thickness ranges from 30 to 60 m, with lithology consisting of clayey sand, fine sand, gravel, and pebble layers. The basement is composed of Neogene deposits, which generally exhibit a disconformity or unconformity contact with the underlying strata. The stratification in the alluvial plain is distinct, and the groundwater flow conditions are good. In the low hills at the edge of the Tengger Desert in the northwest, now reclaimed as the fast-growing forest base of the Meili Paper Group, the lithology mainly consists of fine sand, silt, and sandy clay layers. The surface elevation is approximately 1,300 m, with the Quaternary stratum thickness of about 20 m. Moving southeast toward the Yellow River alluvial plain, the sand layers become gradually thinner, while the pebble and gravel layers thicken. The Quaternary strata here often contain interbedded material from weathered muddy sandstone, coarse and fine sandstone, and mudstone, resulting in a relatively complex stratigraphic structure. In the northeastern part of the Weining Plain, the Quaternary stratum thickness ranges from 20 to 40 m, with lithology mainly composed of fine sand, gravel, and pebble layers. The stratification is distinct, with a sand and pebble thickness of nearly 30 m, providing favorable conditions for unconfined groundwater storage. The cross-section of the Weining Plain is shown in Fig. 2.4, and the location of the cross-section is indicated in Fig. 2.1.

The Nanshantaizi area belongs to Shapotou District, Zhongwei City, and is located to the southwestern part of the Weining Plain, distributed along the northern foothills of the Xiangshan Mountains, west of the Qingshui River. The formation and distribution of groundwater in this area are primarily controlled by tectonic factors.

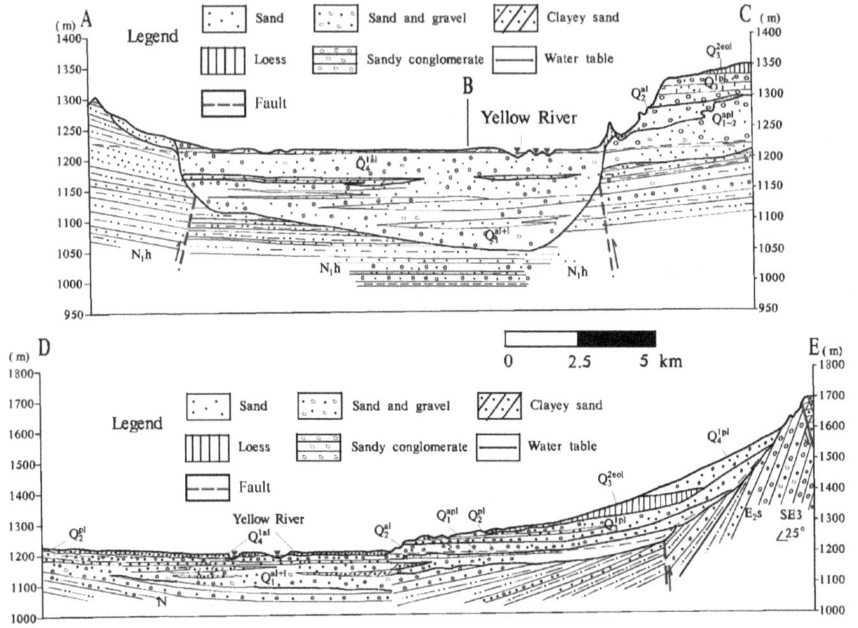

Fig. 2.4 Hydrogeological cross-section of the Weining Plain (Gong et al. 2025; Li 2014)

According to the tectonic system, it belongs to the first-order tectonic component of the Longxi arc tectonic system, the Zhongwei and Chenmajing Cenozoic subsidence zone, which governs the deposition of the Cenozoic strata in this area. The sediments in the subsidence zone, in addition to Quaternary loose deposits, are underlain by Tertiary clastic rock formations, representing alluvial and piedmont deposits. In the lower part of the front edge, the deposits may belong to an alluvial-proluvial transitional facies. In the central and northern parts of Nanshantaizi, the thickness of Quaternary loose deposits reaches 143.5 m, mainly composed of sand, crushed stone, and gravel, interbedded with loess-like clayey sand. The front edge of Nanshantaizi is elevated by over 100 m compared to the adjacent alluvial plain, resulting in generally high groundwater gradient in this area. According to borehole data, the groundwater level depth in the central part of the Nanshantaizi is 39.6 m, representing a rise of about 60 m compared to the water level recorded in the 1970s. The north part of the Nanshantaizi has a dual-aquifer structure, with an unconfined aquifer above 88 m and a confined aquifer below. In the southern part of Nanshantaizi, the Quaternary thickness exceeds 150 m, and the aquifer is a single unconfined aquifer with a water level depth of 51.2 m. The aquifer consists of gravel and coarse sand, interbedded with thin layers of clayey sand.

The storage and distribution of unconfined groundwater in the Weining Plain are controlled by geological structures, lithology, topography, hydrology, and meteorological factors. Based on storage conditions and hydraulic characteristics, the unconfined groundwater in the study area is primarily pore water in loose rocks, and is widely distributed in the Yellow River alluvial plain and the Nanshantaizi area. In the alluvial plain, the thickness of the unconfined aquifer ranges from 6.08 to 47.48 m, showing a decline in aquifer thickness from west to east and from the north and south toward the Yellow River. In the Nanshantaizi area, the unconfined aquifer is generally more than 50 m in thickness. According to borehole data, the unconfined aquifer in the study area consists of several layers of gray fine sand and sand-pebble layers, interbedded with thin layers of sandy clay and, locally, sand layers. From top down, the color deepens, the structure becomes looser, and the grain size becomes more uniform, with individual layer thicknesses ranging from a few meters to tens of meters. In most areas of the Nanshantaizi ground surface is covered by a layer of sandy clay with a thickness of 1.0 to 5.0 m. The lithology of the aquifer bottom is sandy clay, with poor continuity. Except for the Nanshantaizi area, where the groundwater level depth is relatively large, the groundwater level depths in the northern hilly region and the alluvial plain are generally small. In the northern hills, the groundwater depth is mostly between 5.0 and 30 m, while in the alluvial plain, the groundwater level depth is mostly less than 5.0 m. In the Nanshantaizi area, the groundwater depth is generally greater than 30 m, with depths exceeding 90 m at the northern foothills of the Xiangshan Mountains.

Based on pumping tests, the hydraulic conductivity of the unconfined aquifer exhibits significant regional variability due to differences in lithology. The aquifer on the north bank of the Yellow River has strong permeability, with a hydraulic conductivity ranging from 6 to 12 m/d. The aquifer on the south bank of the Yellow River is distributed in a narrow belt, with a hydraulic conductivity generally ranging from 5.5 to 8.5 m/d, decreasing from upstream to downstream. In the Nanshantaizi area, the permeability of the unconfined aquifer is distributed in a banded pattern from south to north.

The water yield of the aquifer is related to the lithology, aquifer thickness, and groundwater recharge. Areas with strong water yield are mainly distributed in the central alluvial plain, where the aquifer is thick, the lithology is relatively loose, and the recharge sources are abundant. Based on exploration results, the single well yield in the Zhongwei part ranges from 199.58 to 4845.31 m^3/d (converted to a pumping well diameter of 209 mm and a water level drawdown of 15 m, the same below), while in the Zhongning part, it ranges from 946 to 6781 m^3/d. The Nanshantaizi area has relatively poor water yield, with a converted yield of 123 to 1451 m^3/d.

(B) **Groundwater Recharge and Discharge**

From a regional hydrogeological and macro-geomorphological perspective, the northwestern part of the study area is mainly composed of low hills, where groundwater flows from north to south. The central part is the Yellow River alluvial plain, where groundwater primarily discharges into the Yellow River. The southern Nanshantaizi is a high terrace formed by alluvial-proluvial forces, and groundwater here flows from southwest to northeast. The groundwater in the study area generally flows from west to east, receiving recharge from lateral inflow through the boundaries, as well as from precipitation infiltration, agricultural irrigation infiltration, and canal leakage. Groundwater discharge primarily occurs through evaporation, drainage ditches, lateral outflow, artificial extraction, and discharge into the Yellow River. The recharge and discharge of groundwater in the study area are influenced not only by geological, geomorphological, structural, lithological, meteorological, and hydrological factors, but also significantly by human activities. In this chapter, the recharge and discharge amounts were calculated using the groundwater balance (groundwater budget) method. The results are shown in Table 2.1.

According to the calculation, the total groundwater recharge in the study area is 394.68 million m^3/year in a hydrological year, with 37.49% from canal leakage recharge, 36.17% from irrigation infiltration recharge, 18.70% from lateral flow recharge, and 7.64% from precipitation infiltration recharge (Table 2.1 and Fig. 2.5). Irrigation and canal leakage are responsible for the most of the recharge to groundwater, while precipitation attributes to the least. During the same period, the total groundwater discharge is 393.58 million m^3/year, with most groundwater discharged through evaporation, artificial extraction, and drainage ditches. Evaporation discharge accounts for 28.03%, artificial extraction accounts for 22.11%, drainage ditch discharge accounts for 32.72%, and lateral flow discharge accounts for 17.14% (Fig. 2.5). Discharge through drainage ditches and evaporation are the two main ways of groundwater discharge followed by the discharge by artificial abstraction and lateral outflow. The water balance difference during the period is 1.10 million m^3/year, with a relative balance difference of 0.28%.

(C) **Groundwater Dynamics**

Groundwater monitoring in Weining Plain is underdeveloped, with limited groundwater monitoring data available. The groundwater system monitoring in the Weining Plain, managed by the Ningxia Environmental Protection Department, began in March 2012. Based on the groundwater level monitoring data, the annual groundwater level dynamics of the Weining Plain are plotted in Fig. 2.6.

Table 2.1 Groundwater balance calculation results

Recharge elements	Recharge (10^6 m^3/year)	Percentages (%)	Discharge elements	Discharge (10^6 m^3/year)	Percentages (%)
Precipitation infiltration	30.16	7.64	Lateral outflow	67.46	17.14
Irrigation infiltration	142.74	36.17	Evaporation	110.31	28.03
Canal leakage recharge	147.96	37.49	Artificial extraction	87.04	22.11
Lateral inflow	73.82	18.70	Drainage ditch discharge	128.78	32.72
Total recharge	394.68	100	**Total discharge**	393.58	100

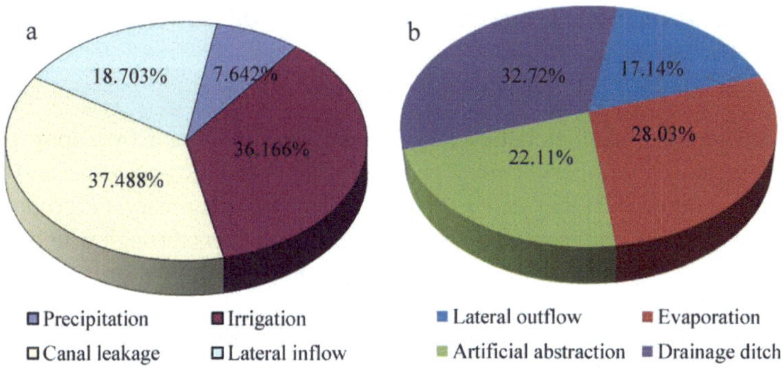

Fig. 2.5 Pie chart of groundwater recharge and discharge in the study area. **a**. Groundwater recharge, **b**. Groundwater discharge

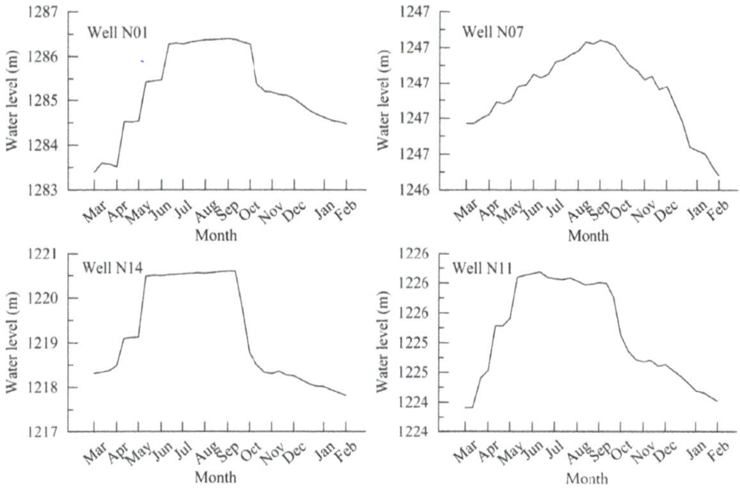

Fig. 2.6 Groundwater level dynamics in the study area

Figure 2.6 shows that the groundwater dynamics in the study area exhibit typical irrigation-related characteristics, with groundwater levels rising during irrigation periods and falling back during non-irrigation periods. In addition to irrigation, precipitation also plays a role in groundwater level changes, but its impact is much smaller than that of irrigation. Influenced by irrigation, groundwater levels begin to rise in mid-April when summer irrigation starts. The summer irrigation continues until August, but due to heavy rainfall in September and the lag effect of irrigation infiltration, the decline in groundwater levels in September is not significant. Winter irrigation begins in mid-October, which causes high groundwater levels until November when the winter irrigation ends. After that the groundwater levels begin to decline, reaching their lowest levels in February to March of the following year.

2.2 Hydrochemical Characteristics of Groundwater in the Weining Plain

The hydrochemical characteristics of groundwater are formed over long periods through a series of hydrogeochemical processes under specific geological and hydrogeological conditions (Qian and Li 2011). The hydrochemistry of groundwater primarily depends on the mineral composition in the aquifers through which it flows. During the movement of groundwater, various hydrogeochemical processes alter its chemical compositions. These processes vary spatially and temporally depending on geological conditions and the chemical characteristics of the aquifer media (Elango and Kannan 2005). Generally, the hydrogeochemical processes controlling groundwater chemistry can be broadly categorized into three types: water–rock interactions, evaporation-concentration, and leaching. These categories can be further subdivided into processes such as rainfall leaching, mineral weathering and dissolution, mineral precipitation, ion exchange, redox reactions, and evaporation (Qian et al. 2012).

The 2000-year history of agricultural irrigation in the Weining Plain has significantly altered its natural hydrogeochemical characteristics. Moreover, recent industrialization and urbanization have had an immeasurable impact on the local groundwater system. A systematic study of the basic hydrochemical characteristics of groundwater in the Weining Plain, exploring the formation mechanisms of groundwater chemistry under the influence of human activities, can not only comprehensively clarify the hydrochemical characteristics and influencing factors of groundwater in the Weining Plain, laying a foundation for groundwater environmental protection, but also provide valuable insights and methodologies that can be applied to other regions, promoting the development of groundwater environmental research under changing conditions.

2.2.1 *Basic Statistical Analysis*

Table 2.2 presents the statistical results of the main chemical indicators of groundwater in the study area. The pH of groundwater in the region ranges from 7.0 to 8.8, indicating neutral to slightly alkaline conditions. The concentrations of various ions vary significantly. The concentration of Na^+ ranges from 35.46 to 3153 mg/L, with an average value of 226.53 mg/L, which exceeds the limit (200 mg/L) specified in drinking water standards. The concentration of K^+ in groundwater is relatively low, with a maximum value of 52.7 mg/L, primarily due to the weak mobility of K^+ caused by its biological activity (Li et al. 2010). The concentrations of Mg^{2+} and Ca^{2+} range from 6.3 to 223.3 mg/L and 17.05 to 574 mg/L, respectively, with average values of 68.61 mg/L and 125.85 mg/L. Among the major anions, SO_4^{2-} exhibits the widest concentration range (30.6–3584 mg/L), while HCO_3^- has the smallest range (119.13–844.22 mg/L). CO_3^{2-} content is minimal and thus not considered. The concentration trends of ions in the groundwater of the study area are as follows:

Table 2.2 Statistics of major ions and chemical indicators in groundwater

Parameters	pH	Na^+ (mg/L)	K^+ (mg/L)	Mg^{2+} (mg/L)	Ca^{2+} (mg/L)	SO_4^{2-} (mg/L)
Max	8.8	3,153	52.7	223.3	574	3,584
Min	7.0	35.46	1.0	6.3	17.05	30.60
Mean	7.76	226.53	8.96	68.61	125.85	409.49
Parameters	Cl^- (mg/L)	HCO_3^- (mg/L)	CO_3^{2-} (mg/L)	F^- (mg/L)	TDS (mg/L)	TH (mg/L)
Max	2,654	844.22	23.95	6.90	7,140	1,979.17
Min	42.25	119.13	0.00	0.11	298.45	93.25
Mean	206.84	413.86	0.33	1.01	1,316.97	591.59

for cations, $Na^+ > Ca^{2+} > Mg^{2+} > K^+$; for anions, $SO_4^{2-} > Cl^- > HCO_3^-$. For total dissolved solids (TDS) in groundwater, the minimum concentration is 298.45 mg/L, meeting the quality standards for drinking water, while the maximum concentration is 7140 mg/L, far exceeding the drinking water standards. The total hardness (TH) of groundwater is also relatively high, reaching up to 1979.17 mg/L. Additionally, F^- concentrations exceed standards in some cases, with a maximum concentration of 6.9 mg/L.

2.2.2 *Hydrochemical Types*

The hydrochemical types of groundwater in the study area are complex, and can be classified into nine categories: SO_4-Na type, SO_4-Na·Ca type, HCO_3·SO_4·Cl-Mg·Na type, HCO_3·Cl-Na·Mg type, HCO_3·Cl-Na·Ca type, HCO_3-Ca·Mg type, HCO_3·SO_4-Ca·Na type, HCO_3·SO_4·Cl-Ca·Na type, and HCO_3·SO_4·Cl-Mg·Ca·Na type. The specific hydrochemical characteristics of the aquifer in the study area are illustrated in Fig. 2.7. Due to the close hydraulic connection between groundwater and the Yellow River, the HCO_3^- content in groundwater on both sides of the Yellow River is significantly higher than that of other ions, as the Yellow River water contains high HCO_3^- concentrations. Additionally, the shallow groundwater depth in the plain area, combined with evaporation and concentration effects, leads to increasing concentrations of SO_4^{2-} and Cl^-. Therefore, the groundwater in the Yellow River alluvial plain is mostly of the HCO_3·SO_4 or HCO_3·SO_4·Cl types. In the Nanshantaizi area, irrigation leaches soluble salts from the vadose zone into the aquifer, increasing the SO_4^{2-} concentration in groundwater, resulting in predominantly SO_4^{2-}-Na type water.

Durov diagram for groundwater in the study area was also generated to show the hydrochemical types of groundwater (Fig. 2.8). Groundwater samples were grouped based on groundwater salinity in Fig. 2.8. It reveals that the pH of groundwater in the area ranges from 7.0 to 8.5, indicating neutral to slightly alkaline conditions. Approximately 50% of the sampling points have TDS values below 1 g/L, classified as freshwater, with HCO_3^- as the dominant anion. Over 40% of the sampling points have TDS values between 1 and 2 g/L with relatively balanced anion concentrations,

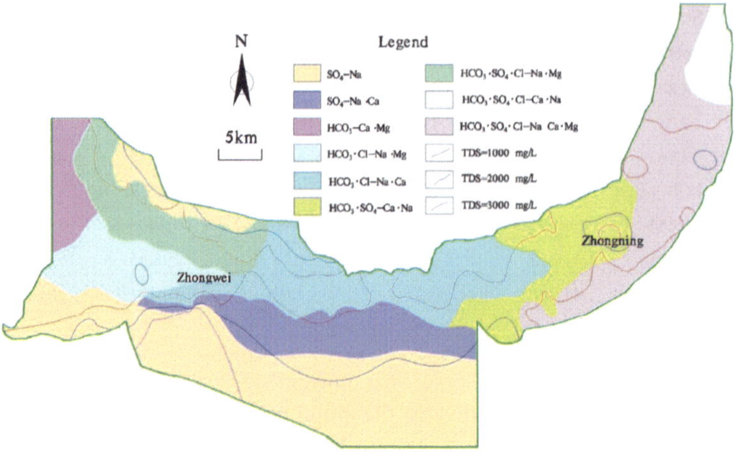

Fig. 2.7 Zonation map of groundwater hydrochemical types

and are classified as slightly saline water. A small portion of samples have TDS values exceeding 2 g/L, classified as saline water, with elevated concentrations of $SO4^{2-}$ and Na^{+}. The TDS content in groundwater is closely related to the strong evaporation conditions and aquifer media in the study area, indicating significant water–rock interactions and evaporation-concentration processes during groundwater flow.

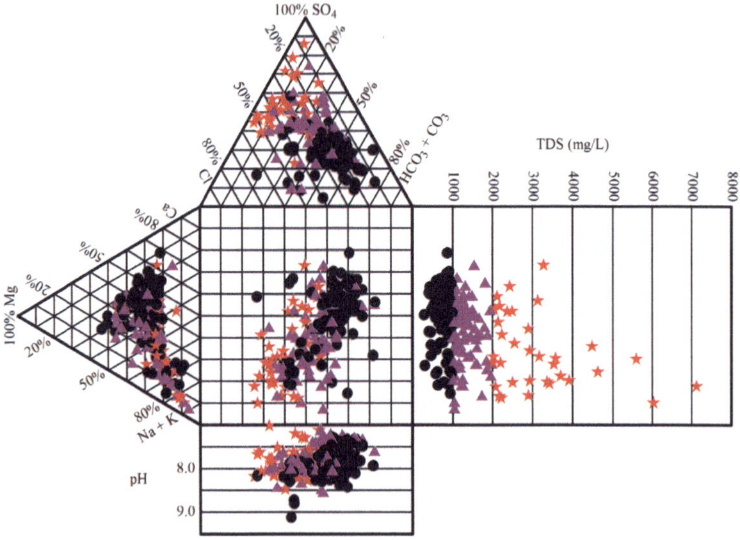

Fig. 2.8 Durov diagram of groundwater chemistry

2.2.3 Hydrochemical Changes Along Groundwater Flow Direction

During groundwater flow from upstream to downstream, a series of water–rock interactions occur, increasing the concentrations of various ions in the groundwater. Additionally, factors such as human interference and evaporation-concentration can cause abnormal increases or decreases in specific ion concentrations. Studying hydrochemical changes along groundwater flow directions is crucial for understanding the influence of water–rock interactions and other factors on groundwater chemistry. Chebotarev (1955) found that the general trend of major anion changes along regional groundwater flow under natural conditions is: $HCO_3^- \rightarrow HCO_3^- + SO_4^{2-} \rightarrow SO_4^{2-} + HCO_3^- \rightarrow SO_4^{2-} + Cl^- \rightarrow Cl^- + SO_4^{2-} \rightarrow Cl^-$.

In this study, three typical groundwater flow paths were selected based on the groundwater flow directions. By examining the changes in groundwater chemical composition along these paths, the primary processes occurring in each flow path were identified. The distribution of flow paths is shown in Fig. 2.9. Flow path A reflects the hydrochemical changes from the fast-growing forest (hilly desert) to the alluvial plain. This path starts from monitoring well N01 in the northern part of the fast-growing forest to monitoring well N08 in the southern part, then shifts direction to monitoring point S016 in the Yellow River alluvial plain. Flow path B represents the hydrochemical changes from Nanshantaizi (high terrace) to the Yellow River alluvial plain, starting from monitoring well NT1 on the southern side of Nanshantaizi to monitoring well NT2-1 on the northern side, and then to monitoring well NW27 in the alluvial plain to the north of Nanshantaizi. Flow path C illustrates the hydrochemical changes within the Yellow River alluvial plain, spanning the wider Zhongning Plain area, from W121 → W054 → W063. Based on water quality analysis data from these monitoring points, Stiff diagrams of groundwater chemistry along the flow paths were plotted (Fig. 2.10).

As shown in Fig. 2.10, the hydrochemical characteristics of groundwater show different variations along different flow paths. In flow path A, the upstream groundwater has relatively high Cl^- content compared to other anions, with Ca^{2+} dominating among cations, resulting in Cl-Ca type water. As groundwater flows downstream, Cl^- content decreases significantly, while HCO_3^- increases, with little change in other ions, transforming the hydrochemical type to $HCO_3{\cdot}SO_4$-Ca type. Further downstream, the concentrations of all ions increase, particularly Na^+, Mg^{2+}, and Ca^{2+} among cations, and HCO_3^- and SO_4^{2-} among anions, changing the hydrochemical type to $HCO_3{\cdot}SO_4$-Ca·Mg type (Fig. 2.11A). Compared to Chebotarev's findings (Chebotarev 1955), the hydrochemical changes in path A do not follow the natural regional groundwater evolution pattern. The significant increase in HCO_3^- from N01 to N08 is primarily due to extensive irrigation in the fast-growing forest, which accelerates groundwater flow and leaches soluble salts, leaving behind slower-dissolving carbonates. Additionally, the irrigation water contains relatively high HCO_3^-, and the weak evaporation in this area results in the notable increase in HCO_3^- at N08.

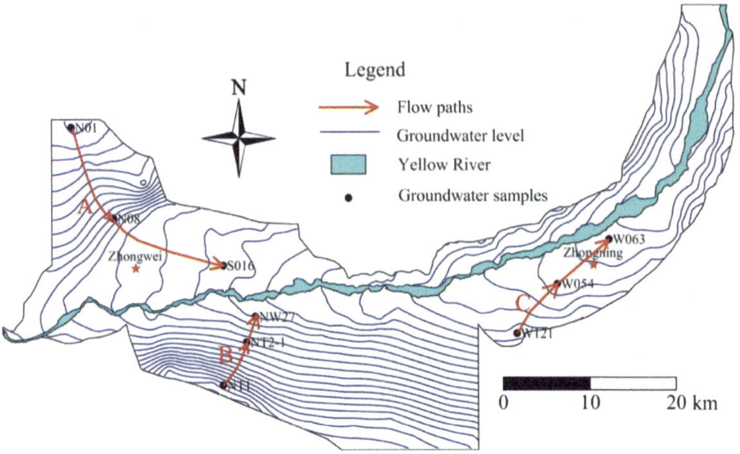

Fig. 2.9 Typical flow path profiles of groundwater hydrochemistry

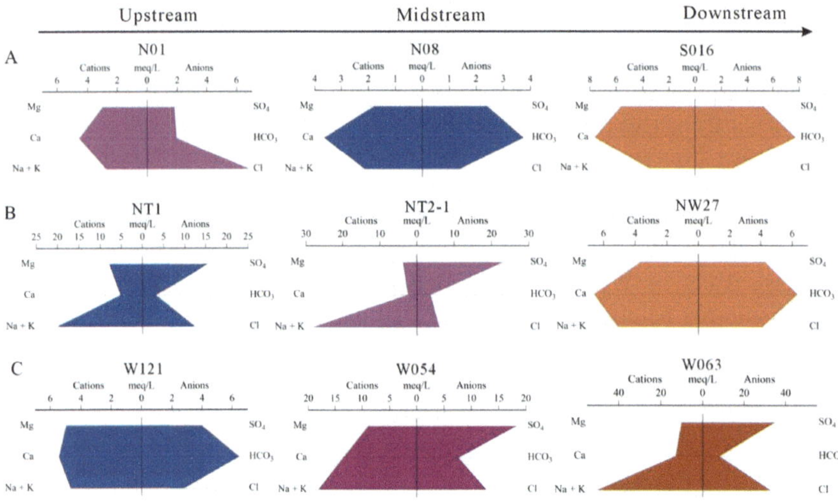

Fig. 2.10 Hydrochemical changes along different flow paths

From N08 to S016, the increase in major ion concentrations is likely due to strong evaporation and concentration effects in the shallow groundwater of the plain area.

In flow path B, the upstream groundwater has high SO_4^{2-} and Cl^- contents among anions, with Na^+ dominating among cations, resulting in SO_4·Cl-Na type water. As groundwater flows, SO_4^{2-} and Na^+ content increases, while Cl^-, Ca^{2+} and Mg^{2+} decrease, changing the hydrochemical type to SO_4-Na type. After passing through the front edge of Nanshantaizi, groundwater flows into the Yellow River alluvial plain. At this point, SO_4^{2-} and Cl^- decrease, while HCO_3^- increases among anions.

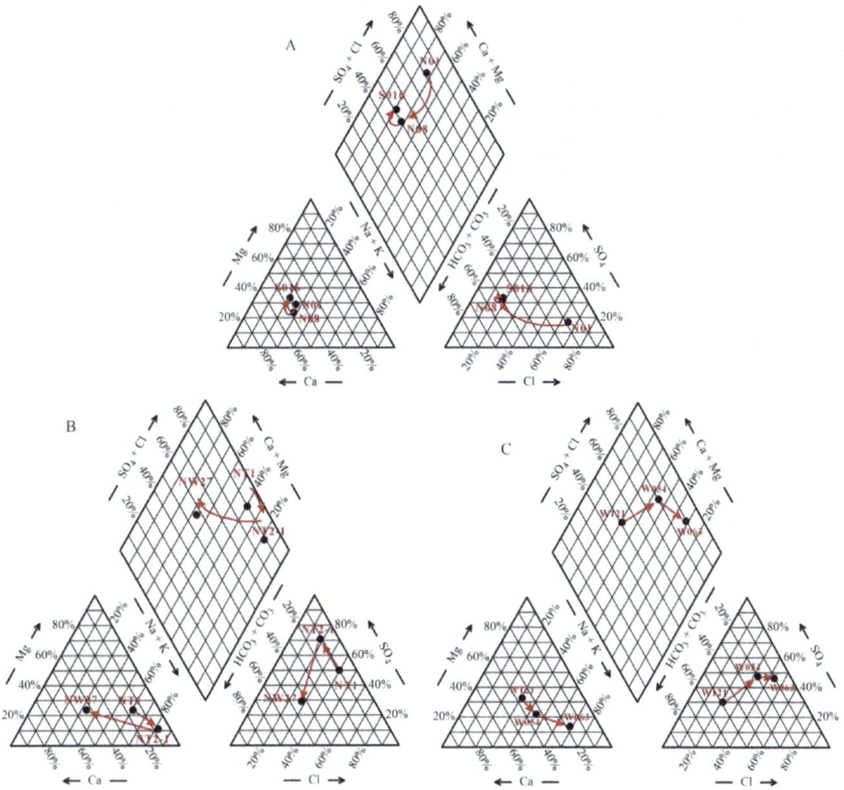

Fig. 2.11 Piper diagrams of hydrochemical evolution along different flow paths

At the same time, Ca^{2+} increases while Na^+ decreases, with little change in Mg^{2+}, resulting in HCO_3-Ca·Na type water (Fig. 2.11B). The anion variation pattern in this path also deviates from Chebotarev's findings (Chebotarev 1955). From NT1 to NT2-1, irrigation water dissolves soluble salts in the vadose zone, which then infiltrate into the aquifer. The primary soluble salt in the Nanshantaizi vadose zone is mirabilite (Na_2SO_4), leading to significant increases in SO_4^{2-} and Na^+. From NT2-1 to NW27, the increase in HCO_3^- and Ca^{2+} is mainly due to recharge from the Yellow River. The thin vadose zone in the plain area limits the leaching effect of irrigation water, but the extensive use of Yellow River water, rich in HCO_3^-, for irrigation increases HCO_3^- in groundwater. The increase in Ca^{2+} and decrease in Na + may result from cation exchange adsorption, while evaporation-concentration effects also influence hydrochemical changes.

In flow path C, the upstream groundwater is dominated by HCO_3^- among anions, with Ca^{2+} having the highest concentration among cations, followed by Mg^{2+} and Na^+, resulting in HCO_3-Ca·Mg·Na type water. From upstream to midstream, the percentage of HCO_3^- in meg/L decreases, while SO_4^{2-} and Cl^- increase rapidly.

Meantime, all cation concentrations increase, with Na_+ showing the largest increase, changing the hydrochemical type to SO_4·Cl-Na·Ca type. Further downstream, the percentage of HCO_3^- continues to decrease, while SO_4^{2-} and Cl^- increase, with the difference between SO_4^{2-} and Cl^- percentages gradually smaller. For cations, Na^+ content continues to increase, while Ca^{2+} and Mg^{2+} remain relatively stable, leading to an increase in the percentage of Na^+ in meq/L and changing the hydrochemical type to SO_4·Cl-Na type (Fig. 2.11C). The anion variation pattern in path C aligns with Chebotarev's findings (Chebotarev 1955), indicating that groundwater in the Zhongning alluvial plain generally follows the natural regional hydrochemical evolution pattern. In the upstream, HCO_3^- dominates among anions. As groundwater flows, water–rock interactions and evaporation-concentration increase SO_4^{2-} and Cl^- while decreasing HCO_3^-, resulting in high salinity water with SO_4 or SO_4·Cl types.

The above studies demonstrate that hydrochemical changes along groundwater flow paths vary in different sections of the study area due to different influencing factors. The long history of irrigation and increasing human activities in the Weining Plain have significantly altered groundwater chemistry, deviating from the natural evolution patterns. This further illustrates that under human interference, the natural attributes of shallow groundwater formation, renewal, and evolution are gradually weakened, which are gradually replaced by the influence of human activities.

2.3 Formation Mechanism of Groundwater Chemistry in the Weining Plain

2.3.1 Groundwater Hydrochemical Formation Based on Gibbs Diagrams

Gibbs (1970) classified the mechanisms controlling the formation of major chemical components in natural surface water bodies into three categories: evaporation-dominated, precipitation-dominated, and rock weathering-dominated. He also developed two diagrams to assess the sources of major components in surface water bodies: one representing the relationship between Na/(Na + Ca) and TDS, and the other representing the relationship between Cl/(Cl + HCO_3) and TDS. These diagrams have been widely used to study the formation mechanisms of groundwater hydrochemistry (Li et al. 2013a, b; Subba Rao 2002, 2006; Naseem et al. 2010; Marghade et al. 2012). However, Marandi and Shand (2018) argued that these diagrams should be used with caution, because groundwater geochemical evolution mechanisms were more complex than surface water, and the Gibbs diagram cannot fully reflect the true evolution mechanisms of groundwater chemistry. In this study, Gibbs diagrams were used to identify the factors controlling the hydrochemical evolution of groundwater in the study area, because we considered phreatic groundwater which is closely connected to surface water in the study area. The major controlling factors are similar to those of surface water. However, we have to admit that the Gibbs diagrams cannot

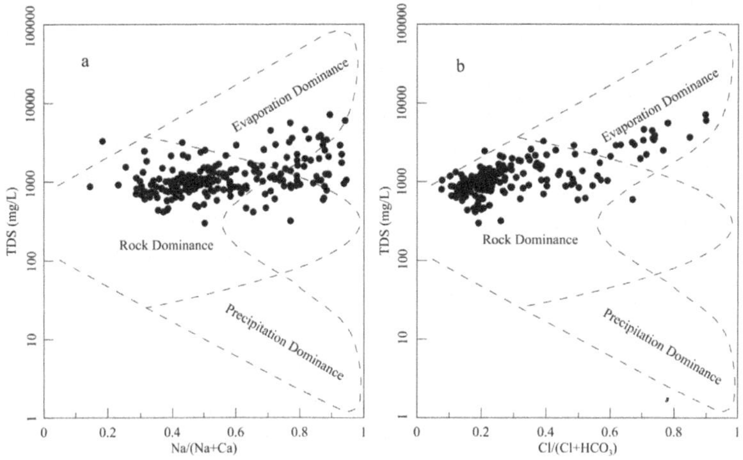

Fig. 2.12 Gibbs diagrams of hydrochemical formation mechanisms

fully reflect the impacts of human activities on groundwater geochemistry, which restrict its wide application in water research (Li et al. 2016a, b). The Gibbs diagrams are shown in Fig. 2.12.

From Fig. 2.12, the majority of water samples fall in the middle section, indicating that the hydrochemical formation of groundwater in the study area is primarily dominated by rock weathering. This process is a long-term and continuous natural process. Additionally, some water samples fall within the evaporation-dominated region, and the overall distribution of samples shows an upward trend, suggesting that evaporation also plays a role in the hydrochemical formation. The high groundwater levels and strong evaporation in the Weining Plain lead to the continuous increase of various components in the water. The Gibbs diagram indicates that rock weathering and evaporation are the primary mechanisms controlling the regional groundwater hydrochemistry.

However, the Gibbs diagram only suggests that the hydrochemical formation in the Weining Plain is influenced by natural processes and does not account for human activities. Human activities can affect groundwater chemistry in two ways: direct and indirect. Direct impacts involve activities that directly alter groundwater chemical compositions, such as wastewater infiltration, which increases Cl^- and SO_4^{2-} concentrations. Indirect impacts involve changes in hydrodynamic conditions that influence water–rock interactions or evaporation, such as irrigation water infiltration and groundwater extraction, which can raise or lower groundwater levels and intensify or weaken evaporation. The increasing influence of human activities makes it difficult to distinguish between natural and anthropogenic contributions using this method, highlighting its limitations.

2.3.2 Correlation Analysis of Ion Sources

The Pearson correlation coefficient matrix is a useful tool in hydrogeochemical studies, as it reveals the relationships between ions or parameters, providing insights into the coexistence of components or the connections between individual indicators (Li et al. 2013b). In this study, SPSS was used to calculate the Pearson correlation coefficient matrix (Table 2.3).

From Table 2.3, it can be seen that Cl^- is significantly positively correlated with Na^+, and SO_4^{2-} is significantly positively correlated with Na^+, indicating that they share common sources or formation processes. Given the high content of halite in the river alluvium of the region, the weathering and dissolution of halite are considered the primary sources of Cl^- and Na^+. In the Nanshantaizi area, the vadose zone contains high levels of mirabilite (Na_2SO_4), which, when leached by irrigation or rainwater, increases the concentrations of SO_4^{2-} and Na^+ in groundwater. These reactions can be expressed as Eqs. 2.1 and 2.2:

$$NaCl \rightarrow Na^+ + Cl^- \tag{2.1}$$

$$Na_2SO_4 \rightarrow 2Na^+ + SO_4^{2-} \tag{2.2}$$

pH is negatively correlated with K^+, Mg^{2+}, Ca^{2+}, and HCO_3^-, and positively correlated with F^-, indicating that under alkaline conditions, the enrichment of K^+, Mg^{2+}, Ca^{2+}, and HCO_3^- is limited, while F^- enrichment is favored. In alkaline conditions, OH^- in water can exchange with F^- in fluorine-containing minerals such as biotite and muscovite (Eqs. 2.3 and 2.4; Guo et al. 2007; Li et al. 2014; Vikas et al. 2013).

$$KMg_3(AlSi_3O_{10})F_{2(Biotite)} + 2OH^- \rightarrow KMg_3(AlSi_3O_{10})(OH)_2 + 2F^- \tag{2.3}$$

$$KAl_2(AlSi_3O_{10})F_{2(Muscovite)} + 2OH^- \rightarrow KAl_2(AlSi_3O_{10})(OH)_2 + 2F^- \tag{2.4}$$

Additionally, F^- is positively correlated with Na^+ and negatively correlated with Ca^{2+}, suggesting that the enrichment of F^- is closely related to the dissolution and precipitation of fluorite (CaF_2). In water with high Ca^{2+} concentrations, Ca^{2+} inhibits the dissolution of fluorite, as shown in the reaction (Eq. 2.5):

$$Ca^{2+} + 2F^- = CaF_2 \tag{2.5}$$

SO_4^{2-} is significantly positively correlated with Ca^{2+}, indicating a common source or formation process of the two ions. Given the presence of gypsum ($CaSO_4{\cdot}2H_2O$) in the region, the weathering and dissolution of gypsum are likely the sources of Ca^{2+} and SO_4^{2-} in groundwater through reactions expressed as Eq. 2.6.

$$CaSO_4 \cdot 2H_2O \rightarrow Ca^{2+} + SO_4^{2-} + 2H_2O \tag{2.6}$$

Table 2.3 Pearson correlation coefficient matrix of water quality indicators

	pH	Na^+	K^+	Mg^{2+}	Ca^{2+}	SO_4^{2-}	Cl^-	HCO_3^-	F^-	TDS	TH	NH_4-N	NO_2-N	NO_3-N
pH	1													
Na^+	0.1	1												
K^+	**–0.22**	*0.14*	1											
Mg^{2+}	**–0.26**	**0.56**	**0.42**	1										
Ca^{2+}	**–0.4**	**0.2**	**0.23**	**0.59**	1									
SO_4^{2-}	–0.09	**0.81**	**0.37**	**0.73**	**0.59**	1								
Cl^-	0.03	**0.94**	**0.19**	**0.63**	**0.29**	**0.8**	1							
HCO_3^-	**–0.45**	**–0.21**	0.09	0.11	*0.16*	**–0.23**	–0.19	1						
F^-	**0.17**	**0.23**	**0.32**	0.07	**–0.24**	**0.21**	**0.18**	**–0.26**	1					
TDS	–0.09	**0.86**	**0.28**	**0.78**	**0.53**	**0.92**	**0.88**	–0.11	*0.15*	1				
TH	**–0.37**	**0.42**	**0.37**	**0.86**	**0.9**	**0.74**	**0.51**	*0.14*	–0.09	**0.73**	1			
NH_4-N	–0.03	0.08	0.09	0.11	0.02	0.1	0.1	0	–0.01	0.12	0.07	1		
NO_2-N	–0.01	–0.08	0.12	0.06	0.04	–0.03	–0.05	0.04	–0.02	–0.06	0.05	0.01	1	
NO_3-N	–0.12	**0.36**	*0.14*	**0.37**	**0.29**	**0.34**	**0.36**	–0.04	0.02	**0.43**	**0.37**	–0.05	0.05	1

Note Bold indicates significance at the 0.01 level, while italic indicates significance at the 0.05 level

Nitrate in groundwater is also correlated with Na^+, Mg^{2+}, Ca^{2+}, SO_4^{2-}, and Cl^-, indicating that these ions are influenced to some extent by human activities. HCO_3^- is weakly correlated with Mg^{2+} and Ca^{2+}, suggesting that the enrichment of HCO_3^-, Mg^{2+}, and Ca^{2+} is influenced by the dissolution and precipitation of minerals such as calcite and dolomite (Eqs. 2.7 and 2.8):

$$CaCO_3 + CO_2 + H_2O \rightarrow Ca^{2+} + 2HCO_3^- \quad (2.7)$$

$$CaMg\,(CO_3)_2 + 2CO_2 + 2H_2O \rightarrow Mg^{2+} + Ca^{2+} + 4HCO_3^- \quad (2.8)$$

Scatter plots of ionic milliequivalent ratios can be used to identify hydrogeochemical processes and formation mechanisms in the aquifer, as shown in Fig. 2.13. The scatter plot of Na^+ and Cl^- forms a nearly straight line, with most points above the line 1:1, indicating that Na^+ content exceeds Cl^- content. This suggests that the dissolution of halite is not the only source of Na^+ and Cl^- in groundwater, and the excess Na^+ may come from the weathering of sodium silicate minerals, such as albite and sodium montmorillonite (Eqs. 2.9 and 2.10). Additionally, cation exchange in the aquifer may also affect Na^+ concentrations.

$$\begin{aligned} 4NaAlSi_3O_8 + 4CO_2 + 22H_2O \rightarrow\ & Al_4(Si_4O_{10})(OH)_8 \\ & + 8H_4SiO_4 + 4Na^+ + 4HCO_3^- \end{aligned} \quad (2.9)$$

$$\begin{aligned} 3Na_{1/3}Al_{7/3}Si_{11/3}O_{10} + 30H_2O + 6OH^- \rightarrow\ & Na^+ + 7Al(OH)_4^- \\ & + 11H_4SiO_4 \end{aligned} \quad (2.10)$$

Ca^{2+}, Mg^{2+}, and HCO_3^- in groundwater generally originate from the dissolution of carbonates, such as calcite and dolomite. In an open system, CO_2 participates in the reactions, resulting in a molar ratio of Ca^{2+} to HCO_3^- between 1:1 and 1:2. In Fig. 2.13b, some water samples fall between the lines 1:1 and 1:2, indicating that calcite dissolution is a source of Ca^{2+} and HCO_3^- in groundwater. However, most samples fall below the line 1:2, suggesting that Ca^{2+} may be reduced due to cation exchange adsorption, while Na^+ increases, further supporting the possibility of cation exchange inferred from Fig. 2.13a. Dolomite dissolution also varies with the amount of CO_2 involved, resulting in a $Ca + Mg/HCO_3^-$ ratio between lines 1:1 and 1:2. From Fig. 2.13c, most samples fall between the lines 1:1 and 1:2, indicating that dolomite dissolution is also a source of Ca^{2+}, Mg^{2+}, and HCO_3^- in groundwater. Some samples fall above the line 1:1, suggesting additional sources of Ca^{2+} and Mg^{2+}, such as gypsum dissolution. The scatter plot of Ca^{2+} and SO_4^{2-} shows a linear trend (Fig. 2.13d), indicating that gypsum dissolution is a source of Ca^{2+} and SO_4^{2-} in groundwater. However, the points fall below the line 1:1 (Fig. 2.13d), suggesting SO_4^{2-} enrichment or Ca^{2+} depletion, possibly due to cation exchange adsorption.

If the dissolution of calcite, dolomite, and gypsum is the primary source of related ions in groundwater, the ratio of $Ca^{2+} + Mg^{2+}$ versus $SO_4^{2-} + HCO_3^-$ should be close

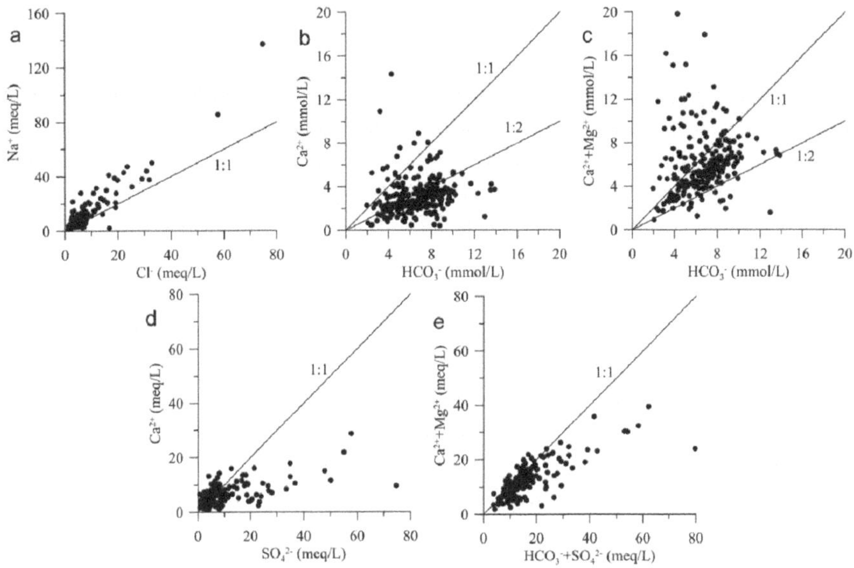

Fig. 2.13 Milliequivalent ratio relationships of ions

to 1:1. From Fig. 2.13e, $Ca^{2+} + Mg^{2+}$ versus $SO_4{}^{2-} + HCO_3{}^{-}$ generally follow the line 1:1, but some samples deviate, indicating that Ca^{2+} and Mg^{2+} in groundwater mainly originate from the dissolution of calcite, dolomite, and gypsum, but may decrease due to cation exchange.

2.3.3 Multivariate Statistical Analysis of Groundwater Quality

Multivariate statistical analysis is a method for handling multi-indicator problems using mathematical statistics, and has been widely applied in geology, hydrology, meteorology, medicine, groundwater quality assessment, hydrochemical analysis, and pollution source identification (Wu et al. 2014). Commonly used methods include principal component analysis (PCA), cluster analysis (CA), factor analysis (FA), and discriminant analysis (DA). In this study, CA and PCA were used to investigate groundwater quality and hydrochemical composition. Cluster analysis, particularly hierarchical cluster analysis, is effective for classifying variables or samples based on numerical characteristics by selecting different distance measures and clustering methods (Tang 2010). The process involves three steps: data transformation, distance coefficient calculation, and hierarchical clustering. Wishart (1969) provided a unified formula for calculating the clustering distances:

$$d_{kr}^2 = \alpha_p d_{kp}^2 + \alpha_q d_{kq}^2 + \beta d_{pq}^2 + \gamma \left| d_{kp}^2 - d_{kq}^2 \right| \quad (2.11)$$

where, class r is formed by merging subclasses p and q, d represents the distance between different classes, and α_p, α_q, β, and γ are relevant coefficients.

Cluster analysis groups water quality indicators into a limited number of clusters based on similarity, helping to identify possible sources or influencing factors. In this study, Z-scores were used to transform the original data (standard normal transformation), and the chi-square distance was used to calculate distance coefficients due to its stronger discriminative power compared to Euclidean distance (Tang 2010). Ward's clustering method was then applied to get the final clusters. All calculations were performed using SPSS 20, and the cluster diagram is shown in Fig. 2.14.

From Fig. 2.14, the 12 selected indicators were grouped into three clusters. Cluster 1 (Group 1) includes pH, F^-, HCO_3^-, Mg^{2+}, TH, and Ca^2 +. This cluster can be further divided into Subgroup 1 (pH, F^-, HCO_3^-) and Subgroup 2 (Mg^{2+}, TH Ca^{2+}). Subgroup 1 represents the dissolution and precipitation of fluorite (CaF_2), as alkaline conditions favor CaF_2 precipitation, and high HCO_3^- indicates the dissolution of carbonates such as calcite and dolomite. The dissolution of carbonates also affects F^- enrichment. Subgroup 2 represents the dissolution and precipitation of carbonate minerals. Cluster 2 (Group 2) includes Na^+ and Cl^-, representing the dissolution of halite. Cluster 3 (Group 3) includes SO_4^{2-}, NH_4^+, TDS, and NO_3^-. Generally, NH_4^+ and NO_3^- originate mainly from human activities such as agriculture. SO_4^{2-} and TDS are primarily influenced by natural processes, such as mineral dissolution

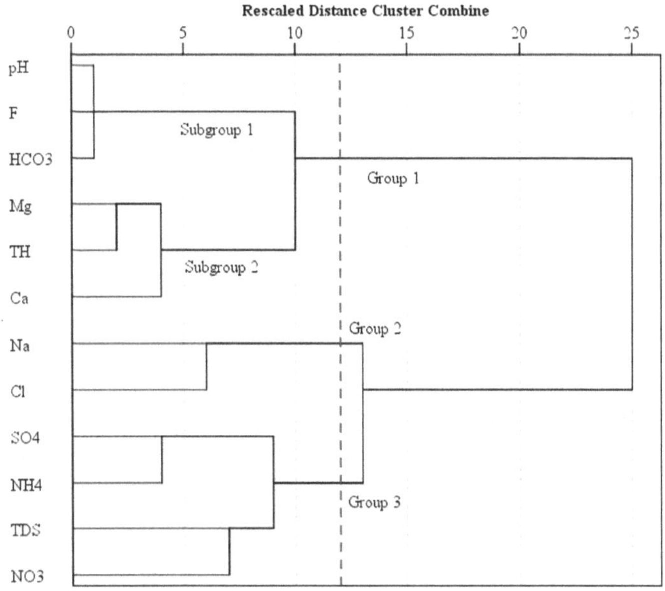

Fig. 2.14 Cluster diagram of major chemical indicators in groundwater

and evaporation, but are also affected by human activities, such as leachate from landfills. Therefore, Cluster 3 represents the combined influence of natural processes and human activities.

PCA is a statistical method that reduces multiple indicators into a few comprehensive indicators, simplifying complex datasets to better analyze their internal structure (Wu et al. 2014). Suppose N samples each have p indicators (1, 2,…,x_1, x_2,…, x_p), PCA transforms them into p composite variables:

$$\begin{cases} y_1 = c_{11}x_1 + c_{12}x_2 + \cdots + c_{1p}x_p \\ y_2 = c_{21}x_1 + c_{22}x_2 + \cdots + c_{2p}x_p \\ \quad\quad \cdots\cdots \\ y_p = c_{p1}x_1 + c_{p2}x_2 + \cdots + c_{pp}x_p \end{cases} \tag{2.12}$$

where, $c^2{}_{k1} + c^2{}_{k2} + \ldots + c^2{}_{kp} = 1$ ($k = 1, 2, \ldots, p$). The composite factors y_1, y_2,…,y_p are called the first, second, …, and pth principal components. In this study, SPSS was used to perform PCA on water quality data. Principal components with eigenvalues greater than 1 were extracted, resulting in three principal components that account for 72.08% of the total variance (Table 2.4).

As shown in Table 2.4, the first principal component accounts for 45.11% of the total variance and is highly positively correlated with TDS, $SO_4{}^{2-}$, Mg^{2+}, TH, Cl^-, $Na^+ + K^+$, Ca^{2+}, and $NO_3{}^-$. The second principal component accounts for 18.34% of the variance and is highly correlated with pH, $HCO_3{}^-$, and F^-. The third principal component accounts for 8.63% of the variance and is positively correlated with $NH_4{}^+$.

Table 2.4 Principal component loading matrix and scores

Parameters	Components		
	1	2	3
TDS	**0.96**	0.14	0.03
$SO_4{}^{2-}$	**0.93**	0.16	0.03
Mg^{2+}	**0.86**	–0.19	0.05
TH	**0.85**	–0.41	–0.03
Cl^-	**0.85**	0.35	0.04
Na + K	**0.82**	0.43	0.02
Ca^{2+}	**0.67**	–0.55	–0.08
NO_3-N	**0.50**	–0.03	–0.37
pH	–0.20	**0.72**	–0.05
$HCO_3{}^-$	–0.06	**–0.70**	0.09
F^-	0.11	**0.57**	–0.01
NH_4-N	0.11	0.01	**0.94**
Eigenvalue	5.41	2.20	1.04
% of variance	45.11	18.34	8.63
Cumulative %	45.11	63.45	72.08

Figure 2.15 shows the relationship between the principal component scores and the major indicators. From Fig. 2.15, the major indicators (TDS, SO_4^{2-}, Mg^{2+}, TH, Cl^-, $Na^+ + K^+$, Ca^{2+}, NO_3^-) are most closely related to the first principal component. Mg^{2+}, Ca^{2+}, and TH (Subgroup 2), Cl^- and $Na^+ + K^+$ (Group 2), are primarily associated with natural processes such as mineral dissolution and evaporation. For SO_4^{2-}, TDS, and NO_3^- (Group 3), both human activities and natural processes have significant impacts. Therefore, the first principal component represents the combined influence of natural processes and human activities on groundwater quality, while the second principal component represents the natural evolution of groundwater, such as the dissolution or precipitation of CaF_2. NH_4^+ is closely related to human activities, so the third principal component represents the impact of human activities on groundwater quality.

Based on the principal component scores of each water sample, the influencing factors were classified, as shown in Fig. 2.16. From Fig. 2.16a, water samples in the fourth quadrant are primarily influenced by the first principal component, while those in the second quadrant are influenced by the second principal component. Samples in the first quadrant are influenced by both the first and second principal components, and those in the third quadrant are not influenced by the first or second principal components. Similarly, from Fig. 2.16b, most water samples fall in the first, second, and fourth quadrants, indicating that hydrochemical composition is primarily influenced by the second and third principal components, the third principal component, and the second principal component, respectively. Figure 2.16c

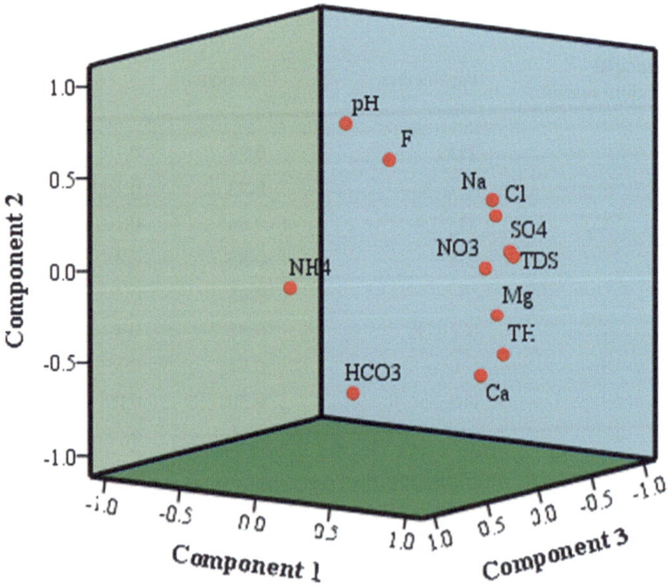

Fig. 2.15 Principal component relationship diagram

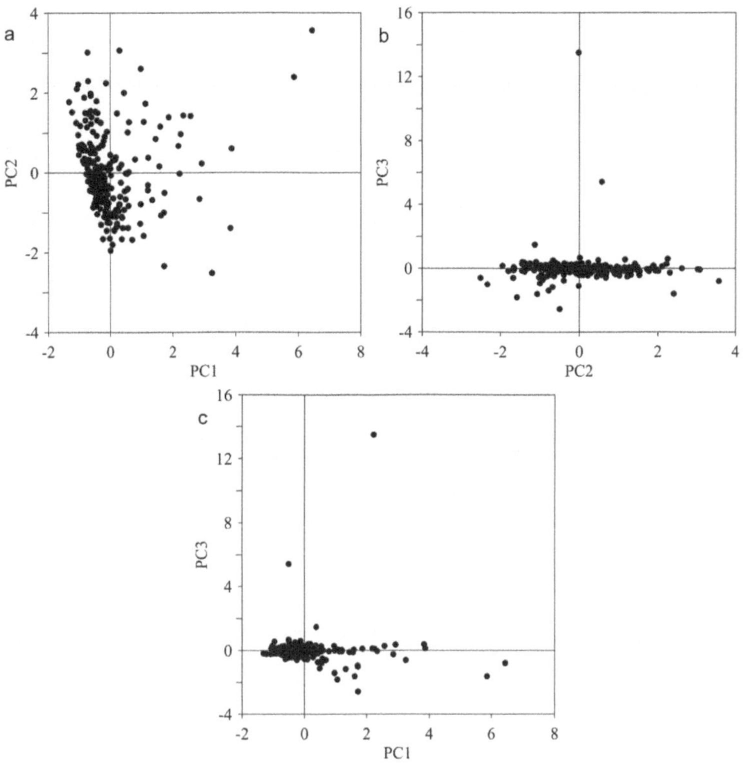

Fig. 2.16 Relationship of principal component score of water samples

shows that some water samples are influenced by a combination of the first and third principal components, while others are influenced by all three principal components.

Based on above analysis, the mechanisms controlling groundwater hydrochemical composition in the study area can be broadly divided into three categories: (1) pure natural factors, such as mineral dissolution, evaporation, and cation exchange, (2) pure human activities, and (3) a combination of both. The term "pure" is used here as a relative concept, indicating that one factor has a much stronger influence than the other. It is not a absolute concept. Statistics show that 23.2% of water samples fall into the first category, 23.2% into the second category, and 53.6% into the third category. This indicates that while groundwater in the Weining Plain is controlled by natural processes, it is also significantly influenced by human activities.

Figure 2.17 shows the distribution of water samples under different influencing factors based on PCA results. Water samples influenced by natural factors are mainly distributed along the foot of the mountain and the northernmost part of the fast-growing forest base. These areas have minimal human activity due to their geomorphology. The second category is mainly distributed around Zhongwei City and

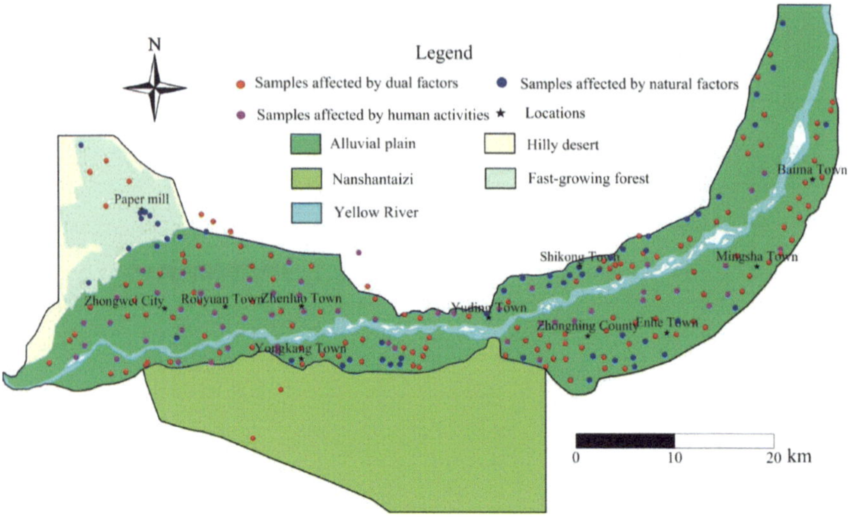

Fig. 2.17 Distribution of water samples under different influencing factors

Zhongning County in the Yellow River alluvial plain. These areas have dense populations and intense human activities, which have become the dominant factor influencing groundwater quality. The third category is mainly distributed in the central part of the fast-growing forest and the vast areas of the Yellow River alluvial plain. In these areas, groundwater is influenced by both natural factors and strong human activities.

2.4 Factors Affecting the Groundwater Chemistry in the Weining Plain

Groundwater in the Weining Plain has evolved greatly under the influence of multiple factors, including geological and hydrogeological conditions, recharge water composition, leaching, rock weathering (including water–rock interactions), cation exchange, evaporation, and human activities. Inthis section, each factor is discussed in detail.

2.4.1 Impacts of Geological and Hydrogeological Conditions

Geological and hydrogeological conditions play a decisive role in shaping groundwater chemistry. Factors such as topography, lithology, groundwater circulation conditions, and groundwater level depth significantly influence the chemical composition of groundwater (Wu et al. 2008). In the Nanshantaizi area, the unconfined aquifer located in the foot of the mountains has higher porosity and permeability, leading to faster groundwater flow and stronger water circulation. As a result, Ca^{2+} and Mg^{2+} concentrations are higher, while Na^+, SO_4^{2-}, and TDS are lower compared to areas with a dual aquifer structure. In the Yellow River alluvial plain, groundwater chemistry varies spatially. In the upstream region of the alluvial plain, coarser sediments and faster water circulation, along with close hydraulic connections to the Yellow River, result in higher HCO_3^- concentrations. In the downstream region, finer sediments and poorer hydrodynamic conditions, combined with lower groundwater extraction and shallower water tables, lead to increased SO_4^{2-} and Cl^- concentrations. Additionally, high TH water is widely distributed in the transition zone between the alluvial plain and mountainous areas, indicating higher Ca^{2+} and Mg^{2+} concentrations in mountainous groundwater, which is closely related to the abundance of carbonate rocks in these areas.

2.4.2 *Influence of Recharge Water Composition*

Groundwater in the Weining Plain has multiple recharge sources, with the primary sources being canal leakage, irrigation infiltration, lateral flow, and precipitation infiltration. Table 2.5 shows the hydrochemical composition of rainwater, Yellow River water, and spring water in the Weining Plain. Rainwater is dominated by HCO_3^- among anions and Mg^{2+} among cations, with a hydrochemical type of HCO_3-Mg·Ca. Yellow River water is dominated by Ca^{2+} among cations and HCO_3^- among anions, with a hydrochemical type of $HCO_3 \cdot SO_4$-Ca. As the primary recharge source, the low TDS Yellow River water primarily dilutes and freshens groundwater. In addition to canal leakage and irrigation infiltration, groundwater in the alluvial plain also receives lateral recharge from the Nanshantaizi area. Groundwater from Nanshantaizi has higher ion concentrations and TDS (see spring water in Table 2.5), leading to higher salinity and ion concentrations in the transition zone between the Nanshantaizi and the alluvial plain. Additionally, high TDS and TH around the alluvial plain are influenced by lateral groundwater inflow from surrounding mountainous areas, where high TDS and TH are common.

2.4.3 *Leaching*

Leaching refers to the process by which soluble substances in rocks and soils are transferred into groundwater through water infiltration (Wang et al. 1995). The intensity of leaching is closely related to the content of soluble salts in the rock/soil layers. To study the relationship between soluble salts and groundwater chemistry, 10 soil

Table 2.5 Hydrochemical composition of rainwater, Yellow River water, and Spring water in the Weining Plain (Units: mg/L, except pH)

Water sample	pH	TDS	Ca^{2+}	Mg^{2+}	K^+	Na^+	Cl^-	SO_4^{2-}	HCO_3^-	Hydrochemical type
Rainwater	7.0	13.8	1.5	1.6	0.2	1.0	0.4	0.9	7.5	HCO_3-Mg·Ca
Yellow river	7.9	326.3	84.0	6.4	1.7	30.9	32.6	73.5	194.5	HCO_3·SO_4-Ca
Yellow river	8.0	358.0	85.2	9.3	2.0	37.0	42.6	84.7	194.5	HCO_3·SO_4-Ca
Yellow river	7.7	334.4	80.2	10.5	1.6	29.6	32.0	78.4	204.3	HCO_3·SO_4-Ca
Yellow river	7.7	324.1	64.0	16.8	1.8	34.1	39.8	72.5	190.2	HCO_3·SO_4-Ca
Mean of yellow river	7.8	335.7	78.3	10.8	1.8	32.9	36.7	77.3	195.9	HCO_3·SO_4-Ca
Spring water	8.1	3716.9	254.9	178.5	11.7	874.4	1059.1	1204.4	267.7	Cl·SO_4-Na

samples were collected from the Nanshantaizi area, and their major ions, pH, and total salt content were analyzed in the laboratory (Table 2.6).

Table 2.6 shows that the Nanshantaizi area has high soil salinity, with a maximum total salt content of 16,276.27 mg/kg. The main chemical components in the soil are Na^+, K^+, Ca^{2+}, Cl^-, SO_4^{2-}, and HCO_3^-. Na^+ is the dominant cation, accounting for about 20% of the total salt content, while SO_4^{2-}, HCO_3^-, and Cl^- are the dominant

Table 2.6 Surface soil salinity in the Nanshantaizi area (Units: mg/kg, except pH)

Sample ID	K^+	Na^+	Ca^{2+}	Mg^{2+}	Cl^-	SO_4^{2-}	HCO3$^-$	CO_3^{2-}	Total Salt	pH
NS2-1	148.00	7.38	355.20	11.64	69.04	101.56	1068.45	61.70	1822.97	9.02
NS2-2	108.40	14.50	144.00	34.92	51.78	184.09	345.68	61.70	945.07	9.01
NS2-3	148.40	65.54	211.20	11.64	34.52	92.71	848.48	61.70	1474.19	9.07
NS2-4	148.80	18.76	211.20	17.46	69.04	41.99	691.35	92.55	1291.15	9.06
NS2-5	146.50	11.22	144.00	17.46	17.26	45.76	534.23	92.55	1008.98	9.06
NS1-1	35.00	1035.52	566.40	145.50	345.20	3595.52	219.98	30.85	5973.97	8.55
NS1-2	35.00	3351.10	316.80	273.54	3969.80	3496.55	251.40	30.85	11,725.04	8.50
NS1-3	32.00	4537.08	441.60	203.70	1950.38	8766.41	314.25	30.85	16,276.27	8.53
NS1-4	35.00	3122.69	240.00	46.56	2036.68	4492.25	377.10	30.85	10,381.13	8.47
NS1-5	42.00	704.78	182.40	69.84	552.32	1201.56	408.53	30.85	3192.28	8.91
Min	32.00	7.38	144.00	11.64	17.26	41.99	219.98	30.85	945.07	8.47
Max	148.80	4537.08	566.40	273.54	3969.80	8766.41	1068.45	92.55	16,276.27	9.07
Mean	87.91	1286.86	281.28	83.23	909.60	2201.84	505.95	52.45	5409.11	8.82

anions, accounting for around 40%, 10%, and 16% of the total salt, respectively. This indicates that the soil contains high levels of Na_2SO_4, NaCl, and $NaHCO_3$. Additionally, the soil pH ranges from 8.47 to 9.07, indicating possible alkalization in the Nanshantaizi area. As a well-known irrigation area by lifting water from the Yellow River, the infiltration of large amounts of Yellow River water continuously dissolves and leaches these soluble salts into the groundwater, altering groundwater quality and chemical composition (Li 2014).

2.4.4 Rock Weathering

Rock weathering, including physical, chemical, and biological weathering, forms the material basis for groundwater hydrochemistry. Water–rock interactions are a common form of chemical weathering that occurs when water interacts with rocks. The impact of water–rock interactions on groundwater chemistry is primarily reflected in the dissolution and precipitation of minerals. Whether a mineral dissolves or precipitates in groundwater depends on its saturation state relative to the water solution. Based on calculations of the equilibrium distribution of water soluble components, the saturation indices (SIs) of groundwater relative to halite, calcite, dolomite, gypsum, and fluorite were calculated using Eq. 2.13.

$$SI = \lg \frac{IAP}{K} \tag{2.13}$$

where, SI represents the saturation index of a mineral, IAP is the ion activity product in the water solution, and K is the equilibrium constant for the mineral dissolution reaction. If $SI > 0$, the water solution is supersaturated relative to the mineral, leading to precipitation. If $SI < 0$, the water solution is undersaturated, leading to dissolution. If $SI = 0$, the water solution is in equilibrium state with the mineral. The relationship between mineral saturation indices and TDS is shown in Fig. 2.18.

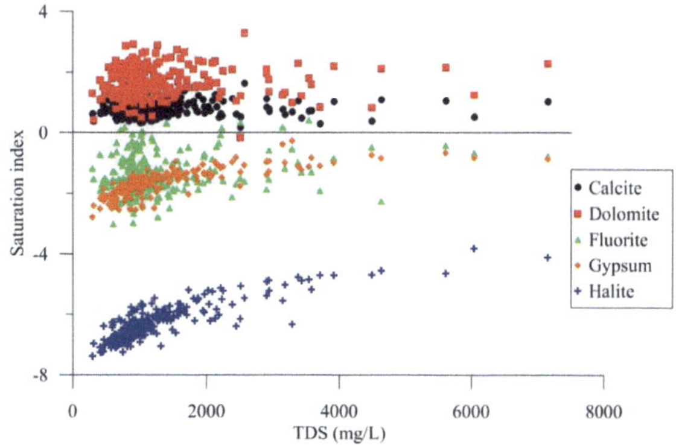

Fig. 2.18 Relationship between mineral saturation indices and TDS

Figure 2.18 shows that in the Weining Plain, except for one sample with a dolomite saturation index below zero (–0.16), all other samples are supersaturated relative to calcite and dolomite, with saturation indices ranging from 0.17 to 1.63 and 0.41 to 3.29, respectively. This indicates that calcite and dolomite are saturated and precipitation may occur in the Weining Plain. For gypsum and halite, all samples are undersaturated, with saturation indices ranging from –2.77 to –0.26 and –7.38 to –3.81, respectively. As TDS increases, the saturation indices for gypsum and halite also increase, indicating ongoing dissolution. For fluorite, most groundwater samples are undersaturated, with only a few samples being saturated, suggesting that fluorite dissolution occurs in most areas, with precipitation occurring only locally.

In contrast to calcite and dolomite, gypsum and halite undergo dissolution throughout the study area. This is also evident from the scatter plots of related ion concentrations (Fig. 2.13), where Ca^{2+} and $SO_4{}^{2-}$, as well as Na^+ and Cl^-, show positive correlations. However, the relationships do not follow theoretical proportions, indicating that other factors, such as cation exchange, influence Ca^{2+} and Na^+ concentrations. The relationships of Ca^{2+} and Mg^{2+} with $HCO_3{}^-$ are more complex and cannot be explained solely by calcite and dolomite dissolution. For example, gypsum dissolution may lead to calcite and dolomite precipitation, and CO_2 dissolution or release may affect $HCO_3{}^-$ concentrations. These dissolution and precipitation processes cause the changes in groundwater chemistry along the flow paths.

2.4.5 *Cation Exchange*

Cation exchange is a common chemical reaction in groundwater systems, particularly in freshwater-saltwater interfaces. Cation exchange can be studied using the chloro-alkaline indices (CAI-I and CAI-II) proposed by Schoeller, as shown in Eqs. 2.14 and 2.15 (Marghade et al. 2012; Li et al. 2013a). Negative values of CAI-I and CAI-II indicate cation exchange as shown in Eq. 2.16, while positive values indicate reverse cation exchange as shown in Eq. 2.17. The results of CAI-I and CAI-II calculations are plotted in Fig. 2.19a.

$$\mathrm{CAI-I} = \frac{\mathrm{Cl^-} - (\mathrm{Na^+} + \mathrm{K^+})}{\mathrm{Cl^-}} \tag{2.14}$$

$$\mathrm{CAI-II} = \frac{\mathrm{Cl^-} - (\mathrm{Na^+} + \mathrm{K^+})}{\mathrm{HCO_3^-} + \mathrm{SO_4^{2-}} + \mathrm{CO_3^{2-}} + \mathrm{NO_3^-}} \tag{2.15}$$

$$\mathrm{Na_2X} + \mathrm{Ca^{2+}} \rightarrow 2\mathrm{Na^+} + \mathrm{CaX} \tag{2.16}$$

$$2\mathrm{Na^+} + \mathrm{CaX} \rightarrow \mathrm{Ca^{2+}} + \mathrm{Na_2X} \tag{2.17}$$

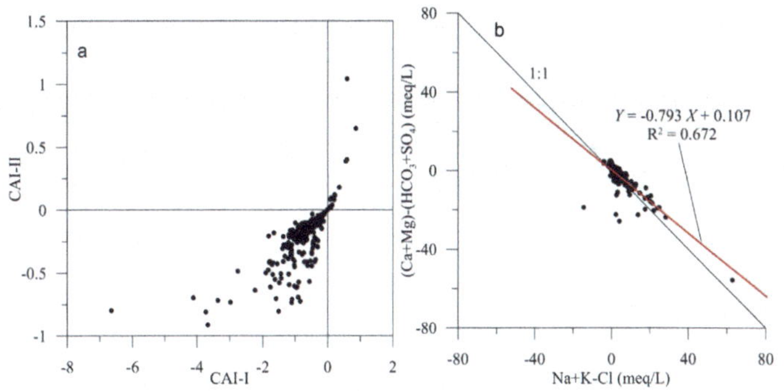

Fig. 2.19 Distribution of CAI-I and CAI-II and ion relationship diagram

The relationship between [(Na + K)-Cl] and [(Ca + Mg) − (HCO_3 + SO_4)] is often used to study the possibility of cation exchange.[(Na + K) − Cl] represents the increase or decrease in Na + due to factors other than halite dissolution, while [(Ca + Mg) − (HCO_3 + SO_4)] represents the increase or decrease in Ca^{2+} and Mg^{2+} due to factors other than gypsum, calcite, and dolomite dissolution (Farid et al. 2013). If cation exchange is the primary process affecting groundwater mineralization, the relationship should be linear with a slope close to –1. The relationship for all water samples is shown in Fig. 2.19b.

Most water samples in the study area fall in the lower-left corner of Fig. 2.19a, indicating that cation exchange as shown in Eq. 2.16 occurs in most areas, increasing Na^+ and decreasing Ca^{2+} concentrations. Figure 2.19b shows that the relationship between [(Na + K) − Cl] and [(Ca + Mg) − (HCO_3 + SO_4)] follows the statistical relationship (Eq. 2.18).

$$Y = -0.793X + 0.107 \text{ with } R^2 = 0.672 \tag{2.18}$$

The slope of –0.793 is close to the theoretical value of -1, indicating that Na^+, Ca^{2+}, and Mg^{2+} indeed participate in cation exchange. However, the deviation from the theoretical value suggests that cation exchange is not the only process influencing Na^+, Ca^{2+}, and Mg^{2+} concentrations.

2.4.6 *Evaporation*

Evaporation can cause solution concentration which plays a significant role in shaping groundwater chemistry in the Weining Plain. Located in an arid to semi-arid region, the area experiences low rainfall and intense evaporation. Evaporation and concentration directly affects the chemical composition and salinity of surface water and shallow groundwater. Groundwater evaporation is a major discharge mechanism in the Weining Plain, accounting for 28.03% of total groundwater discharge, primarily in areas with shallow water tables in the alluvial plain. During evaporation, water is evaporated to the atmosphere, while salts remain in the residual water and soil. This process leads to salt accumulation, deteriorating water quality and causing soil salinization. Therefore, shallow water tables and intense evaporation are major factors contributing to increased TDS and poor water quality in the alluvial plain.

2.4.7 Human Activities

Rapid economic development in the Weining Plain has intensified human activities, posing serious impacts on groundwater. Human activities can influence groundwater chemistry in several ways.

Fast-Growing Forest Area: In the fast-growing forest area, some regions are irrigated with Yellow River water, while others use a mixture of Yellow River water and paper mill wastewater. Irrigation with Yellow River water does not strongly affect groundwater chemistry directly, because the Yellow River water is basically fresh water. However, irrigation with paper mill wastewater introduces high concentrations of pollutants, such as Cl^- (up to 600 mg/L) and Na^+ (up to 700 mg/L), into the groundwater. Additionally, irrigation raises groundwater levels, altering hydrodynamic conditions. Elevated water levels accelerate groundwater flow from the forest area to the alluvial plain, facilitating the transport of pollutants. The rise of groundwater level also increases evaporation, enhancing evaporation and potentially causing soil salinization.

Agricultural Activities in the Alluvial Plain: Intensive and extensive Yellow River irrigation can raise groundwater levels and increase the risk of soil salinization. Fertilizer application is a major source of nitrate and phosphate in groundwater. In rural areas, inadequate waste management systems lead to the indiscriminate disposal of household waste and livestock manure, which contain pollutants such as Na^+, Cl^-, and NO_3^-. These pollutants can be transported into aquifers with rainwater and irrigation water, increasing groundwater contamination.

Nanshantaizi Area: Agricultural irrigation in the Nanshantaizi area raises groundwater levels, increasing lateral flow to the alluvial plain and enhancing evaporation. This also leads to soil salinization in the alluvial plain. Additionally, soluble salts in the vadose zone of the Nanshantaizi are leached into the groundwater by irrigation water, increasing ion concentrations in the groundwater recharging the alluvial plain, altering its hydrochemical characteristics.

The above studies demonstrate that the factors influencing groundwater hydrochemical composition are highly complex. The chemical reactions occurring in groundwater are interconnected and mutually influential, continuously altering the chemical composition of groundwater. When simulating and predicting water quality, it is essential to fully take into account these chemical reactions. However, most existing mathematical models primarily focus on the decay of radioactive elements and rarely consider these common reactions. In recent years, reactive transport modelling has gained significant attention from researchers. It is believed that advancements in reactive transport modeling will bring new insights and vitality to hydrogeochemical research and groundwater pollution prevention.

2.5 Isotopic Evidence of Human Impacts on the Groundwater Environment

Isotopes have been widely applied in groundwater studies. It can be used to identify sources of groundwater pollution and analyze the extent of human impact on groundwater environments. The Weining Plain, a typical Yellow River irrigation area. On one hand, the extensive use of Yellow River water for irrigation has raised groundwater levels, enhanced reactions between groundwater and aquifer media, and induced issues such as secondary soil salinization. On the other hand, the application of fertilizers and pesticides in agriculture has directly led to the deterioration of groundwater quality. This section explores the use of isotopes to identify the impact of human activities on the groundwater environment.

2.5.1 Isotope Sampling and Analysis

To identify the impact of human activities on the groundwater environment in the Weining Plain, systematic isotope sampling and analysis were conducted in August 2013. Samples were collected from shallow private wells, groundwater monitoring wells, the Yellow River, and drainage ditches. A total of 67 groundwater samples, 2 Yellow River water samples, and 6 drainage ditch water samples were collected. The sampling locations of isotope samples are shown in Fig. 2.20. During sampling, coordinates were recorded using a portable GPS. Sampling bottles were 500 mL disposable groundwater sampling bottles. After sampling, addictive was added promptly, the bottles were sealed, labeled, and records were maintained. On-site measurements of dissolved oxygen, TDS, temperature, and pH were conducted. Isotope testing was performed by the State Key Laboratory of Environmental Geochemistry at the Institute of Geochemistry, Chinese Academy of Sciences. The isotope analysis included $\delta^{18}O$ and $\delta^{2}H$. $\delta^{18}O$ was prepared using the CO_2 equilibrium method, and $\delta^{2}H$ was prepared using the zinc reduction method, followed by measurement using a Finnigan MAT isotope mass spectrometer (MAT 252). The results were expressed in δ notation, with precisions of $\pm$ 0.9‰ for $\delta^{18}O$ and $\pm$ 0.2‰ for $\delta^{2}H$.

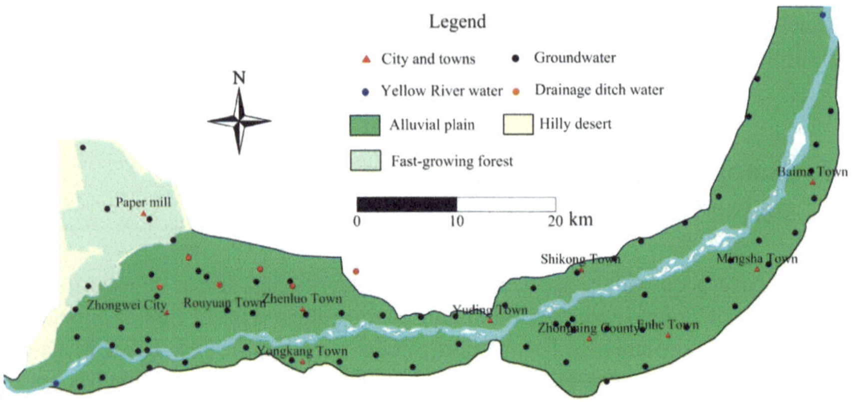

Fig. 2.20 Sampling locations of water isotopes in the Weining Plain

2.5.2 Hydrogen and Oxygen Stable Isotopes and Their Indication of Human Activities

The statistical results of isotopes in different water bodies are shown in Table 2.7. In groundwater, δ^2H ranges from –80.24‰ to –53.09‰, with an average of –67.96‰, and δ^{18}O ranges from –11.13‰ to –5.79‰, with an average of –9.28‰. In Yellow River water, δ^2H ranges from –69.61‰ to –60.98‰, with an average of –65.12‰, and δ^{18}O ranges from –9.60‰ to –8.03‰, with an average of –8.99‰. In drainage ditch water, δ^2H ranges from –66.70‰ to –58.10‰, with an average of –64.55‰, and δ^{18}O ranges from –9.52‰ to –7.83‰, with an average of –8.86‰. Comparison of the isotope contents in different water bodies reveals that the hydrogen and oxygen stable isotope contents are similar in these water bodies, indicating that groundwater, Yellow River water, and drainage ditch water share similar sources or have undergone similar processes. This is consistent with the fact that groundwater in the plain is primarily recharged by Yellow River water, while a significant portion of groundwater is discharged through drainage ditches.

The relationship between hydrogen and oxygen isotopes in different water bodies is graphed in Fig. 2.21. The global meteoric water line (GMWL, Eq. 2.19, Craig

Table 2.7 Statistics of isotopes in different water bodies (‰)

Statistics	Groundwater		Yellow River water		Drainage ditch water	
	δD	δ^{18}O	δD	δ^{18}O	δD	δ^{18}O
Max	–53.09	–5.79	–60.98	–8.03	–58.10	–7.83
Min	–80.24	–11.13	–69.61	–9.60	–66.70	–9.52
Mean	–67.96	–9.28	–65.12	–8.99	–64.55	–8.86
SD	4.33	0.85	2.73	0.52	2.95	0.51

Fig. 2.21 Relationship between hydrogen and oxygen isotopes in different water bodies in the Weining Plain

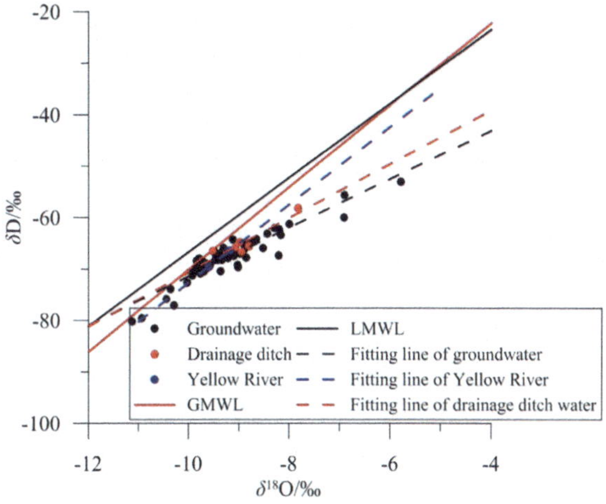

1961) and the local meteoric water line (LMWL) for Yinchuan (Eq. 2.20, Qian et al. 2013), an adjacent alluvial plain with similar climate and hydrogeolgical conditions to the present study area are also plotted in Fig. 2.21.

$$\delta D = 8\delta^{18}O + 10 \tag{2.19}$$

$$\delta D = 7.22\delta^{18}O + 5.5 \tag{2.20}$$

Figure 2.21 shows that the LMWL deviates from the GMWL, indicating secondary evaporation during precipitation. Groundwater, Yellow River water, and drainage ditch water further deviate from the LMWL, suggesting strong evaporation during the flow of Yellow River water. As Yellow River water infiltrates into groundwater, evaporation continues, causing the hydrogen and oxygen isotope composition of groundwater to deviate further to the lower right of the LMWL. The relationships between hydrogen and oxygen stable isotopes in groundwater, Yellow River water, and drainage ditch water can be expressed by Eqs. 2.21–2.23, respectively. Groundwater has the smallest fitting slope of 4.73, Yellow River water has the largest fitting slope of 7.59, and drainage ditch water has an intermediate fitting slope of 5.27. This indicates that drainage ditch water is a mixture of discharged groundwater and irrigation return flow from the Yellow River.

$$\delta D = 4.73\delta^{18}O - 24.13\ r = 0.93 \tag{2.21}$$

$$\delta D = 7.59\delta^{18}O + 3.27\ r = 1.00 \tag{2.22}$$

$$\delta D = 5.27\delta^{18}O - 17.85\ r = 0.92 \tag{2.23}$$

The similar hydrogen and oxygen stable isotope compositions of groundwater and Yellow River water in the Weining Plain indicate that groundwater is recharged by the infiltration of Yellow River water. If this similarity were only observed near the Yellow River, it would be difficult to attribute it to human activities. However, the consistent relationship across the entire plain suggests that agricultural irrigation has significantly altered the local groundwater environment, making the hydrogen and oxygen isotope composition of groundwater closely resemble that of Yellow River water. This provides evidence that human agricultural activities have impacted the groundwater environment in the Weining Plain. In fact, the use of $\delta^{15}N$ and $\delta^{18}O$ in nitrate can further demonstrate the impact of agricultural activities on groundwater nitrate pollution, which has been well reported in literature (He et al. 2022; Li et al. 2024).

2.6 Conclusions

This chapter provides a comprehensive investigation of the hydrogeochemical characteristics of groundwater in the Weining Plain, a region with over 2000 years of irrigation history, using hydrogeochemical and isotopic techniques. The study reveals the formation, distribution, and evolution of groundwater chemistry under the influence of agricultural activities and explores the impacts of human activities on groundwater quality. The main conclusions are as follows:

- The groundwater in the Weining Plain exhibits neutral to slightly alkaline conditions, with pH values ranging from 7.0 to 8.8. The major ion concentrations vary significantly, with Na^+ concentrations ranging from 35.46 to 3153 mg/L and SO_4^{2-} concentrations from 30.6 to 3584 mg/L. The hydrochemical types of groundwater are complex and diverse, including SO_4-Na type, SO_4-Na·Ca type, and HCO_3·SO_4·Cl-Mg·Na type, among others. The hydrochemical characteristics of groundwater show specific variation patterns along different flow paths. For example, in the flow path from the fast-growing forest to the alluvial plain, the Cl^- content decreases while HCO_3^- increases, transitioning from Cl-Ca type to HCO_3·SO_4-Ca type and then to HCO_3·SO_4-Ca·Mg type. In the flow path from Nanshantaizi to the alluvial plain, the groundwater evolves from SO_4·Cl-Na type to SO_4-Na type and then to HCO_3-Ca·Na type.
- The hydrochemical formation of groundwater in the Weining Plain is primarily controlled by rock weathering and evaporation. The Gibbs diagram indicates that most water samples fall in the rock weathering-dominated region, while some samples are in the evaporation-dominated region. Correlation analysis of ions suggests that Cl^- is significantly positively correlated with Na^+, indicating that halite weathering and dissolution are the primary sources of Cl^- and Na^+. The weathering and dissolution of gypsum are considered the main sources of Ca^{2+} and SO_4^{2-} in groundwater. Multivariate statistical analysis reveals that groundwater hydrochemical composition is influenced by three types of mechanisms: pure

natural factors (such as mineral dissolution, evaporation, and cation exchange), pure human activities, and a combination of both. Statistics show that 23.2% of water samples are influenced by natural factors alone, 23.2% by human activities alone, and 53.6% by a combination of both.

- Geological and hydrogeological conditions, recharge water composition, leaching, rock weathering, cation exchange, evaporation, and human activities collectively shape the hydrogeochemical characteristics of groundwater in the Weining Plain. The groundwater in the Nanshantaizi area, located at the foot of the mountains, has higher porosity and permeability in the unconfined aquifer, leading to faster groundwater flow, stronger water circulation, and relatively higher Ca^{2+} and Mg^{2+} concentrations. In the Yellow River alluvial plain, groundwater chemistry varies spatially. In the upstream region of the alluvial plain, coarser sediments, faster water circulation, and close hydraulic connections with the Yellow River result in higher HCO_3^- concentrations. In the downstream region, finer sediments, poorer hydrodynamic conditions, lower groundwater extraction rates, and shallower water tables lead to increased SO_4^{2-} and Cl^- concentrations. Agricultural irrigation activities in the Weining Plain have significantly altered the natural hydrogeochemical characteristics of the region. Irrigation with paper wastewater in the fast-growing forest area introduces high concentrations of pollutants (e.g., Cl^- and Na^+) into groundwater, raises groundwater levels, alters hydrodynamic conditions, and accelerates groundwater flow from the forest area to the alluvial plain, facilitating pollutant transport. Agricultural irrigation in the alluvial plain increases groundwater levels, raising the risk of soil salinization. Fertilizer application is a major source of nitrate and phosphate in groundwater. In rural areas, improper waste management systems lead to the random disposal of household waste and livestock manure, which contain pollutants such as Na^+, Cl^-, and NO_3^-. These pollutants can infiltrate aquifers with rainwater and irrigation water, increasing groundwater contamination.
- The hydrogen and oxygen stable isotope compositions of groundwater in the Weining Plain are similar to those of Yellow River water and drainage ditch water, indicating that groundwater is primarily recharged by the Yellow River water. Agricultural irrigation has significantly altered the local groundwater environment, making the hydrogen and oxygen isotope composition of groundwater closely resemble that of Yellow River water. This provides evidence of human agricultural activities impacting the groundwater environment in the Weining Plain.

References

Chebotarev II (1955) Metamorphism of natural waters in the crust of weathering-1. Geochim Cosmochim Acta 8(1–2):22–48

Craig H (1961) Isotope variations in meteoric waters. Science 133:1702–1703. https://doi.org/10.1126/science.133.3465.1702

Elango L, Kannan R (2005) Mass balance modelling and its concepts: A case study. In: Elango L (ed) Numerical simulation of groundwater flow and solute transport. Allied Publishers Pvt. Ltd., Chennai, pp 227–235

Farid I, Trabelsi R, Zouari K, Abid K, Ayachi M (2013) Hydrogeochemical processes affecting groundwater in an irrigated land in Central Tunisia. Environ Earth Sci 68:1215–1231. https://doi.org/10.1007/s12665-012-1788-7

Gibbs RJ (1970) Mechanisms controlling world water chemistry. Science 17:1088–1090. https://doi.org/10.1126/science.170.3962.1088

Gong M, Li P, Kong J (2025) Estimation of groundwater recharge and assessment of groundwater quality in the Weining Plain, China. Water 17(5):704. https://doi.org/10.3390/w17050704

Guo QH, Wang YX, Ma T, Ma R (2007) Geochemical processes controlling the elevated fluoride concentrations in groundwaters of the Taiyuan Basin, Northern China. J Geochem Explor 93:1–12. https://doi.org/10.1016/j.gexplo.2006.07.001

He S, Li P, Su F, Wang D, Ren X (2022) Identification and apportionment of shallow groundwater nitrate pollution in Weining Plain, northwest China, using hydrochemical indices, nitrate stable isotopes, and the new Bayesian stable isotope mixing model (MixSIAR). Environ Pollut 298:118852. https://doi.org/10.1016/j.envpol.2022.118852

Li P (2014) Research on groundwater environment under human interferences: A case study from Weining Plain, Northwest China. Ph.D. thesis, Chang'an University

Li L, Li P, Tian Y, Kou X, He S (2024) Nitrate sources and transformation in surface water and groundwater in Huazhou District, Shaanxi, China: integrated research using hydrochemistry, isotopes and MixSIAR model. Environ Res 263:120052. https://doi.org/10.1016/j.envres.2024.120052

Li P, He S, He X, Tian R (2018) Seasonal hydrochemical characterization and groundwater quality delineation based on matter element extension analysis in a paper wastewater irrigation area, Northwest China. Expo Health 10(4):241–258. https://doi.org/10.1007/s12403-017-0258-6

Li P, He X, Wu J, Elumalai V (2025) Sustainable groundwater and environment: Challenges and solutions. Springer Cham, Switzerland. https://doi.org/10.1007/978-3-031-82194-3

Li PY, Qian H, Wu JH, Chen J, Zhang YQ, Zhang HB (2014) Occurrence and hydrogeochemistry of fluoride in shallow alluvial aquifer of Weihe River, China. Environ Earth Sci 71(7):3133–3145. https://doi.org/10.1007/s12665-013-2691-6

Li PY, Qian H, Wu JH, Zhang YQ, Zhang HB (2013b) Major ion chemistry of shallow groundwater in the Dongsheng Coalfield, Ordos Basin, China. Mine Water Environ 32(3):195–206. https://doi.org/10.1007/s10230-013-0234-8

Li P, Wu J, Qian H (2013a) Assessment of groundwater quality for irrigation purposes and identification of hydrogeochemical evolution mechanisms in Pengyang County, China. Environ Earth Sci 69:2211–2225. https://doi.org/10.1007/s12665-012-2049-5

Li P, Wu J, Qian H, Zhang Y, Yang N, Jing L, Yu P (2016a) Hydrogeochemical characterization of groundwater in and around a wastewater irrigated forest in the southeastern edge of the Tengger Desert, Northwest China. Expo Health 8:331–348. https://doi.org/10.1007/s12403-016-0193-y

Li P, Zhang Y, Yang N, Jing L, Yu P (2016b) Major ion chemistry and quality assessment of groundwater in and around a mountainous tourist town of China. Expo Health 8(2):239–252. https://doi.org/10.1007/s12403-016-0198-6

Li X, Sun Z, Liu J (2010) Hydrogeochemistry, 3rd edn. Atomic Energy Publishing House, Beijing (in Chinese)

Marandi A, Shand P (2018) Groundwater chemistry and the Gibbs diagram. Appl Geochem 9:209–212. https://doi.org/10.1016/j.apgeochem.2018.07.009

Marghade D, Malpe DB, Zade AB (2012) Major ion chemistry of shallow groundwater of a fast growing city of Central India. Environ Monit Assess 184:2405–2418. https://doi.org/10.1007/s10661-011-2126-3

Naseem S, Rafique T, Bashir E, Bhanger MI, Laghari A, Usmani TH (2010) Lithological influences on occurrence of high-fluoride groundwater in Nagar Parkar area, Thar Desert, Pakistan. Chemosphere 78:1313–1321. https://doi.org/10.1016/j.chemosphere.2010.01.010

Qian H, Li PY (2011) Hydrochemical characteristics of groundwater in Yinchuan plain and their control factors. Asian J Chem 23(7):2927–2938

Qian H, Li P, Wu J, Zhou Y (2013) Isotopic characteristics of precipitation, surface and ground waters in the Yinchuan plain, Northwest China. Environ Earth Sci 70(1):57–70. https://doi.org/10.1007/s12665-012-2103-3

Qian H, Ma Z, Li P (2012) Hydrogeochemistry, 2nd edn. Geological Publishing House, Beijing (in Chinese)

Subba Rao N (2002) Geochemistry of groundwater in parts of Guntur district, Andhra Pradesh, India. Environ Geol 41(5):552–562. https://doi.org/10.1007/s002540100431

Subba Rao N (2006) Seasonal variation of groundwater quality in a part of Guntur district, Andhra Pradesh, India. Environ Geol 49(3):413–429. https://doi.org/10.1007/s00254-005-0089-9

Tang Q (2010) DPS data processing system: Experimental design, statistical analysis, and data mining. Science Press, Beijing (in Chinese)

Vikas C, Kushwaha R, Ahmad W, Prasannakumar V, Reghunath R (2013) Genesis and geochemistry of high fluoride bearing groundwater from a semi-arid terrain of NW India. Environ Earth Sci 68:289–305. https://doi.org/10.1007/s12665-012-1739-3

Wang D, Zhang R, Shi Y et al (1995) Introduction to hydrogeology. Geological Publishing House, Beijing (in Chinese)

Wishart D (1969) An algorithm for hierarchical classifications. Biometrics 25(1):165–170

Wu JH, Li PY, Qian H, Duan Z, Zhang XD (2013) Using correlation and multivariate statistical analysis to identify hydrogeochemical processes affecting the major ion chemistry of waters: a case study in Laoheba phosphorite mine in Sichuan, China. Arab J Geosci 7(10):3973–3982. https://doi.org/10.1007/s12517-013-1057-4

Wu J, Li P, Qian H, Fang Y (2014) Assessment of soil salinization based on a low-cost method and its influencing factors in a semi-arid agricultural area, Northwest China. Environ Earth Sci 71(8):3465–3475. https://doi.org/10.1007/s12665-013-2736-x

Wu X, Qian H, Yu D et al (2008) Investigation and evaluation of rational allocation of groundwater resources in Yinchuan Plain. Geological Publishing House, Beijing (in Chinese)

Xu F, Li (2024) Biogeochemical mechanisms of iron (Fe) and manganese (Mn) in groundwater and soil profiles in the Zhongning section of the Weining Plain (northwest China). Sci Total Environ, 939:173506. https://doi.org/10.1016/j.scitotenv.2024.173506

Chapter 3
Groundwater Quality and Main Groundwater Quality Issues in the Weining Plain

Groundwater environmental investigation and assessment serve as the foundation for groundwater pollution prevention and prediction. The primary objectives of it are to identify the types of groundwater pollution sources, measure the concentrations of pollutants, understand the migration and transformation characteristics of pollutants within the groundwater system, and determine the main factors influencing pollutant concentrations, based on which, groundwater quality assessment can be carried out (Zheng 2009). Groundwater quality assessment is a complex systematic task that must be built upon a foundation of hydrogeological surveys, monitoring, and experimental work. The reliability of the assessment results depends on two key factors: the accuracy of the monitoring data and the scientific rigour of the evaluation methods. Currently, there are over 60 methods available for groundwater quality assessment (Li et al. 2010a, 2011a, 2012), but each of these methods has its own advantages and disadvantages. For instance, the water quality index method provides an intuitive reflection of how much groundwater indicators exceed standard limits and is computationally straightforward. However, its major drawback lies in the subjective nature of water quality classification during comprehensive evaluation, as different individuals or regions may adopt varying classification standards, making cross-comparisons challenging. Methods such as the fuzzy comprehensive evaluation method, set pair analysis method, and grey theory evaluation method consider the fuzziness and uncertainty in indicator grading, but they usually involve more complex computational processes. The TOPSIS (Technique for Order Preference by Similarity to Ideal Solution) method treats groundwater quality standards as hypothetical samples in the evaluation process, then rank the results uniformly based on calculated values to determine quality levels. Due to their clear computational principles and simplicity, these methods are increasingly being applied in groundwater quality assessments.

In this chapter, the TOPSIS method was selected as the fundamental evaluation approach. Building upon this, the information entropy theory was employed

P. Li and V. Elumalai, *Groundwater Quality Under Agricultural Activities—Cases from China and South Africa*, Professional Practice in Earth Sciences,
https://doi.org/10.1007/978-3-031-98993-3_3

to determine the weights of various indicators, thereby formulating the entropy-weighted TOPSIS method to achieve a more comprehensive and accurate assessment of water quality. This integrated approach enhances the robustness and reliability of the evaluation, ensuring that the results are both scientifically sound and practically applicable. In addition to groundwater quality assessment using the entropy-weighted TOPSIS method, this chapter also summarized the main environmental issues related to groundwater in the Weining Plain, and discussed the evolution of groundwater system under human activities in this plain.

3.1 Groundwater Quality Assessment

3.1.1 Sample Collection and Analysis

A total of 223 groundwater samples were collected to systematically evaluate the groundwater quality in the Weining Plain with the sampling locations illustrated in Fig. 3.1. Notably, the Nanshantaizi area had two groundwater monitoring wells built in 2012. Groundwater samples in the Yellow River alluvial plain were mainly collected from private wells and those in the fast-growing forest were collected from groundwater monitoring wells built between 2011 and 2012. During sampling, sampling locations were recorded using portable GPS. After collection, samples were promptly treated with preservatives, sealed, labeled, and documented before being sent to the Ningxia Environmental Monitoring Center for analysis. The water quality analysis included over 30 parameters, such as pH, common ions, TH, TDS, volatile phenols, nitrogen (ammonia, nitrite, and nitrate), metasilicic acid, cyanide, total phosphorus, total copper, total aluminum, total manganese, total zinc, total chromium, cadmium, lead, arsenic, and fluoride.

3.1.2 Entropy Weight TOPSIS Method for Water Quality Assessment

(1) **Entropy Weight**

The information entropy theory has been widely applied to determine the weights of water quality indicators. The entropy weight method is an objective approach to determine indicator weights, relying solely on the selected samples and their evaluation criteria. The process of determining weights using information entropy theory involves the following steps (Li et al. 2011b, 2013a):

(1) Establishment of the Initial Evaluation Matrix

Suppose there are m samples to be evaluated, each with n evaluation indicators. The initial evaluation matrix $\mathbf{X}$ can be represented as Eq. 3.1, where x_{ij} is the initial value

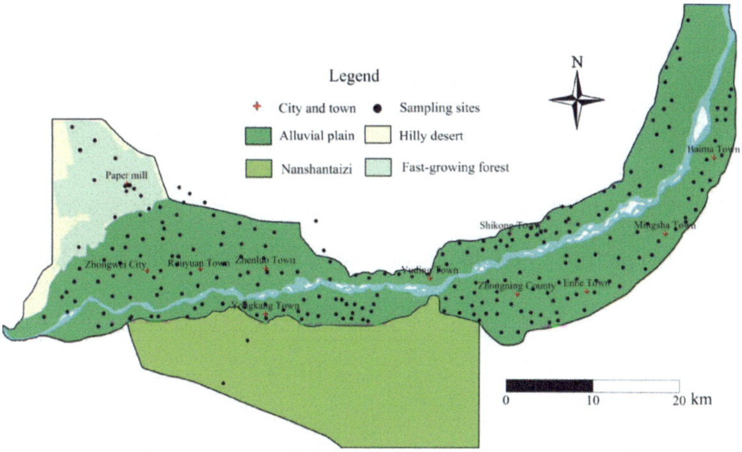

Fig. 3.1 Distribution of groundwater sampling points for water quality assessment

of the jth indicator for the ith sample, with $i = 1, 2, \ldots, m$, and $j = 1, 2, \ldots, n$.

$$\mathbf{X} = \begin{bmatrix} x_{11} & x_{12} & \cdots & x_{1n} \\ x_{21} & x_{22} & \cdots & x_{2n} \\ \vdots & \vdots & \ddots & \vdots \\ x_{m1} & x_{m2} & \cdots & x_{mn} \end{bmatrix} \tag{3.1}$$

(B) Standardization of Initial Data

Typically, different indicators have varying units and scales. If these differences are not addressed, the calculated weights may be significantly skewed, failing to reflect the actual situation. Therefore, before calculating weights, the initial data must be standardized to eliminate the influence of differing units and scales. The standardization of initial data is generally performed using Eq. 3.2.

$$y_{ij} = \begin{cases} \dfrac{x_{ij} - (x_{ij})^j_{\min}}{(x_{ij})^j_{\max} - (x_{ij})^j_{\min}} \text{efficiency type} \\ \dfrac{(x_{ij})^j_{\max} - x_{ij}}{(x_{ij})^j_{\max} - (x_{ij})^j_{\min}} \text{ cost type} \end{cases} \tag{3.2}$$

where, y_{ij} is the standardized value of the indicators, and $(x_{ij})^j{}_{\max}$和$(x_{ij})^j{}_{\min}$ are the maximum and minimum values of the jth indicator in the initial evaluation matrix, respectively. After standardization, the standardized evaluation matrix **Y** can be obtained, as shown in Eq. 3.3.

$$\mathbf{Y} = \begin{bmatrix} y_{11} & y_{12} & \cdots & y_{1n} \\ y_{21} & y_{22} & \cdots & y_{2n} \\ \vdots & \vdots & \ddots & \vdots \\ y_{m1} & y_{m2} & \cdots & y_{mn} \end{bmatrix} \tag{3.3}$$

(C) Calculation of Information Entropy

The information entropy of the indicators can be calculated using Eqs. 3.4 and 3.5.

$$P_{ij} = \frac{y_{ij} + 10^{-4}}{\sum_{i=1}^{m} \left(y_{ij} + 10^{-4}\right)} \tag{3.4}$$

$$e_j = -\frac{1}{\ln m} \sum_{i=1}^{m} \left(P_{ij} \ln P_{ij}\right) \tag{3.5}$$

where, P_{ij} is the ratio of the standardized value of a column in the standardized evaluation matrix to the sum of the values in that column, e_j is the information entropy of the indicator, and 10^{-4} is a correction factor introduced to avoid mathematical issues when P_{ij} is zero. Zhang and Ren (2011) demonstrated that using 10^{-4} in Eq. 3.4 ensures that $\ln P_{ij}$ remains mathematically meaningful while keeping its impact on the entropy value e_j within a reasonable range.

(D) Calculation of Entropy Weight

The entropy weights can be calculated using Eq. 3.6, where w_j is the weight of the indicator.

$$w_j = \frac{1 - e_j}{\sum_{j=1}^{n} \left(1 - e_j\right)} \tag{3.6}$$

(B) **TOPSIS Method**

The TOPSIS method, initially proposed by Hwang and Yoon (1981), is a classic approach for solving multi-objective decision-making problems. When applied to groundwater quality assessment, the TOPSIS method involves the following steps:

(1) Establishment of the Initial Decision Matrix for Water Quality Assessment

Assuming there are m water samples, each with n evaluation indicators, the initial decision matrix $\mathbf{C}$ can be established based on the water quality analysis results as Eq. 3.7.

$$\mathbf{C} = \begin{bmatrix} c_{11} & c_{12} & \cdots & c_{1n} \\ c_{21} & c_{22} & \cdots & c_{2n} \\ \vdots & \vdots & \ddots & \vdots \\ c_{m1} & c_{m2} & \cdots & c_{mn} \end{bmatrix} \tag{3.7}$$

where, **C** is the initial decision matrix, c_{ij} is the concentration value of the jth indicator for the ith sample, with $i = 1, 2, 3 \ldots m$, and $j = 1, 2, 3 \ldots n$.

(B) Standardization of the Initial Decision Matrix

Since the indicators in the initial decision matrix have different units and scales, the matrix must be standardized. The standardization of the initial matrix can be performed using Eq. 3.8.

$$r_{ij} = \begin{cases} \dfrac{c_{ij}}{\left(\sum\limits_{i=1}^{m} x_{ij}^2\right)^{\frac{1}{2}}} \text{efficiency type} \\ \dfrac{-c_{ij}}{\left(\sum\limits_{i=1}^{m} x_{ij}^2\right)^{\frac{1}{2}}} \text{ cost type} \end{cases} \tag{3.8}$$

After standardization, the standardized decision matrix **R** can be obtained (Eq. 3.9).

$$\mathbf{R} = \begin{bmatrix} r_{11} & r_{12} & \cdots & r_{1n} \\ r_{21} & r_{22} & \cdots & r_{2n} \\ \vdots & \vdots & \ddots & \vdots \\ r_{m1} & r_{m2} & \cdots & r_{mn} \end{bmatrix} \tag{3.9}$$

where, **R** is the standardized decision matrix, and r_{ij} is the standardized value of the indicators.

(C) Determination of Weights for Evaluation Indicators

The weights of the indicators are crucial for ensuring the objectivity and rationality of the evaluation results. The entropy weight method, as described above, was used to determine the weights in this chapter. Then the standardized decision matrix is multiplied by the entropy weights to obtain the weighted standardized decision matrix according to Eq. 3.10. The weighted standardized decision matrix is expressed as $\mathbf{F} = (f_{ij})_{m\times n}$.

$$f_{ij} = r_{ij} \times w_j \tag{3.10}$$

(D) Determination of the Positive and Negative Ideal Solutions

The positive and negative ideal solutions are virtual points representing the best and worst water quality, respectively, based on the weighted standardized water quality indicator values.

Let $\begin{cases} f_j^+ = \max(f_{1j}, f_{2j}, \cdots, f_{mj}) \\ f_j^- = \min(f_{1j}, f_{2j}, \cdots, f_{mj}) \end{cases}$, $j = 1, 2, \ldots, n$, then the positive and negative ideal solutions can be obtained. The positive ideal solution is $G = (f_1^+, f_2^+, \ldots, f_n^+)$, and the negative ideal solution is $B = (f_1^-, f_2^-, \ldots, f_n^-)$.

(E) Calculation of Distances to the Positive and Negative Ideal Solutions

The distances of each water sample to the positive and negative ideal solutions can be calculated using Eq. 3.11.

$$\begin{cases} d^+ = \sqrt{\sum_{j=1}^{n} \left[f_{ij} - (f_{ij})_G\right]^2} \\ d^- = \sqrt{\sum_{j=1}^{n} \left[f_{ij} - (f_{ij})_B\right]^2} \end{cases} \tag{3.11}$$

where, $(f_{ij})_G$ and $(f_{ij})_B$ are the indicator values in the positive and negative ideal solutions, respectively, d^+ is the distance to the positive ideal solution, and d^- is the distance to the negative ideal solution.

(F) Calculation of Closeness Coefficient for Water Quality Assessment

The relative distance of a water sample to the positive ideal solution is called the closeness coefficient (*CC*), and can be calculated using Eq. 3.12.

$$CC = \frac{d^-}{d^+ + d^-} \tag{3.12}$$

where, *CC* is the closeness coefficient, representing the relative distance of the water sample to the positive ideal solution. A higher *CC* value indicates better water quality, with $CC = 1$ when the water quality matches the positive ideal solution. Conversely, a lower *CC* value indicates poorer water quality, with $CC = 0$ when the water quality matches the negative ideal solution. By considering the groundwater quality standards as virtual samples and sorting them together with the real water samples based on the *CC* values, the water quality grades of the samples can be determined.

3.1.3 Statistical Analysis of Water Quality

The drinking water for urban residents in the Weining Plain is entirely sourced from groundwater, making the assessment of drinking water quality crucial for public health. In this chapter, a statistical analysis was conducted to understand the basic groundwater pollution status in the study area. The statistical results are presented in Table 3.1

The statistical results reveal that groundwater in the study area is significantly contaminated, with major contaminants being SO_4^{2-}, Cl^-, F^-, TDS, TH, NH_4-N,

Table 3.1 Statistical results of major Groundwater Pollution indicators and Exceedance Rates (Unit: mg/L except pH)

Indicators	pH	SO_4^{2-}	Cl^-	F^-	TDS	TH	NH_4-N	NO_2-N
Min	7	30.6	42.25	0.11	298.45	93.25	0.01	0
Max	8.8	3584	2654	6.9	7140	1979.17	37	0.52
Mean	7.76	409.49	206.84	1.01	1316.97	591.59	0.39	0.03
Standard limits	6.5–8.5	250	250	1	1000	450.00	0.20	0.02
Exceedance count	2	123	38	67	120	159	44	54
Exceedance rate (%)	0.90	55.16	17.04	30.04	53.81	71.30	19.73	24.22
Indicators	NO_3-N	COD_{Mn}	Mn	Zn	Volatile phenols	Ni	Pb	Cr
Min	0.08	0.44	0	0	0	0	0	0
Max	137	28.4	6.46	1.963	0.004	0.022	0.091	0.16
Mean	14.56	1.85	0.31	0.02	0.00	0.00	0.003	0.0028
Standard limits	20	3	0.1	1	0.002	0.05	0.05	0.05
Exceedance count	42	10	110	1	4	0	1	1
Exceedance rate (%)	18.83	4.48	49.33	0.45	1.79	0.00	0.45	0.45

NO_2-N, NO_3-N, and Mn. Especially, NO_3-N and Mn pollution have already been widely reported in groundwater as well in sediments in previous research (He et al. 2022; Liu et al. 2022; Li et al. 2022; Xu and Li 2024). The exceedance rates for these contaminants all exceeded 15%, with TH, TDS, and SO_4^{2-} exceeding 50%. COD_{Mn}, pH, volatile phenols, zinc, lead, and chromium showed slight contamination in some samples, but the rates were below 5%, with pH, zinc, lead, and chromium all below 1%.

3.1.4 Comprehensive Water Quality Assessment

Based on the statistical results of pollutant concentrations, this study selected 10 indicators including SO_4^{2-}, Cl^-, F^-, TDS, TH, NH_4-N, NO_2-N, NO_3-N, Mn, and COD_{Mn} for a comprehensive assessment using the entropy weight TOPSIS method (Li et al. 2013a, b). The assessment results are presented in Table 3.2, and a comprehensive groundwater quality zoning map of the study area was created based on these results, as shown in Fig. 3.2.

Table 3.2 Groundwater quality assessment results based on entropy weighted TOPSIS

Sample	d^+	d^-	CC	Grade	Water quality
N15	0.04	0.139	0.778	V	Unsuitable for drinking
N16	0.014	0.155	0.915	IV	Suitable for drinking after treatment
N17	0.119	0.103	0.463	V	Unsuitable for drinking
N18	0.007	0.162	0.956	III	Generally suitable for drinking
N19	0.005	0.163	0.968	II	Suitable for drinking
N20	0.008	0.16	0.95	III	Generally suitable for drinking
N21	0.011	0.158	0.934	IV	Suitable for drinking after treatment
GC1	0.005	0.162	0.971	II	Suitable for drinking
GC2	0.051	0.154	0.752	V	Unsuitable for drinking
NT1	0.015	0.158	0.912	IV	Suitable for drinking after treatment
NT2-1	0.029	0.156	0.843	IV	Suitable for drinking after treatment
NT3	0.009	0.159	0.944	III	Generally suitable for drinking
N23	0.036	0.152	0.811	V	Unsuitable for drinking
N06	0.03	0.158	0.84	IV	Suitable for drinking after treatment
N07	0.005	0.164	0.973	II	Suitable for drinking
N08	0.004	0.163	0.974	II	Suitable for drinking
N11	0.007	0.162	0.957	III	Generally suitable for drinking
N14	0.006	0.162	0.962	III	Generally suitable for drinking
O1	0.007	0.162	0.961	III	Generally suitable for drinking
O2	0.007	0.162	0.961	III	Generally suitable for drinking
O3	0.007	0.162	0.961	III	Generally suitable for drinking
N05	0.024	0.149	0.861	IV	Suitable for drinking after treatment
N12	0.01	0.157	0.939	III	Generally suitable for drinking
N02	0.008	0.16	0.951	III	Generally suitable for drinking
G1	0.027	0.152	0.851	IV	Suitable for drinking after treatment
G2	0.064	0.146	0.694	V	Unsuitable for drinking
G3	0.04	0.148	0.787	V	Unsuitable for drinking
N01	0.006	0.162	0.963	III	Generally suitable for drinking
N09	0.014	0.159	0.919	IV	Suitable for drinking after treatment
N13	0.019	0.161	0.897	IV	Suitable for drinking after treatment
N10	0.037	0.15	0.804	V	Unsuitable for drinking
N03	0.007	0.161	0.957	III	Generally suitable for drinking
N04	0.01	0.161	0.94	III	Generally suitable for drinking
W001	0.012	0.159	0.929	IV	Suitable for drinking after treatment
W003	0.072	0.143	0.666	V	Unsuitable for drinking
W004	0.01	0.16	0.941	III	Generally suitable for drinking
W005	0.009	0.162	0.948	III	Generally suitable for drinking

(continued)

Table 3.2 (continued)

Sample	d^+	d^-	CC	Grade	Water quality
W006	0.012	0.16	0.928	IV	Suitable for drinking after treatment
W007	0.01	0.161	0.939	III	Generally suitable for drinking
W008	0.013	0.16	0.926	IV	Suitable for drinking after treatment
W009	0.013	0.16	0.925	IV	Suitable for drinking after treatment
W010	0.01	0.161	0.942	III	Generally suitable for drinking
W011	0.014	0.159	0.92	IV	Suitable for drinking after treatment
W012	0.009	0.159	0.948	III	Generally suitable for drinking
W013	0.011	0.159	0.936	III	Generally suitable for drinking
W014	0.008	0.162	0.953	III	Generally suitable for drinking
W015	0.008	0.161	0.952	III	Generally suitable for drinking
W016	0.009	0.16	0.945	III	Generally suitable for drinking
W017	0.007	0.162	0.96	III	Generally suitable for drinking
W018	0.021	0.156	0.879	IV	Suitable for drinking after treatment
W019	0.006	0.163	0.963	III	Generally suitable for drinking
W020	0.012	0.158	0.927	IV	Suitable for drinking after treatment
W021	0.012	0.16	0.932	IV	Suitable for drinking after treatment
W022	0.006	0.161	0.962	III	Generally suitable for drinking
W023	0.008	0.16	0.953	III	Generally suitable for drinking
W024	0.009	0.161	0.949	III	Generally suitable for drinking
W025	0.012	0.159	0.93	IV	Suitable for drinking after treatment
W026	0.011	0.159	0.938	III	Generally suitable for drinking
W027	0.027	0.152	0.851	IV	Suitable for drinking after treatment
W028	0.008	0.161	0.955	III	Generally suitable for drinking
W029	0.01	0.162	0.941	III	Generally suitable for drinking
W030	0.005	0.163	0.97	II	Suitable for drinking
W031	0.007	0.162	0.958	III	Generally suitable for drinking
W032	0.006	0.163	0.967	II	Suitable for drinking
W033	0.009	0.161	0.945	III	Generally suitable for drinking
W034	0.006	0.163	0.965	III	Generally suitable for drinking
W035	0.009	0.162	0.949	III	Generally suitable for drinking
W036	0.004	0.163	0.974	II	Suitable for drinking
W037	0.007	0.161	0.961	III	Generally suitable for drinking
W038	0.014	0.157	0.919	IV	Suitable for drinking after treatment
W039	0.008	0.16	0.952	III	Generally suitable for drinking
W040	0.004	0.163	0.976	II	Suitable for drinking
W041	0.027	0.154	0.852	IV	Suitable for drinking after treatment
W042	0.006	0.162	0.966	II	Suitable for drinking

(continued)

Table 3.2 (continued)

Sample	d^+	d^-	CC	Grade	Water quality
W043	0.003	0.164	0.982	II	Suitable for drinking
W044	0.006	0.163	0.966	III	Generally suitable for drinking
W045	0.019	0.155	0.892	IV	Suitable for drinking after treatment
W046	0.014	0.157	0.917	IV	Suitable for drinking after treatment
W047	0.018	0.154	0.894	IV	Suitable for drinking after treatment
W048	0.008	0.16	0.952	III	Generally suitable for drinking
W050	0.017	0.156	0.9	IV	Suitable for drinking after treatment
W051	0.041	0.15	0.787	V	Unsuitable for drinking
W052	0.012	0.158	0.931	IV	Suitable for drinking after treatment
W053	0.012	0.158	0.932	IV	Suitable for drinking after treatment
W054	0.028	0.149	0.843	IV	Suitable for drinking after treatment
W055	0.01	0.159	0.94	III	Generally suitable for drinking
W056	0.009	0.16	0.947	III	Generally suitable for drinking
W057	0.007	0.163	0.961	III	Generally suitable for drinking
W058	0.007	0.162	0.961	III	Generally suitable for drinking
W059	0.021	0.156	0.884	IV	Suitable for drinking after treatment
W060	0.011	0.16	0.934	IV	Suitable for drinking after treatment
W061	0.008	0.162	0.952	III	Generally suitable for drinking
W062	0.014	0.156	0.919	IV	Suitable for drinking after treatment
W063	0.04	0.143	0.781	V	Unsuitable for drinking
W064	0.013	0.158	0.926	IV	Suitable for drinking after treatment
W065	0.013	0.158	0.926	IV	Suitable for drinking after treatment
W066	0.013	0.158	0.927	IV	Suitable for drinking after treatment
W067	0.012	0.158	0.929	IV	Suitable for drinking after treatment
W068	0.017	0.157	0.904	IV	Suitable for drinking after treatment
W069	0.012	0.158	0.927	IV	Suitable for drinking after treatment
W070	0.009	0.159	0.944	III	Generally suitable for drinking
W071	0.011	0.16	0.935	III	Generally suitable for drinking
W072	0.009	0.161	0.945	III	Generally suitable for drinking
W073	0.01	0.161	0.942	III	Generally suitable for drinking
W074	0.007	0.163	0.959	III	Generally suitable for drinking
W075	0.01	0.162	0.944	III	Generally suitable for drinking
W076	0.008	0.16	0.954	III	Generally suitable for drinking
W077	0.008	0.161	0.95	III	Generally suitable for drinking
W078	0.013	0.158	0.921	IV	Suitable for drinking after treatment
W079	0.019	0.154	0.891	IV	Suitable for drinking after treatment
W080	0.01	0.16	0.941	III	Generally suitable for drinking

(continued)

Table 3.2 (continued)

Sample	d^+	d^-	CC	Grade	Water quality
W081	0.021	0.155	0.882	IV	Suitable for drinking after treatment
W082	0.014	0.161	0.921	IV	Suitable for drinking after treatment
W083	0.021	0.155	0.88	IV	Suitable for drinking after treatment
W084	0.012	0.158	0.927	IV	Suitable for drinking after treatment
W085	0.021	0.155	0.881	IV	Suitable for drinking after treatment
W086	0.066	0.142	0.683	V	Unsuitable for drinking
W087	0.032	0.152	0.826	IV	Suitable for drinking after treatment
W089	0.016	0.157	0.907	IV	Suitable for drinking after treatment
W090	0.01	0.16	0.943	III	Generally suitable for drinking
W091	0.008	0.161	0.951	III	Generally suitable for drinking
W092	0.014	0.161	0.92	IV	Suitable for drinking after treatment
W093	0.026	0.157	0.858	IV	Suitable for drinking after treatment
W094	0.011	0.159	0.936	III	Generally suitable for drinking
W095	0.016	0.158	0.907	IV	Suitable for drinking after treatment
W096	0.014	0.156	0.92	IV	Suitable for drinking after treatment
W097	0.009	0.16	0.944	III	Generally suitable for drinking
W098	0.012	0.16	0.932	IV	Suitable for drinking after treatment
W099	0.013	0.159	0.926	IV	Suitable for drinking after treatment
W100	0.013	0.158	0.926	IV	Suitable for drinking after treatment
W101	0.015	0.157	0.912	IV	Suitable for drinking after treatment
W102	0.012	0.158	0.93	IV	Suitable for drinking after treatment
W104	0.019	0.154	0.891	IV	Suitable for drinking after treatment
W105	0.036	0.152	0.81	V	Unsuitable for drinking
W107	0.021	0.155	0.881	IV	Suitable for drinking after treatment
W108	0.018	0.156	0.898	IV	Suitable for drinking after treatment
W110	0.008	0.161	0.951	III	Generally suitable for drinking
W111	0.012	0.159	0.928	IV	Suitable for drinking after treatment
W112	0.011	0.159	0.935	III	Generally suitable for drinking
W113	0.025	0.156	0.863	IV	Suitable for drinking after treatment
W114	0.01	0.161	0.942	III	Generally suitable for drinking
W115	0.017	0.158	0.901	IV	Suitable for drinking after treatment
W116	0.006	0.163	0.962	III	Generally suitable for drinking
W117	0.028	0.16	0.853	IV	Suitable for drinking after treatment
W118	0.029	0.155	0.844	IV	Suitable for drinking after treatment
W119	0.01	0.161	0.943	III	Generally suitable for drinking
W120	0.029	0.153	0.842	IV	Suitable for drinking after treatment
W121	0.01	0.161	0.944	III	Generally suitable for drinking

(continued)

Table 3.2 (continued)

Sample	d^+	d^-	CC	Grade	Water quality
W123	0.004	0.164	0.976	II	Suitable for drinking
W125	0.025	0.154	0.859	IV	Suitable for drinking after treatment
W126	0.016	0.158	0.909	IV	Suitable for drinking after treatment
W127	0.009	0.16	0.949	III	Generally suitable for drinking
W128	0.013	0.158	0.922	IV	Suitable for drinking after treatment
W130	0.008	0.161	0.953	III	Generally suitable for drinking
W131	0.009	0.16	0.947	III	Generally suitable for drinking
W132	0.014	0.157	0.917	IV	Suitable for drinking after treatment
W134	0.011	0.158	0.935	III	Generally suitable for drinking
W135	0.011	0.159	0.935	III	Generally suitable for drinking
W137	0.008	0.16	0.951	III	Generally suitable for drinking
W138	0.029	0.152	0.841	IV	Suitable for drinking after treatment
W139	0.018	0.154	0.894	IV	Suitable for drinking after treatment
W140	0.009	0.159	0.944	III	Generally suitable for drinking
W141	0.023	0.154	0.872	IV	Suitable for drinking after treatment
W142	0.015	0.155	0.91	IV	Suitable for drinking after treatment
W143	0.005	0.163	0.969	II	Suitable for drinking
W144	0.025	0.156	0.861	IV	Suitable for drinking after treatment
W145	0.016	0.156	0.908	IV	Suitable for drinking after treatment
S001	0.009	0.163	0.95	III	Generally suitable for drinking
S002	0.009	0.161	0.945	III	Generally suitable for drinking
S005	0.038	0.153	0.803	V	Unsuitable for drinking
S006	0.012	0.16	0.931	IV	Suitable for drinking after treatment
S007	0.011	0.159	0.937	III	Generally suitable for drinking
S008	0.006	0.162	0.964	III	Generally suitable for drinking
S009	0.03	0.15	0.832	IV	Suitable for drinking after treatment
S010	0.008	0.162	0.95	III	Generally suitable for drinking
S011	0.009	0.161	0.947	III	Generally suitable for drinking
S012	0.034	0.151	0.818	V	Unsuitable for drinking
S013	0.009	0.161	0.949	III	Generally suitable for drinking
S014	0.009	0.16	0.948	III	Generally suitable for drinking
S015	0.007	0.162	0.959	III	Generally suitable for drinking
S016	0.008	0.162	0.952	III	Generally suitable for drinking
S017	0.009	0.16	0.946	III	Generally suitable for drinking
S018	0.032	0.144	0.821	V	Unsuitable for drinking
S019	0.024	0.153	0.864	IV	Suitable for drinking after treatment
S020	0.036	0.15	0.805	V	Unsuitable for drinking

(continued)

Table 3.2 (continued)

Sample	d^+	d^-	CC	Grade	Water quality
S021	0.042	0.153	0.786	V	Unsuitable for drinking
S022	0.008	0.161	0.951	III	Generally suitable for drinking
S023	0.007	0.163	0.957	III	Generally suitable for drinking
S024	0.023	0.158	0.873	IV	Suitable for drinking after treatment
S025	0.024	0.156	0.866	IV	Suitable for drinking after treatment
S026	0.009	0.161	0.948	III	Generally suitable for drinking
S027	0.009	0.161	0.947	III	Generally suitable for drinking
S028	0.026	0.154	0.858	IV	Suitable for drinking after treatment
S029	0.019	0.157	0.891	IV	Suitable for drinking after treatment
S030	0.009	0.16	0.947	III	Generally suitable for drinking
S031	0.018	0.158	0.897	IV	Suitable for drinking after treatment
S032	0.021	0.153	0.881	IV	Suitable for drinking after treatment
S033	0.013	0.159	0.926	IV	Suitable for drinking after treatment
S034	0.019	0.159	0.894	IV	Suitable for drinking after treatment
S035	0.015	0.16	0.915	IV	Suitable for drinking after treatment
S036	0.014	0.159	0.921	IV	Suitable for drinking after treatment
S037	0.036	0.151	0.81	V	Unsuitable for drinking
S038	0.007	0.163	0.959	III	Generally suitable for drinking
S039	0.008	0.162	0.954	III	Generally suitable for drinking
S040	0.005	0.163	0.969	II	Suitable for drinking
S041	0.011	0.159	0.937	III	Generally suitable for drinking
S043	0.006	0.162	0.962	III	Generally suitable for drinking
S044	0.013	0.16	0.926	IV	Suitable for drinking after treatment
S045	0.011	0.16	0.938	III	Generally suitable for drinking
S046	0.012	0.158	0.93	IV	Suitable for drinking after treatment
S047	0.01	0.161	0.944	III	Generally suitable for drinking
S048	0.009	0.161	0.95	III	Generally suitable for drinking
S049	0.027	0.156	0.852	IV	Suitable for drinking after treatment
S050	0.022	0.157	0.878	IV	Suitable for drinking after treatment
S051	0.009	0.161	0.949	III	Generally suitable for drinking
S052	0.006	0.163	0.966	II	Suitable for drinking
S053	0.021	0.153	0.88	IV	Suitable for drinking after treatment
S054	0.007	0.162	0.959	III	Generally suitable for drinking
S086	0.015	0.161	0.917	IV	Suitable for drinking after treatment
S087	0.013	0.161	0.925	IV	Suitable for drinking after treatment
S090	0.01	0.162	0.94	III	Generally suitable for drinking
S091	0.011	0.161	0.936	III	Generally suitable for drinking
S093	0.003	0.165	0.981	II	Suitable for drinking

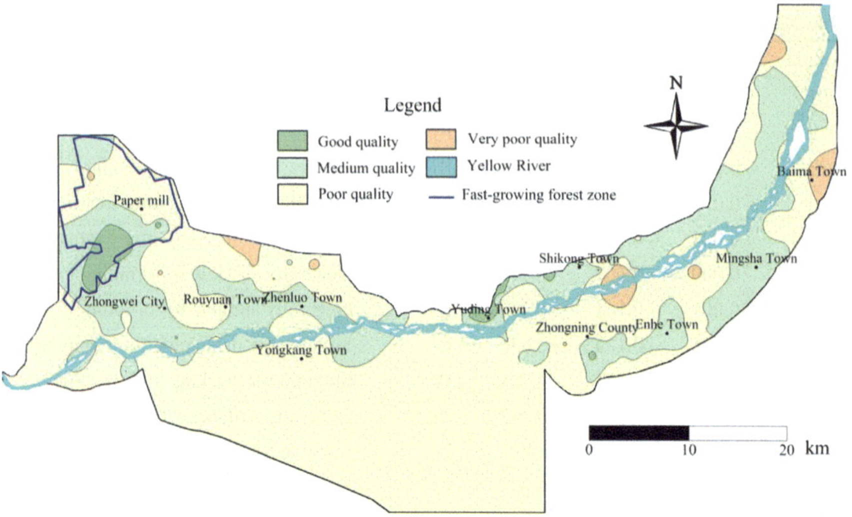

Fig. 3.2 Comprehensive Groundwater Quality zoning map of the study area

The assessment results indicate that out of the 223 water samples evaluated, 15 were classified as grade II, indicating suitability for drinking, 96 as grade III, which suggests they are generally suitable for drinking, 94 as grade IV, requiring treatment before use, and 18 as grade V, unsuitable for drinking. The proportions of each class were 6.73%, 43.01%, 42.15%, and 8.07%, respectively. As shown in Fig. 3.2, most areas in the study region have poor water quality, classified as grades IV or V. Only in the western part of the study area, near the Zhongwei City, Zhenluo Town, and Mingsha Town and Shikong Town in Zhongning County, grades II or III waters are found. The alluvial plain area, heavily influenced by human activities, usually exhibits poor water quality. Under the strong influence of human activities, the groundwater level in the plain area has been continuously rising, leading to increased evaporation and concentration effects, ultimately resulting in elevated concentrations of various ions. Additionally, industrial wastewater discharge has further deteriorated groundwater quality. The Nanshantaizi area generally has poor water quality, classified as grade IV, primarily due to the high salt content in the vadose zone soil. Under the influence of extensive irrigation, these salts are continuously leached into the aquifer, exacerbating water quality issues.

3.1.5 Irrigation Water Quality Assessment

The groundwater in the study area is also used for agricultural irrigation as supplementary of Yellow River water, primarily in rural regions. The quality of irrigation water significantly impacts soil structure and plant growth, making its assessment essential. Currently, the primary methods or indicators for evaluating agricultural irrigation water quality include the Sodium Adsorption Ratio (SAR), Residual Sodium Carbonate (RSC), Permeability Index (PI), and Sodium Percentage (%Na).

SAR represents the soil's ability to adsorb sodium ions. A higher SAR indicates a stronger capacity to adsorb sodium ions, which can affect the plant ability to absorb water (Singh and Singh 2008). The SAR is calculated as per Eq. 3.13 (Hakim et al. 2009).

$$\mathrm{SAR} = \frac{\mathrm{Na}^{+}}{\sqrt{\frac{\mathrm{Ca}^{2+} + \mathrm{Mg}^{2+}}{2}}} \tag{3.13}$$

RSC indicates the alkalinity hazard of groundwater. A negative RSC value suggests that the content of CO_3^{2-} and HCO_3^- is less than that of Ca^{2+} and Mg^{2+}, meaning there is no excess carbonate to react with Na^+, thus irrigation with this water will not increase alkalinity. Conversely, a positive RSC value indicates a potential alkalinity hazard, with higher values posing greater risks (Almeida et al. 2008). The RSC is calculated using Eq. 3.14 (Li et al. 2010b).

$$\mathrm{RSC} = \left(\mathrm{CO}_3^{2-} + \mathrm{HCO}_3^{-}\right) - \left(\mathrm{Ca}^{2+} + \mathrm{Mg}^{2+}\right) \tag{3.14}$$

%Na also represents the risk of alkalinity hazard from irrigation water, with higher values indicating greater risk. The %Na is calculated using Eq. 3.15.

$$\%\ \mathrm{Na} = \frac{\mathrm{Na}}{\mathrm{Ca} + \mathrm{Mg} + \mathrm{Na} + \mathrm{K}} \times 100\% \tag{3.15}$$

PI is influenced by long-term irrigation and the content of Na^+, Ca^{2+}, Mg^{2+}, and HCO_3^- in the soil (Sivasankar et al. 2014; Wang 20,123). Ragunath (1987) and Doneen (1964) provided evaluation standards for irrigation water quality based on PI, calculated using Eq. 3.16. (Aghazadeh and Mogaddam 2010).

$$\mathrm{PI} = \frac{\mathrm{Na} + \sqrt{\mathrm{HCO}_3}}{\mathrm{Ca} + \mathrm{Mg} + \mathrm{Na}} \times 100\% \tag{3.16}$$

In these equations, ion concentrations are all expressed in meq/L. While these indicators primarily reflect the alkalinity hazard of irrigation water, the salinity hazard must also be considered when assessing irrigation water quality. Salinity hazards are often represented by TDS or electrical conductivity (EC). When TDS or EC exceeds certain thresholds, irrigation water can cause salinity hazards. The evaluation standards for alkalinity and salinity hazards are presented in Table 3.3, with the evaluation results shown in Table 3.4.

Based on TDS, 192 out of the 223 groundwater samples were rated as excellent, 17 as good, 9 as moderate, and 5 as poor, accounting for 86.10%, 7.62%, 4.04%, and 2.24%, respectively. Groundwater rated as excellent or good can be used for irrigation

Table 3.3 Irrigation water quality evaluation standards

SAR	TDS	EC	Water quality	RSC	%Na	PI	Water quality
< 10	< 2000	< 250	Excellent	< 1.25	< 30%	> 75%	Suitable
10–18	< 3000	250–750	Good	1.25–2.5	30–60%	25–75%	Marginally suitable
18–26	< 4000	750–2250	Moderate	> 2.5	> 60%	< 25%	Unsuitable
> 26	> 4000	> 2250	Poor				

without causing soil salinization. Groundwater rated as moderate may be used for irrigation under specific conditions, such as appropriate irrigation patterns, irrigation amounts, and crop types. Groundwater rated as poor is unsuitable for irrigation. According to the irrigation water quality zoning map (Fig. 3.3a), most areas in the study region have excellent or good groundwater quality. Only in local areas, such as parts of Baima Town and Qingtongxia Town, water quality is slightly poorer and unsuitable for irrigation.

SAR, RSC, %Na, and PI all reflect the risk of alkalinity hazards from irrigation water. Based on SAR evaluation results, 1 sample was rated as poor, 1 as moderate, 13 as good, and the remaining 208 as excellent, accounting for 0.45%, 0.45%, 5.83%, and 93.27%, respectively. According to RSC evaluation results, 214 samples were suitable for irrigation, 3 were generally suitable, and 6 were unsuitable. Based on %Na, 40.81% were suitable, 43.05% were generally suitable, and 16.14% were unsuitable. According to PI evaluation results, 32 samples were suitable, 190 were generally suitable, and only 1 was unsuitable. Overall, the groundwater in the study area is suitable for irrigation, with only local areas witnessing a risk of alkalinity hazards. The irrigation water quality zoning map (Fig. 3.3) shows that these unsuitable areas are mainly located at the interface between the plain and hilly regions, where groundwater quality is inherently poorer, with higher levels of soluble salts leading to elevated TDS and Na^+ concentration.

The suitability of groundwater for irrigation also depends on groundwater availability and the feasibility of groundwater extraction. If groundwater availability is limited or extraction is difficult, even if the water quality meets irrigation standards, the area should be classified as unsuitable for irrigation. Nanshantaizi is a typical example of this case. Although the groundwater quality in Nanshantaizi generally meets irrigation standards, the deep water table and high extraction costs make it economically unfeasible for irrigation. Additionally, the single well yield in this area is much lower than that in the plain region. Therefore, based on RSC, %Na, and PI evaluations criteria, the groundwater in Nanshantaizi has been classified as unsuitable for irrigation, even though the water quality for NT1 was marginally suitable for irrigation (Fig. 3.3c, d, e).

The above analysis evaluates the quality of groundwater for irrigation purposes based on individual factors, considering both salinity and alkalinity hazards. Wilcox (1948) and the US Salinity Laboratory (1954) proposed a classification method for

Table 3.4 Assessment results for irrigation water quality

Sample	TDS	Quality	SAR	Quality	RSC	Quality	%Na	Quality	PI	Quality
W001	1266	Excellent	3.77	Excellent	–6.31	Suitable	43.91	Marginally suitable	55.14	Marginally suitable
W003	6040	Poor	39.57	Poor	–19.05	Suitable	84.89	Unsuitable	86.48	Suitable
W004	1348	Excellent	4.67	Excellent	–6.46	Suitable	49.26	Marginally suitable	59.43	Marginally suitable
W005	777	Excellent	2.39	Excellent	–3.22	Suitable	37.08	Marginally suitable	54.45	Marginally suitable
W006	1005	Excellent	2.33	Excellent	–5.58	Suitable	33.25	Marginally suitable	47.45	Marginally suitable
W007	818	Excellent	2.32	Excellent	–3.60	Suitable	35.76	Marginally suitable	52.63	Marginally suitable
W008	1044	Excellent	4.87	Excellent	–2.67	Suitable	54.90	Marginally suitable	68.07	Marginally suitable
W009	1210	Excellent	2.20	Excellent	–8.19	Suitable	29.95	Suitable	41.89	Marginally suitable
W010	950	Excellent	2.04	Excellent	–6.37	Suitable	29.51	Suitable	43.55	Marginally suitable
W011	1606	Excellent	4.70	Excellent	–7.18	Suitable	47.10	Marginally suitable	57.09	Marginally suitable
W012	1116	Excellent	2.16	Excellent	–7.76	Suitable	27.72	Suitable	40.85	Marginally suitable
W013	1138	Excellent	2.99	Excellent	–4.30	Suitable	36.42	Marginally suitable	50.96	Marginally suitable
W014	793	Excellent	2.30	Excellent	–3.86	Suitable	35.09	Marginally suitable	51.58	Marginally suitable

(continued)

Table 3.4 (continued)

Sample	TDS	Quality	SAR	Quality	RSC	Quality	%Na	Quality	PI	Quality
W015	859	Excellent	1.66	Excellent	–4.70	Suitable	24.93	Suitable	41.85	Marginally suitable
W016	1256	Excellent	5.58	Excellent	–5.43	Suitable	55.98	Marginally suitable	65.58	Marginally suitable
W017	932	Excellent	1.67	Excellent	–5.01	Suitable	24.46	Suitable	40.98	Marginally suitable
W018	2943	Good	7.71	Excellent	–14.77	Suitable	54.94	Marginally suitable	60.28	Marginally suitable
W019	758	Excellent	2.31	Excellent	–3.88	Suitable	36.02	Marginally suitable	52.42	Marginally suitable
W020	1324	Excellent	3.22	Excellent	–6.72	Suitable	35.66	Marginally suitable	48.25	Marginally suitable
W021	1542	Excellent	1.52	Excellent	–13.15	Suitable	19.44	Suitable	29.93	Marginally suitable
W022	976	Excellent	1.80	Excellent	–3.64	Suitable	26.50	Suitable	44.18	Marginally suitable
W023	989	Excellent	1.79	Excellent	–5.26	Suitable	26.03	Suitable	42.04	Marginally suitable
W024	1075	Excellent	3.25	Excellent	–6.92	Suitable	42.28	Marginally suitable	52.37	Marginally suitable
W025	1576	Excellent	3.15	Excellent	–9.93	Suitable	36.03	Marginally suitable	45.91	Marginally suitable
W026	1128	Excellent	2.41	Excellent	–5.19	Suitable	31.87	Marginally suitable	46.66	Marginally suitable

(continued)

Table 3.4 (continued)

Sample	TDS	Quality	SAR	Quality	RSC	Quality	%Na	Quality	PI	Quality
W027	891	Excellent	2.07	Excellent	−2.81	Suitable	30.44	Marginally suitable	48.66	Marginally suitable
W028	1027	Excellent	1.64	Excellent	−4.80	Suitable	23.47	Suitable	40.09	Marginally suitable
W029	414	Excellent	1.59	Excellent	−1.42	Suitable	32.31	Marginally suitable	57.73	Marginally suitable
W030	575	Excellent	1.16	Excellent	−2.89	Suitable	21.86	Suitable	43.80	Marginally suitable
W031	762	Excellent	3.23	Excellent	−2.97	Suitable	45.45	Marginally suitable	61.25	Marginally suitable
W032	415	Excellent	1.06	Excellent	−2.12	Suitable	23.44	Suitable	48.82	Marginally suitable
W033	820	Excellent	2.08	Excellent	−5.27	Suitable	31.95	Marginally suitable	46.92	Marginally suitable
W034	710	Excellent	1.33	Excellent	−3.06	Suitable	22.75	Suitable	43.13	Marginally suitable
W035	778	Excellent	1.43	Excellent	−5.75	Suitable	23.29	Suitable	39.45	Marginally suitable
W036	484	Excellent	1.57	Excellent	−1.92	Suitable	30.68	Marginally suitable	54.20	Marginally suitable
W037	517	Excellent	1.66	Excellent	−2.06	Suitable	30.61	Marginally suitable	52.92	Marginally suitable
W038	1040	Excellent	2.08	Excellent	0.19	Suitable	28.22	Suitable	47.96	Marginally suitable

(continued)

Table 3.4 (continued)

Sample	TDS	Quality	SAR	Quality	RSC	Quality	%Na	Quality	PI	Quality
W039	628	Excellent	1.58	Excellent	−3.37	Suitable	28.11	Suitable	47.62	Marginally suitable
W040	501	Excellent	1.57	Excellent	−2.45	Suitable	30.23	Marginally suitable	52.15	Marginally suitable
W041	838	Excellent	1.50	Excellent	−3.85	Suitable	23.71	Suitable	42.17	Marginally suitable
W042	746	Excellent	1.68	Excellent	−2.58	Suitable	27.49	Suitable	47.63	Marginally suitable
W043	438	Excellent	1.13	Excellent	−1.51	Suitable	24.59	Suitable	51.56	Marginally suitable
W044	690	Excellent	1.16	Excellent	−2.53	Suitable	20.29	Suitable	42.13	Marginally suitable
W045	924	Excellent	2.30	Excellent	−3.96	Suitable	31.90	Marginally suitable	48.26	Marginally suitable
W046	462	Excellent	3.16	Excellent	−1.53	Suitable	43.16	Marginally suitable	61.01	Marginally suitable
W047	1452	Excellent	6.16	Excellent	−3.60	Suitable	57.29	Marginally suitable	68.37	Marginally suitable
W048	442	Excellent	1.14	Excellent	−2.03	Suitable	24.88	Suitable	50.27	Marginally suitable
W050	2128	Good	2.31	Excellent	−13.63	Suitable	25.48	Suitable	35.54	Marginally suitable
W051	5610	Poor	11.29	Good	−25.28	Suitable	58.80	Marginally suitable	62.20	Marginally suitable

(continued)

Table 3.4 (continued)

Sample	TDS	Quality	SAR	Quality	RSC	Quality	%Na	Quality	PI	Quality
W052	1248	Excellent	3.08	Excellent	–4.13	Suitable	37.21	Marginally suitable	51.81	Marginally suitable
W053	964	Excellent	2.36	Excellent	–3.80	Suitable	32.97	Marginally suitable	49.40	Marginally suitable
W054	2903	Good	5.27	Excellent	–14.67	Suitable	43.52	Marginally suitable	50.97	Marginally suitable
W055	971	Excellent	1.93	Excellent	–3.95	Suitable	29.19	Suitable	46.51	Marginally suitable
W056	880	Excellent	1.86	Excellent	–3.17	Suitable	28.27	Suitable	46.73	Marginally suitable
W057	798	Excellent	1.69	Excellent	0.51	Suitable	27.88	Suitable	52.13	Marginally suitable
W058	731	Excellent	1.96	Excellent	–2.10	Suitable	32.50	Marginally suitable	52.91	Marginally suitable
W059	863	Excellent	1.87	Excellent	–0.72	Suitable	28.98	Suitable	50.42	Marginally suitable
W060	1260	Excellent	1.88	Excellent	–7.81	Suitable	26.47	Suitable	39.55	Marginally suitable
W061	895	Excellent	1.66	Excellent	–3.36	Suitable	25.57	Suitable	44.16	Marginally suitable
W062	1612	Excellent	5.61	Excellent	–6.15	Suitable	52.06	Marginally suitable	62.02	Marginally suitable
W063	4630	Poor	14.69	Good	–15.13	Suitable	68.05	Unsuitable	72.23	Marginally suitable

(continued)

Table 3.4 (continued)

Sample	TDS	Quality	SAR	Quality	RSC	Quality	%Na	Quality	PI	Quality
W064	984	Excellent	2.18	Excellent	–3.52	Suitable	31.16	Marginally suitable	48.31	Marginally suitable
W065	1452	Excellent	4.34	Excellent	–1.87	Suitable	44.62	Marginally suitable	58.52	Marginally suitable
W066	1706	Excellent	4.75	Excellent	–8.21	Suitable	46.00	Marginally suitable	55.63	Marginally suitable
W067	1264	Excellent	3.16	Excellent	–5.03	Suitable	38.22	Marginally suitable	51.88	Marginally suitable
W068	1274	Excellent	3.19	Excellent	–3.87	Suitable	38.32	Marginally suitable	52.87	Marginally suitable
W069	1099	Excellent	2.47	Excellent	–3.38	Suitable	33.12	Marginally suitable	49.73	Marginally suitable
W070	1077	Excellent	2.63	Excellent	–3.29	Suitable	35.26	Marginally suitable	51.56	Marginally suitable
W071	1574	Excellent	7.23	Excellent	–4.46	Suitable	60.64	Unsuitable	70.05	Marginally suitable
W072	1174	Excellent	2.03	Excellent	–1.15	Suitable	26.97	Suitable	45.49	Marginally suitable
W073	1352	Excellent	2.44	Excellent	–5.31	Suitable	30.32	Marginally suitable	44.82	Marginally suitable
W074	869	Excellent	1.34	Excellent	–2.69	Suitable	20.97	Suitable	40.82	Marginally suitable
W075	1010	Excellent	1.40	Excellent	–7.38	Suitable	22.54	Suitable	36.30	Marginally suitable

(continued)

Table 3.4 (continued)

Sample	TDS	Quality	SAR	Quality	RSC	Quality	%Na	Quality	PI	Quality
W076	881	Excellent	3.83	Excellent	–0.84	Suitable	45.80	Marginally suitable	62.78	Marginally suitable
W077	1130	Excellent	4.33	Excellent	–3.86	Suitable	48.87	Marginally suitable	61.66	Marginally suitable
W078	1854	Excellent	7.74	Excellent	–3.62	Suitable	59.91	Marginally suitable	69.38	Marginally suitable
W079	1120	Excellent	1.51	Excellent	–13.05	Suitable	18.51	Suitable	29.65	Marginally suitable
W080	1194	Excellent	4.67	Excellent	–2.41	Suitable	50.75	Marginally suitable	64.44	Marginally suitable
W081	1868	Excellent	8.78	Excellent	–2.12	Suitable	64.32	Unsuitable	73.93	Marginally suitable
W082	660	Excellent	1.83	Excellent	–2.67	Suitable	31.69	Marginally suitable	51.76	Marginally suitable
W083	2904	Good	17.95	Good	–4.56	Suitable	79.33	Unsuitable	84.37	Suitable
W084	1598	Excellent	4.94	Excellent	–4.03	Suitable	48.12	Marginally suitable	59.90	Marginally suitable
W085	2936	Good	13.53	Good	–9.74	Suitable	70.69	Unsuitable	75.45	Suitable
W086	7140	Poor	22.03	Moderate	–26.32	Suitable	73.64	Unsuitable	75.62	Suitable
W087	1354	Excellent	2.87	Excellent	–7.24	Suitable	33.94	Marginally suitable	46.29	Marginally suitable
W089	1088	Excellent	2.10	Excellent	–4.75	Suitable	29.25	Suitable	45.03	Marginally suitable

(continued)

Table 3.4 (continued)

Sample	TDS	Quality	SAR	Quality	RSC	Quality	%Na	Quality	PI	Quality
W090	1378	Excellent	5.99	Excellent	–6.03	Suitable	57.61	Marginally suitable	66.23	Marginally suitable
W091	1114	Excellent	4.45	Excellent	–2.24	Suitable	49.86	Marginally suitable	64.16	Marginally suitable
W092	1962	Excellent	9.61	Excellent	–6.15	Suitable	67.79	Unsuitable	74.40	Marginally suitable
W093	3580	Moderate	9.23	Excellent	–21.12	Suitable	57.09	Marginally suitable	60.19	Marginally suitable
W094	631	Excellent	1.20	Excellent	–1.84	Suitable	21.83	Suitable	45.24	Marginally suitable
W095	1502	Excellent	3.00	Excellent	–8.26	Suitable	34.23	Marginally suitable	45.79	Marginally suitable
W096	1299	Excellent	3.03	Excellent	–6.34	Suitable	36.33	Marginally suitable	49.08	Marginally suitable
W097	1036	Excellent	2.14	Excellent	–3.97	Suitable	29.76	Suitable	46.33	Marginally suitable
W098	1466	Excellent	2.69	Excellent	–8.23	Suitable	31.70	Marginally suitable	43.81	Marginally suitable
W099	810	Excellent	1.47	Excellent	–3.66	Suitable	23.88	Suitable	42.80	Marginally suitable
W100	958	Excellent	2.25	Excellent	–4.72	Suitable	31.20	Marginally suitable	46.81	Marginally suitable
W101	916	Excellent	1.82	Excellent	–2.43	Suitable	26.84	Suitable	45.79	Marginally suitable

(continued)

Table 3.4 (continued)

Sample	TDS	Quality	SAR	Quality	RSC	Quality	%Na	Quality	PI	Quality
W102	1045	Excellent	2.34	Excellent	-4.21	Suitable	31.92	Marginally suitable	47.86	Marginally suitable
W104	1670	Excellent	3.69	Excellent	-6.55	Suitable	38.09	Marginally suitable	50.12	Marginally suitable
W105	1824	Excellent	1.64	Excellent	−12.26	Suitable	20.50	Suitable	31.65	Marginally suitable
W107	1052	Excellent	2.11	Excellent	-6.49	Suitable	28.09	Suitable	42.25	Marginally suitable
W108	2560	Good	6.38	Excellent	−10.34	Suitable	49.48	Marginally suitable	57.76	Marginally suitable
W110	908	Excellent	2.31	Excellent	-5.02	Suitable	32.04	Marginally suitable	47.18	Marginally suitable
W111	1218	Excellent	2.10	Excellent	-0.54	Suitable	28.20	Suitable	47.07	Marginally suitable
W112	1606	Excellent	8.01	Excellent	-3.65	Suitable	63.95	Unsuitable	73.29	Marginally suitable
W113	2148	Good	4.07	Excellent	−14.93	Suitable	38.45	Marginally suitable	45.78	Marginally suitable
W114	1071	Excellent	4.27	Excellent	−2.34	Suitable	49.91	Marginally suitable	64.54	Marginally suitable
W115	2036	Good	6.92	Excellent	−2.98	Suitable	56.96	Marginally suitable	67.70	Marginally suitable
W116	581	Excellent	2.26	Excellent	−1.57	Suitable	38.48	Marginally suitable	59.80	Marginally suitable

(continued)

Table 3.4 (continued)

Sample	TDS	Quality	SAR	Quality	RSC	Quality	%Na	Quality	PI	Quality
W117	778	Excellent	7.11	Excellent	1.35	Marginally suitable	70.85	Unsuitable	87.53	Suitable
W118	2504	Good	12.38	Good	-9.00	Suitable	71.58	Unsuitable	75.82	Suitable
W119	1005	Excellent	2.91	Excellent	-3.91	Suitable	38.86	Marginally suitable	54.19	Marginally suitable
W120	3920	Moderate	15.31	Good	-14.19	Suitable	71.00	Unsuitable	74.60	Marginally suitable
W121	917	Excellent	1.99	Excellent	-3.88	Suitable	30.13	Marginally suitable	47.42	Marginally suitable
W123	468	Excellent	3.04	Excellent	-0.07	Suitable	51.22	Marginally suitable	75.88	Suitable
W125	3374	Moderate	14.79	Good	-11.50	Suitable	71.49	Unsuitable	75.65	Suitable
W126	2088	Good	12.91	Good	-5.11	Suitable	74.64	Unsuitable	80.46	Suitable
W127	1280	Excellent	4.78	Excellent	-2.67	Suitable	49.96	Marginally suitable	63.28	Marginally suitable
W128	1090	Excellent	1.64	Excellent	-7.20	Suitable	22.55	Suitable	37.05	Marginally suitable
W130	1260	Excellent	9.68	Excellent	0.68	Suitable	75.22	Unsuitable	87.38	Suitable
W131	878	Excellent	2.03	Excellent	-3.64	Suitable	30.62	Marginally suitable	48.11	Marginally suitable
W132	1045	Excellent	2.14	Excellent	-4.85	Suitable	30.12	Marginally suitable	45.83	Marginally suitable
W134	1130	Excellent	2.31	Excellent	-3.01	Suitable	32.54	Marginally suitable	50.06	Marginally suitable

(continued)

Table 3.4 (continued)

Sample	TDS	Quality	SAR	Quality	RSC	Quality	%Na	Quality	PI	Quality
W135	1512	Excellent	3.25	Excellent	–5.92	Suitable	37.26	Marginally suitable	50.15	Marginally suitable
W137	922	Excellent	2.45	Excellent	–2.14	Suitable	35.43	Marginally suitable	53.96	Marginally suitable
W138	3706	Moderate	12.53	Good	–15.53	Suitable	66.95	Unsuitable	70.36	Marginally suitable
W139	1178	Excellent	2.39	Excellent	–4.05	Suitable	32.01	Marginally suitable	47.98	Marginally suitable
W140	766	Excellent	1.46	Excellent	–2.62	Suitable	23.53	Suitable	43.88	Marginally suitable
W141	3428	Moderate	14.21	Good	–11.01	Suitable	71.83	Unsuitable	75.92	Suitable
W142	1428	Excellent	2.95	Excellent	–7.37	Suitable	35.12	Marginally suitable	47.27	Marginally suitable
W143	643	Excellent	1.09	Excellent	–2.48	Suitable	20.62	Suitable	43.77	Marginally suitable
W144	2440	Good	1.75	Excellent	–18.58	Suitable	19.10	Suitable	27.96	Marginally suitable
W145	702	Excellent	1.09	Excellent	–4.05	Suitable	18.59	Suitable	38.11	Marginally suitable
S001	872	Excellent	0.66	Excellent	–8.99	Suitable	10.04	Suitable	25.07	Marginally suitable
S002	802	Excellent	1.73	Excellent	–3.07	Suitable	27.39	Suitable	46.75	Marginally suitable
S005	3288	Moderate	1.26	Excellent	–35.30	Suitable	12.05	Suitable	16.99	Unsuitable

(continued)

Table 3.4 (continued)

Sample	TDS	Quality	SAR	Quality	RSC	Quality	%Na	Quality	PI	Quality
S006	658	Excellent	1.25	Excellent	–0.78	Suitable	22.40	Suitable	47.84	Marginally suitable
S007	1006	Excellent	1.58	Excellent	–4.43	Suitable	22.62	Suitable	40.21	Marginally suitable
S008	740	Excellent	1.31	Excellent	–3.22	Suitable	21.74	Suitable	42.10	Marginally suitable
S009	1412	Excellent	2.04	Excellent	–7.78	Suitable	25.79	Suitable	39.42	Marginally suitable
S010	1126	Excellent	1.63	Excellent	–5.77	Suitable	22.65	Suitable	38.62	Marginally suitable
S011	990	Excellent	1.82	Excellent	–3.94	Suitable	25.65	Suitable	43.12	Marginally suitable
S012	2182	Good	1.98	Excellent	–14.88	Suitable	22.68	Suitable	32.39	Marginally suitable
S013	1072	Excellent	1.84	Excellent	–3.99	Suitable	26.73	Suitable	44.65	Marginally suitable
S014	1038	Excellent	2.57	Excellent	–3.24	Suitable	33.84	Marginally suitable	50.71	Marginally suitable
S015	828	Excellent	1.13	Excellent	–3.69	Suitable	19.37	Suitable	39.54	Marginally suitable
S016	1040	Excellent	1.32	Excellent	–5.02	Suitable	20.27	Suitable	37.62	Marginally suitable
S017	882	Excellent	1.48	Excellent	–4.46	Suitable	22.17	Suitable	40.16	Marginally suitable

(continued)

Table 3.4 (continued)

Sample	TDS	Quality	SAR	Quality	RSC	Quality	%Na	Quality	PI	Quality
S018	1396	Excellent	2.70	Excellent	–5.15	Suitable	30.80	Marginally suitable	45.63	Marginally suitable
S019	2214	Good	7.85	Excellent	–9.65	Suitable	58.43	Marginally suitable	65.68	Marginally suitable
S020	3144	Moderate	3.59	Excellent	–29.20	Suitable	30.37	Marginally suitable	34.64	Marginally suitable
S021	2382	Good	3.44	Excellent	–16.64	Suitable	33.71	Marginally suitable	41.14	Marginally suitable
S022	936	Excellent	1.60	Excellent	–4.80	Suitable	23.39	Suitable	40.86	Marginally suitable
S023	882	Excellent	1.15	Excellent	–5.95	Suitable	18.49	Suitable	35.36	Marginally suitable
S024	740	Excellent	1.57	Excellent	–2.77	Suitable	26.34	Suitable	47.40	Marginally suitable
S025	914	Excellent	1.02	Excellent	–10.07	Suitable	14.73	Suitable	28.04	Marginally suitable
S026	1090	Excellent	1.66	Excellent	–3.83	Suitable	24.00	Suitable	42.45	Marginally suitable
S027	1008	Excellent	2.27	Excellent	–2.15	Suitable	30.72	Marginally suitable	51.35	Marginally suitable
S028	1408	Excellent	2.33	Excellent	–7.70	Suitable	28.14	Suitable	41.62	Marginally suitable
S029	900	Excellent	4.53	Excellent	–1.99	Suitable	55.27	Marginally suitable	70.67	Marginally suitable

(continued)

Table 3.4 (continued)

Sample	TDS	Quality	SAR	Quality	RSC	Quality	%Na	Quality	PI	Quality
S030	1020	Excellent	1.22	Excellent	−5.94	Suitable	17.93	Suitable	34.58	Marginally suitable
S031	918	Excellent	1.46	Excellent	−5.20	Suitable	21.74	Suitable	38.89	Marginally suitable
S032	1104	Excellent	1.47	Excellent	−6.97	Suitable	20.25	Suitable	35.46	Marginally suitable
S033	990	Excellent	1.70	Excellent	−5.22	Suitable	23.87	Suitable	40.80	Marginally suitable
S034	1034	Excellent	3.65	Excellent	−1.99	Suitable	43.12	Marginally suitable	61.29	Marginally suitable
S035	1052	Excellent	7.07	Excellent	−1.71	Suitable	67.30	Unsuitable	79.64	Suitable
S036	894	Excellent	5.44	Excellent	−1.96	Suitable	61.56	Unsuitable	75.56	Suitable
S037	720	Excellent	1.91	Excellent	−7.15	Suitable	23.74	Suitable	39.00	Marginally suitable
S038	1322	Excellent	1.04	Excellent	−3.43	Suitable	17.79	Suitable	38.48	Marginally suitable
S039	748	Excellent	1.26	Excellent	−3.98	Suitable	21.05	Suitable	40.65	Marginally suitable
S040	604	Excellent	1.21	Excellent	−4.28	Suitable	21.92	Suitable	41.01	Marginally suitable
S041	726	Excellent	1.18	Excellent	−3.21	Suitable	20.32	Suitable	40.97	Marginally suitable
S043	748	Excellent	1.38	Excellent	−3.06	Suitable	23.35	Suitable	43.92	Marginally suitable

(continued)

Table 3.4 (continued)

Sample	TDS	Quality	SAR	Quality	RSC	Quality	%Na	Quality	PI	Quality
S044	1488	Excellent	1.47	Excellent	−9.38	Suitable	19.36	Suitable	32.76	Marginally suitable
S045	978	Excellent	2.26	Excellent	−3.22	Suitable	31.47	Marginally suitable	50.04	Marginally suitable
S046	910	Excellent	1.69	Excellent	−3.17	Suitable	25.21	Suitable	44.62	Marginally suitable
S047	950	Excellent	1.66	Excellent	−3.43	Suitable	24.73	Suitable	44.15	Marginally suitable
S048	1160	Excellent	2.44	Excellent	−3.39	Suitable	32.02	Marginally suitable	49.19	Marginally suitable
S049	2052	Good	1.73	Excellent	−10.98	Suitable	21.97	Suitable	33.89	Marginally suitable
S050	1702	Excellent	3.26	Excellent	−7.24	Suitable	37.61	Marginally suitable	50.23	Marginally suitable
S051	820	Excellent	1.07	Excellent	−3.24	Suitable	18.83	Suitable	39.95	Marginally suitable
S052	842	Excellent	1.13	Excellent	−3.14	Suitable	19.19	Suitable	40.15	Marginally suitable
S053	2500	Good	1.74	Excellent	9.75	Unsuitable	40.59	Marginally suitable	107.46	Suitable
S054	750	Excellent	0.82	Excellent	−3.63	Suitable	15.66	Suitable	38.10	Marginally suitable
S086	1030	Excellent	7.43	Excellent	−2.25	Suitable	67.28	Unsuitable	78.83	Suitable
S087	1186	Excellent	7.90	Excellent	−3.10	Suitable	66.83	Unsuitable	77.22	Suitable

(continued)

Table 3.4 (continued)

Sample	TDS	Quality	SAR	Quality	RSC	Quality	%Na	Quality	PI	Quality
S090	840	Excellent	5.61	Excellent	–1.77	Suitable	63.56	Unsuitable	77.40	Suitable
S091	924	Excellent	7.30	Excellent	–1.65	Suitable	68.17	Unsuitable	80.54	Suitable
S093	318	Excellent	4.05	Excellent	0.19	Suitable	66.66	Unsuitable	92.53	Suitable
N03	877.09	Excellent	4.47	Excellent	–0.50	Suitable	54.24	Marginally suitable	71.89	Marginally suitable
N04	876.53	Excellent	7.81	Excellent	1.54	Marginally suitable	73.09	Unsuitable	89.71	Suitable
N06	1049.82	Excellent	13.07	Good	3.81	Unsuitable	84.65	Unsuitable	99.56	Suitable
N07	298.45	Excellent	1.16	Excellent	–0.58	Suitable	29.60	Suitable	64.43	Marginally suitable
N08	458.01	Excellent	1.21	Excellent	–1.70	Suitable	26.25	Suitable	52.75	Marginally suitable
N11	1042.96	Excellent	6.40	Excellent	3.12	Unsuitable	62.67	Unsuitable	82.33	Suitable
N14	806.74	Excellent	1.48	Excellent	–2.41	Suitable	23.94	Suitable	44.78	Marginally suitable
O1	807.29	Excellent	1.45	Excellent	–2.25	Suitable	23.48	Suitable	44.65	Marginally suitable
O2	811.36	Excellent	1.46	Excellent	–2.37	Suitable	23.62	Suitable	44.56	Marginally suitable
O3	818.87	Excellent	1.51	Excellent	–2.23	Suitable	24.24	Suitable	45.22	Marginally suitable
N05	1733.50	Excellent	13.76	Good	1.45	Marginally suitable	79.89	Unsuitable	89.69	Suitable

(continued)

Table 3.4 (continued)

Sample	TDS	Quality	SAR	Quality	RSC	Quality	%Na	Quality	PI	Quality
N12	783.27	Excellent	1.73	Excellent	−3.14	Suitable	27.84	Suitable	47.08	Marginally suitable
N02	692.94	Excellent	3.26	Excellent	−0.84	Suitable	47.35	Marginally suitable	67.43	Marginally suitable
G1	1038.94	Excellent	5.81	Excellent	0.98	Suitable	60.51	Unsuitable	77.01	Suitable
G2	1227.04	Excellent	9.47	Excellent	3.18	Unsuitable	73.85	Unsuitable	88.38	Suitable
G3	606.34	Excellent	3.48	Excellent	0.03	Suitable	52.01	Marginally suitable	74.59	Marginally suitable
N01	595.36	Excellent	1.35	Excellent	−5.65	Suitable	25.32	Suitable	39.34	Marginally suitable
N09	962.66	Excellent	6.54	Excellent	−1.10	Suitable	66.11	Unsuitable	79.31	Suitable
N13	944.88	Excellent	7.98	Excellent	5.27	Unsuitable	68.55	Unsuitable	93.73	Suitable
N10	4488.75	Poor	8.88	Excellent	−29.03	Suitable	50.98	Marginally suitable	54.75	Marginally suitable
N15	1168.01	Excellent	3.27	Excellent	3.06	Unsuitable	47.25	Marginally suitable	77.78	Suitable
N16	1252.78	Excellent	8.10	Excellent	0.13	Suitable	69.21	Unsuitable	81.76	Suitable
N17	3179.41	Moderate	8.45	Excellent	−15.91	Suitable	55.95	Marginally suitable	61.11	Marginally suitable
N18	837.72	Excellent	2.25	Excellent	−3.29	Suitable	34.28	Marginally suitable	51.82	Marginally suitable
N19	643.19	Excellent	1.46	Excellent	−2.03	Suitable	26.44	Suitable	48.99	Marginally suitable

(continued)

Table 3.4 (continued)

Sample	TDS	Quality	SAR	Quality	RSC	Quality	%Na	Quality	PI	Quality
N20	853.47	Excellent	3.04	Excellent	–3.06	Suitable	42.38	Marginally suitable	58.41	Marginally suitable
N21	855.27	Excellent	2.81	Excellent	–2.24	Suitable	39.93	Marginally suitable	57.69	Marginally suitable
GC1	620.64	Excellent	4.39	Excellent	–0.88	Suitable	60.43	Unsuitable	75.80	Suitable
GC2	589.37	Excellent	4.54	Excellent	–0.63	Suitable	62.21	Unsuitable	78.79	Suitable
NT1	1969.10	Excellent	7.77	Excellent	–9.95	Suitable	59.93	Marginally suitable	65.67	Marginally suitable
NT2–1	2227.84	Good	15.83	Good	–2.61	Suitable	81.58	Unsuitable	87.33	Suitable
NT3	838.21	Excellent	1.60	Excellent	–4.66	Suitable	25.73	Suitable	42.96	Marginally suitable
N23	3538.14	Moderate	9.28	Excellent	–19.38	Suitable	56.44	Marginally suitable	60.93	Marginally suitable

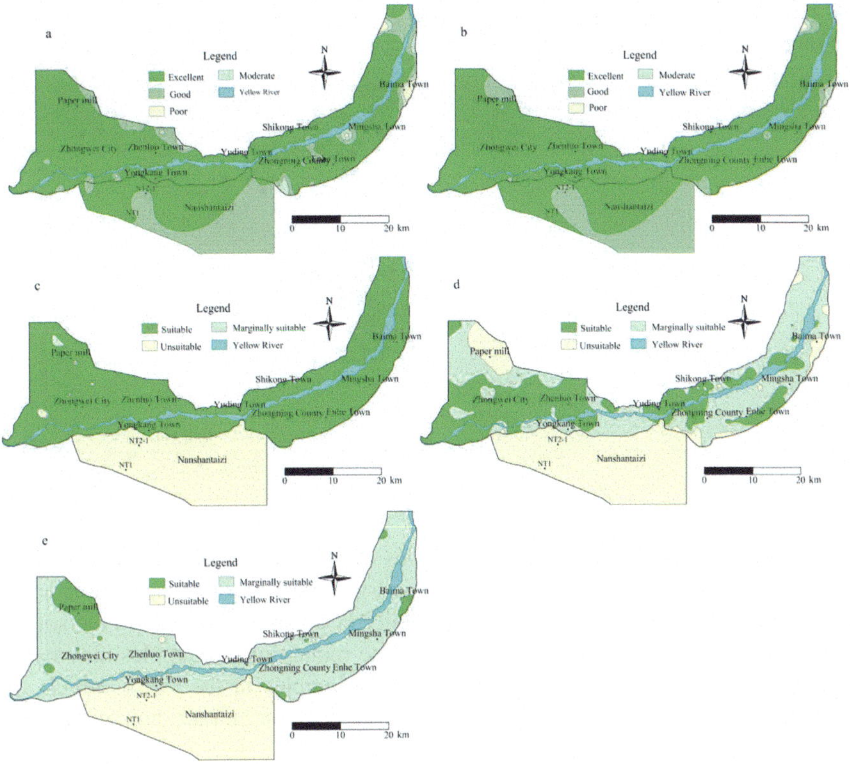

Fig. 3.3 Irrigation water quality zoning map based on **a** TDS, **b** SAR, **c** RSC, **d** %Na and **e** PI

groundwater irrigation water quality, considering simultaneously both salinity and alkalinity hazards (Singh et al. 2010). The Wilcox diagram (Fig. 3.4) simultaneously represents %Na and EC, dividing the classification into five zones. Zone I represents excellent irrigation water quality, Zone II represents good quality, Zone III represents acceptable quality, Zone IV represents marginal quality, and Zone V represents unsuitable quality (Li et al. 2013c). Water samples falling in Zones I and II are suitable for agricultural irrigation without causing salinity or alkalinity hazards. Water in Zone III carries a minor risk of alkalinity hazards, but with appropriate irrigation practices and controlled irrigation volumes, adverse effects can generally be avoided. Water in Zone IV poses risks of both salinity and alkalinity hazards and should only be used for irrigating specific salt-tolerant and alkali-tolerant crops. Water in Zone V is unsuitable for irrigation and may cause severe salinization and alkalization.

As shown in Fig. 3.4, there are 155 water samples in Zones I and II, accounting for 69.51% of the total water samples, 26 samples in Zone III (11.66%), 26 samples in Zone IV (11.66%), and 16 samples in Zone V (7.17%). The Wilcox diagram indicates that most of the groundwater in the area meets the requirements for irrigation water and will not cause severe salinization or alkalization. Similarly, the US Salinity

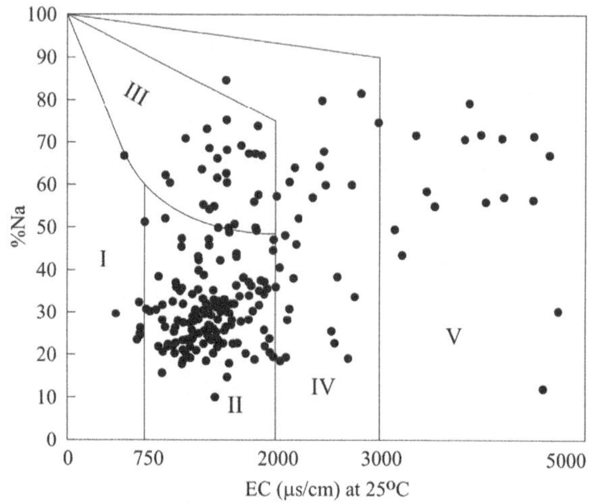

Fig. 3.4 Wilcox diagram of irrigation water quality

Laboratory (USSL) diagram (Fig. 3.5) shows that 190 water samples fall into the C2S1, C3S1, and C3S2 zones. These samples generally meet the irrigation water quality requirements and will not cause salinity or alkalinity hazards to the soil or crops under reasonable irrigation practices. The remaining 33 samples may cause salinity or alkalinity hazards, especially the 10 samples in the C4S4, which are associated with more severe salinization and alkalization and are not suitable for irrigation.

Overall, the groundwater in the study area generally meets the irrigation water quality standards. With appropriate irrigation practices and irrigation methods, it will not harm crops or cause soil salinization and alkalization. However, considering both the volume and the feasibility of groundwater extraction, the groundwater in Nanshantaizi is not suitable for agricultural irrigation due to its great depth and high extraction difficulty.

3.2 Main Environmental Issues Related to Groundwater

The Weining Plain is a typical irrigation area that relies on water diversion from the Yellow River for irrigation, which has significant impacts on groundwater environment in the plain. Intense human activities in the region have significantly disrupted the groundwater environment, leading to dramatic changes. The primary groundwater-related environmental issues in the Weining Plain include pollution caused by industrial production, agricultural activities and residential use, secondary soil salinization caused by improper irrigation practices, and continuous groundwater level decline due to long-term over-extraction.

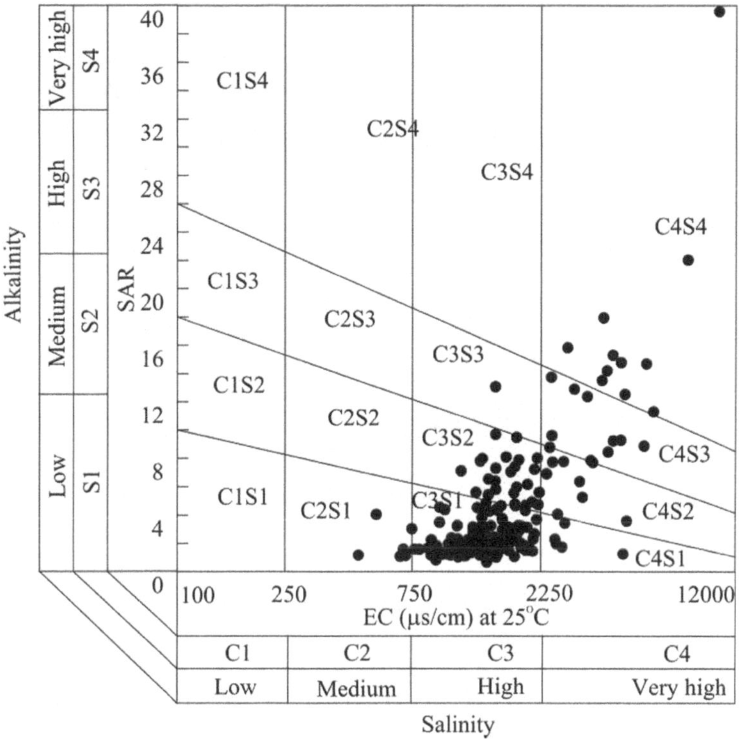

Fig. 3.5 USSL diagram of irrigation water quality

3.2.1 Groundwater Pollution

The Weining Plain has two major industrial parks: the Meili Industrial Park in Shapotou District, Zhongwei City, and the Shikong Industrial Park in Shikong Town, Zhongning County. The Meili Industrial Park was established along the edge of the Tengger Desert to support the expansion of Meili Paper Industry (Fig. 3.6). Prior to the construction of this industrial park, the area was a barren desert. To treat wastewater from the papermaking process, Meili Paper Industry established a big fast-growing forest within and around the industrial park. The forest can be divided into eastern and western sections: the western section is irrigated with pure Yellow River water, while the eastern section is irrigated with a mixture of papermaking wastewater and Yellow River water at a 1:1 ratio. Before mixing, the wastewater undergoes natural oxidation in four oxidation tanks. However, sampling results show that even after this treatment, the wastewater still contains high levels of pollutants, such as TDS (1884 mg/L), COD (252 mg/L), and Cl^- (563 mg/L). After mixing with Yellow River water at a 1:1 ratio, these values are reduced to TDS (1102 mg/L), COD (127 mg/L), and Cl^- (300 mg/L). Despite this dilution, COD levels remain high, leading to severe

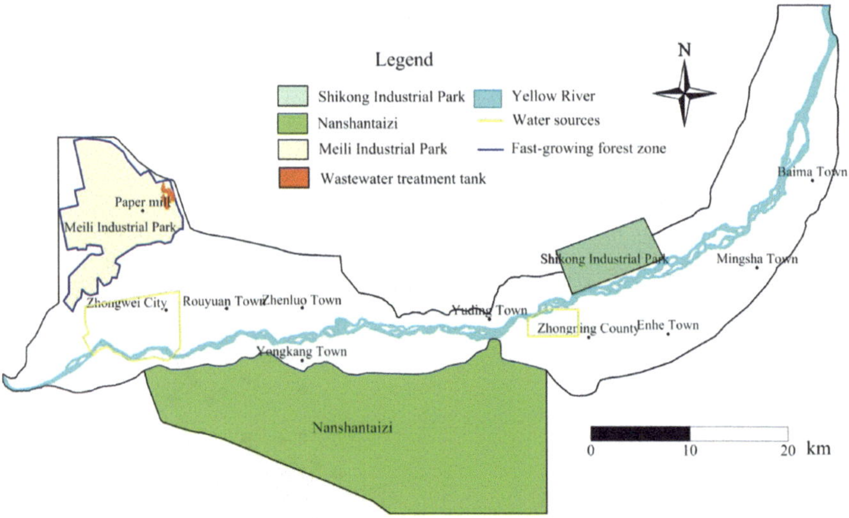

Fig. 3.6 Distribution of major industrial parks in the Weining plain

groundwater contamination in the irrigation zone. Comprehensive evaluations indicate that groundwater quality in most areas of the region falls into Category IV (Wu et al. 2013a). Residents have expressed strong dissatisfaction for two main reasons. First, farmers believe that excessive irrigation in the forest has caused downstream water tables to rise, disrupting normal agricultural production. Second, residents are concerned that contaminants in the irrigation water pose a threat to urban water sources and scattered residential wells downstream, thereby affecting their daily lives.

The Shikong Industrial Park primarily engages in metallurgy, chemical production, deep coal processing, building materials, and agricultural product processing. It is home to Tianyuan Manganese Industry, the world's largest electrolytic manganese producer. Due to the nature of its industries, the primary groundwater environmental issue in this area is heavy metal pollution in both soil and groundwater (Wu et al. 2013b; Li et al. 2014). Studies show that Mn concentrations in soil around the Tianyuan Manganese Industry exceed background values by more than four times (Li et al. 2014), and Mn concentrations in groundwater are also four times higher than the acceptable limit (Wu et al. 2013b).

Apart from industrial pollution posed by the two industrial parks, agricultural activities also significantly contribute to groundwater contamination. With an irrigation history spanning over 2000 years, the impact of agriculture on groundwater quality in the Weining Plain has long been established. The region suffers from extensive groundwater nitrate pollution, which has become a major constraint on normal production and daily life (Gong et al. 2025).

3.2.2 Secondary Soil Salinization

The Weining Plain is a typical Yellow River irrigation zone with a long history of irrigation. However, improper irrigation practices have led to widespread secondary soil salinization. This problem is particularly severe in the Nanshantaizi region to the southern bank of the Yellow River. Nanshantaizi is the third-order terrace of the Yellow River formed by alluvial and pluvial deposits from the northern foothills of Xiangshan Mountain. This area is an important grain production base that relies on lifted Yellow River water for irrigation.

Due to its higher elevation, which is several dozen to over a hundred meters higher than the alluvial plain, excess irrigation water infiltrates downward, raising the water table in the alluvial plain and transporting poor quality groundwater from Nanshantaizi to the lower alluvial plain. As a result, severe secondary soil salinization has developed in the alluvial plain downstream of the Nanshantaizi. According to Wu et al. (2014), 53.62% of the 79.28 km^2 downstream of the Nanshantaizi is classified as moderately saline soil, while 27.98% is heavily saline soil. In these regions, groundwater levels are typically shallower than 2 m. The expansion of saline soils has progressively reduced the amount of arable land, restricting local agricultural production.

3.2.3 Groundwater Level Decline

Groundwater is a crucial water supply source for both residential and industrial use in the Weining Plain. Surveys indicate that in 2012, total groundwater extraction in the region reached 870.35 million m^3/year. Of this, 83.7% was extracted from centralized sources, primarily for industrial production and drinking water, while the remaining 16.3% was extracted from private wells and irrigation wells, mainly for agricultural irrigation (Gong et al. 2025).

Nearly all centralized groundwater extraction takes place in the Zhongwei water source site and surrounding industrial enterprises. Years of excessive extraction have led to continuous groundwater level declines in the Zhongwei City. During investigation, many residents reported that many rural wells had dried up due to over extraction of groundwater in this area. In response, farmers had been forced to drill new wells at their own expense, leading to widespread dissatisfaction.

3.2.4 Outdated Water Resource Management

As an arid to semi-arid region, the Weining Plain experiences high evaporation and low precipitation, resulting in scarcity of surface water. However, since the Yellow River flows from west to east through the entire plain, the region is not inherently water-deficient. The persistence of groundwater environmental issues mentioned above is largely due to outdated water resource management, particularly the lack of coordinated allocation between groundwater and surface water.

To date, no comprehensive water resource allocation studies have been conducted for the Weining Plain. Additionally, outdated irrigation infrastructure and inefficient irrigation practices have led to significant wastage of water resources and worsening soil salinization. The lack of groundwater protection research further exacerbates these issues. Except for a regional hydrogeological survey conducted in the 1970s–80 s and in the 2010s, no systematic groundwater studies have been carried out. Consequently, there is little understanding of the region's exploitable groundwater resources or groundwater quality.

Furthermore, in the past, the Zhongwei City government prioritized rapid economic development over sustainable groundwater management, introducing numerous polluting industries such as Tianyuan Manganese Industry without fully considering the carrying capacity of groundwater resources. This has resulted in severe groundwater contamination. Although environmental awareness has grown in recent years, and local authorities have begun to pay more attention to environmental protection, efforts remain constrained by a lack of specialized research institutions and environmental monitoring teams, making groundwater protection efforts difficult to implement effectively.

3.3 Development of Irrigation Zone and Environmental Changes

The Weining Plain and Yinchuan Plain are two significant parts of the Ningxia Plain. During the Han Dynasty, the Weining Plain belonged to Anding Commandery (Wang 2005). The current geographical landscape and agricultural productivity of the Weining Plain are the results of extensive human intervention. The impact of human activities on this region has been so profound that it is now almost impossible to find areas untouched by anthropogenic influence.

Agricultural cultivation in the Weining Plain began during the Qin Dynasty, with large-scale agricultural infrastructure development commencing in the Han Dynasty. According to Wang (2005), the historical agricultural development of the Weining Plain can be categorized into three major phases: the first large-scale development during the Han Dynasty, the second wave of expansion during the Tang Dynasty, and the third major development during the Qing Dynasty. Before the Han Dynasty, Qin military forces stationed in the region for border defense initiated some agricultural activities, but animal husbandry remained the dominant agricultural economic activity. It was not until the reign of Emperor Wu of Han, particularly around 118 BC, that agriculture experienced rapid growth. Two large-scale migration events in 119 BC and 118 BC brought an influx of agricultural laborers to this region. These efforts transformed the original grasslands into extensive farmlands through artificial irrigation, shifting the economy from a pastoral-based system to a mixed agro-pastoral model, which persisted until the Ming Dynasty (Wang 2005).

During the Tang Dynasty, agricultural production in the plain further expanded. Building upon the irrigation canals developed during the Han Dynasty, agricultural activities extended into the lower-lying areas of northern Yinchuan, an adjacent plain to the north of the Weining Plain. These areas were originally characterized by harsh ecological conditions and severe soil salinization. The expansion of agriculture into these lowland areas indicates a continuous improvement in productivity. By the Qing Dynasty, the irrigation system in the Weining Plain had become even more refined and extensive. During the Qianlong period of the Qing Dynasty, the population of the entire Ningxia irrigation district reached approximately 1.35 million people (Wang 2005). The Qing government also established specialized water conservancy officials, and involved non-water conservancy officials in water management to enhance governance efficiency (Yue 2014).

Following the establishment of the People's Republic of China, the irrigation system of the Weining Plain underwent further development. Major agricultural projects, including the renovation of old canal systems, the excavation of drainage channels, and land consolidation, contributed to the formation of a more comprehensive irrigation and drainage network.

However, the rapid expansion of irrigated farmland has led to a series of environmental challenges. Historically, periods of significant agricultural expansion were also marked by extensive deforestation and severe soil erosion. The development of agriculture in the Weining Plain came at the cost of destroying the region's forest vegetation, significantly impacting the local ecological environment (Zhang 2006). Continuous hydraulic engineering and farmland expansion led to diversification in land use, alterations in population distribution patterns, changes in the spatial distribution of vegetation, and a severe imbalance among agricultural, pastoral, and forestry resources. This disruption of ecological equilibrium exacerbated desertification and intensified secondary soil salinization (Zhang 2010).

Additionally, increased water cycling and the emergence of an artificial hydrological system made the Weining Plain one of the most groundwater-rich areas in northwest China. However, human activities have also led to significant environmental changes, causing the disappearance of springs in the piedmont areas, intensified wind and water erosion, worsening soil degradation, and a drastic decline in both the number and variety of wild animal species (Wang 1983).

The over 2000-year history of agricultural development in the Weining Plain demonstrates that water has played a pivotal role in shaping the region. The expansion of irrigation in the Weining Plain has been closely linked to the exploitation of Yellow River water resources. Future development and ecological restoration in this region will largely depend on the sustainable management and efficient utilization of water resources. However, relying solely on Yellow River water is insufficient; groundwater must also be incorporated into water resource management. A comprehensive approach that integrates surface water and groundwater resources through coordinated allocation and joint regulation is necessary.

Currently, the Weining Plain is facing a series of water-related environmental challenges due to the unsustainable use of water resources, with groundwater degradation being particularly severe. Protecting these limited water resources, especially groundwater, is critical for the sustainable development of the plain and the well-being of its inhabitants. Thus, urgent attention must be given to this issue.

3.4 Groundwater System Changes Under Human Activities

The impact of human activities on the groundwater environment of the Weining Plain can be divided into two temporal scales: long-term and short-term. The long-term impact refers to the influence of agricultural activities on the Weining Plain, while the short-term impact is attributed to industrialization. The long-term influence of human agricultural activities on the Weining Plain can be traced back over 2000 years, whereas the short-term impact has been concentrated in the past few decades since the founding of the People's Republic of China, especially in the last 40 years following China's reform and opening-up policy. Under the combined influence of long-term and short-term human activities, the formation and occurrence conditions of groundwater in the Weining Plain have undergone unprecedented changes.

3.4.1 Evolution of the Groundwater Flow System

The groundwater-bearing system is a rock mass that stores groundwater, bounded by aquicludes or relatively impermeable boundaries. It consists of aquifers and relatively impermeable layers, and has a unified hydraulic connection within it. In contrast, the groundwater flow system is a system with a unified flow of water. The flow of water moves in a hierarchical and orderly manner, with regular spatiotemporal changes in water volume, salinity, and heat, forming a spatiotemporally ordered structure. Unlike the three-dimensional groundwater-bearing system, the groundwater flow system is a system with a four-dimensional concept (Zhang et al. 2011). Under natural conditions, the groundwater level fluctuates around a certain average value, and the water quality steadily tends to freshening or salinization. However, under the influence of human activities, the flow system undergoes significant changes.

In its original state, the groundwater flow system of the Weining Plain was confined within the range of the Yellow River alluvial plain and exhibited seasonal dynamic changes in response to rainfall. However, under the strong influence of human activities, the original groundwater flow system has undergone tremendous changes. For example, over the past 2000 years of irrigation development and agricultural activities, the dynamic changes in groundwater have shifted from rainfall infiltration type to irrigation infiltration type. In groundwater-bearing systems dominated by rainfall infiltration, the groundwater level generally rises only during the rainy season and then declines afterward. In contrast, in groundwater-bearing systems dominated by irrigation infiltration, the annual dynamic changes in groundwater level mainly depend on the infiltration of irrigation water during the irrigation period. In the Weining Plain, there are two irrigation periods: summer irrigation and winter irrigation. The groundwater level typically experiences two rising periods and two declining periods corresponding to the irrigation seasons.

Moreover, the extensive extraction of groundwater has also led to changes in the groundwater flow system. For example, near the water sources of Zhongwei City and Zhongning County, the concentrated extraction of groundwater has caused the water level to continuously decline in the source areas and their surroundings. The original groundwater flow system has disappeared, and a new groundwater flow system has formed, with groundwater flow directed toward the extraction center. The irrigation of the Meili Paper Industry's fast-growing forest is also a major cause to the groundwater flow system changes of the Weining Plain. The area where the Meili Paper Industry's fast-growing forestis located is on the edge of the Tengger Desert, with an original topography of hilly desert. With the construction of the fast-growing forest, a large oasis has been formed. In its original state, the groundwater level in the hilly desert area was deeply buried, with weak hydraulic connection to the Yellow River alluvial plain. It could be considered an independent groundwater flow system. However, with continuous irrigation of the fast-growing forest, the groundwater level in the forest area has risen, strengthening its hydraulic connection with the groundwater of the Yellow River alluvial plain and increasing the lateral recharge to the alluvial plain. To date, this lateral inflow accounts for 7.80% of the total recharge to the alluvial plain. The groundwater flow system of the hilly desert area has become closely integrated with the groundwater flow system of the Yellow River alluvial plain, forming a new unified groundwater flow system.

Under the long-term influence of human activities, the groundwater flow subsystem of the Nanshantaizi has also undergone significant changes. Due to human activities such as Yellow River irrigation and land reclamation, the surface vegetation types of Nanshantaizi have changed, and the groundwater level has been continuously rising. In some areas with intensive irrigation, the groundwater level has risen rapidly, while in areas with low-intensity irrigation or no irrigation, the groundwater level has risen slowly. This has increased the water level difference between high-intensity and low-intensity irrigation areas, promoting the exchange between them and forming new local groundwater flow microsystems from high-water-level areas to low-water-level areas. In its original state, the groundwater flow system of Nanshantaizi had a certain hydraulic connection with the groundwater flow system of the Yellow River alluvial plain, but the connection was not as close as it is today. Human activities have significantly raised the groundwater level of Nanshantaizi, increasing the hydraulic gradient between Nanshantaizi and the groundwater of the Yellow River alluvial plain, allowing more groundwater to flow laterally into the alluvial plain aquifer, forming a strong groundwater flow system from Nanshantaizi to the alluvial plain.

Human activities have not only altered the flow conditions of the groundwater flow system, but also changed its hydrochemical characteristics. In its original state, the recharge sources of the groundwater in the alluvial plain mainly included atmospheric precipitation, lateral inflow from the surrounding areas, and infiltration from the Yellow River in some sections. The hydrochemical characteristics of the groundwater in the Yellow River alluvial plain showed a process of increasing salinization from upstream to downstream (from west to east), while the infiltration of Yellow River water produced a process of groundwater freshening occurred from the mountainous areas to the direction of the Yellow River. However, under the strong influence of human activities, especially the extensive use of Yellow River water for irrigation, the groundwater in the alluvial plain has received a large amount of freshwater recharge, on the one hand, resulting in freshening in most areas. On the other hand, irrigation with Yellow River water has also increased the evaporation and concentration of groundwater, leading to deteriorating water quality in some areas. The combined effects have caused significant changes in the hydrochemical characteristics of the groundwater flow system in the alluvial plain. The original regular horizontal zoning of groundwater hydrochemistry under natural conditions has been disrupted, resulting in a chaotic and complex distribution of groundwater hydrochemistry.

Unlike the Yinchuan Plain, which has a multi-layered aquifer structure, the Weining Plain has only a single unconfined aquifer structure. This means that any changes in the groundwater flow system of the Weining Plain can have significant impacts on the local human survival and economic development. In the Yinchuan Plain, the main groundwater resources used are from confined aquifers, while the easily polluted unconfined aquifers are mainly used for irrigation. Therefore, even if the unconfined aquifer system is significantly affected by external factors, such as groundwater pollution, it will not have a substantial impact on the deeper confined water in the short term. The situation in the Weining Plain is different. In the Weining Plain, the unconfined aquifer is the only source of groundwater for various uses. The phreatic aquifer system can be considered the lifeline for the survival of the people in the Weining Plain. Any changes or pollution to this aquifer system would cause unimaginable consequences to the local population.

3.4.2 Changes in Groundwater Recharge Sources

Under natural conditions, the main sources of groundwater recharge in the Weining Plain are atmospheric precipitation and lateral inflow. Over 2000 years ago, large-scale human irrigation activities began to significantly alter the groundwater environment in this region, introducing new recharge sources such as irrigation infiltration and canal leakage. This disrupted the original groundwater balance, forcing the system to seek a new balance through changes in water levels. Population growth and industrial development have brought a large amount of domestic sewage and industrial wastewater. During the discharge of these wastewaters, a significant portion has infiltrated into the aquifer, becoming one of the sources of groundwater recharge.

The sources of groundwater recharge have thus increased from the original atmospheric precipitation and lateral inflow to include irrigation infiltration, canal leakage, atmospheric precipitation, lateral inflow, and infiltration of wastewater.

Human activities have not only increased the sources of recharge but also the volume of recharge. The development and utilization of groundwater in the Weining Plain have been vigorous. Both human life and industrial development have relied heavily on groundwater. The development of groundwater has caused changes in the groundwater level, further stimulating lateral inflow and increasing the areal recharge. For example, in the vicinity of Zhongwei City, centralized extraction has led to the formation of small-scale drawdown cones. The creation of these cones has caused groundwater from the surrounding areas to flow towards the center of the cone, further promoting the occurrence of lateral inflow from the periphery of the plain. Human activities have altered the hydraulic connection between the alluvial plain and the surrounding bedrock, directing a large amount of lateral flow towards the interior of the plain. For instance, the irrigation of the Meili Paper Industry's fast-growing forest and the water-lifting irrigation of Nanshantaizi have not only increased the sources of recharge, but also significantly changed the intensity of lateral inflow.

Human activities have also had a significant impact on the quality of recharge water. Firstly, as one of the new sources of groundwater recharge, domestic sewage and industrial wastewater have poorer quality compared to the natural recharge water. For example, the irrigation with wastewater from the Meili Paper Industry has poor water quality. Secondly, human activities have changed the groundwater flow dynamics, thereby altering the quality of recharge water. In the Nanshantaizi, for example, the lifting of Yellow River water for irrigation has caused a large amount of soluble salts from the vadose zone of Nanshantaizi to leach into the groundwater, increasing the concentration of various substances in the groundwater of Nanshantaizi. As Nanshantaizi is one of the important recharge areas for the groundwater of the alluvial plain, changes in the groundwater quality of Nanshantaizi will inevitably lead to the changes in the quality of groundwater in the alluvial plain. Thirdly, human activities can directly alter the hydrochemical composition of recharge sources. Acid rain caused by human activities is a good example of this.

3.4.3 Changes in Groundwater Discharge Pathways

Under natural conditions, the main discharge pathways of groundwater in the Weining Plain were evaporation and lateral outflow. However, with the intensification of human activities, not only have new discharge pathways been added, but the discharge volumes through these pathways have also been altered, allowing groundwater to find a new relative balance between recharge and discharge. First, human activities have increased the discharge pathways of groundwater. Under natural conditions, the Weining Plain mainly had evaporation and lateral outflow as discharge pathways. Extensive extraction by humans has gradually made artificial extraction a new discharge pathway for groundwater in the Weining Plain. Currently, the volume of artificial extraction accounts for 22.11% of the total groundwater discharge. To ensure normal agricultural production, many drainage ditches have been excavated in this area to discharge excess groundwater and prevent soil salinization. At present, the volume of groundwater discharged through drainage ditches accounts for 32.72% of the total groundwater discharge, making it the largest discharge pathway for groundwater in the Weining Plain. Groundwater discharge relying mainly on evaporation and lateral outflow has been replaced by a new pattern dominated by drainage ditch discharge and artificial extraction, supplemented by evaporation and lateral outflow.

Secondly, human activities have also changed the discharge volumes of groundwater through each pathway. For example, the irrigation has caused the groundwater level to rise continuously in some areas, increasing the evaporation discharge of groundwater in these areas. In contrast, in areas with concentrated groundwater extraction, the water level has continuously declined, resulting in a decrease in groundwater evaporation compared to natural conditions. Changes in the water level have also affected the discharge volume of lateral outflow. For example, in some areas, groundwater was originally recharged by the Yellow River, but due to human activities that raised the groundwater level, groundwater now discharges into the Yellow River. Changes in the groundwater level are also the main reason for changes in the discharge volume through drainage ditches. When the groundwater level is high, the discharge volume through drainage ditches increases, and vice versa. The temporal variation of drainage ditch discharge in the Weining Plain corresponds to the agricultural irrigation. During the irrigation period, the water level rises, and the discharge volume through drainage ditches increases. After irrigation stops, the groundwater level drops, and the discharge volume through drainage ditches sharply decreases. In fact, the most direct way human activities affect groundwater discharge volume is through groundwater extraction. With the continuous increase in population, economic development, and industrial growth in the Weining Plain, the volume of groundwater extraction has also been increasing year by year.

In summary, the impact of human activities on groundwater discharge pathways is mainly reflected in two aspects. One is the addition of new discharge pathways for groundwater, and the other is the alteration of discharge volumes through each pathway. This has gradually changed the natural groundwater discharge pattern, which was primarily based on evaporation and lateral flow, to a new pattern dominated by drainage ditch discharge and artificial extraction, supplemented by evaporation and lateral outflow.

3.5 Conclusions

This chapter provides an in-depth assessment of groundwater quality in the Weining Plain using the entropy-weighted TOPSIS method. It also outlines the primary environmental issues related to groundwater in the region and explores the evolution of the groundwater system under human activities. The main conclusions are as follows:

The entropy-weighted TOPSIS method was employed to evaluate groundwater quality in the Weining Plain based on 223 groundwater samples. The results indicate that only 15 of the samples met Grade II standards (suitable for drinking), 96 met Grade III standards (generally suitable for drinking), 94 required treatment before use (Grade IV), and 18 were unsuitable for drinking (Grade V). Poor water quality is prevalent in most areas of the Weining Plain, particularly in the alluvial plain region influenced by human activities, where groundwater quality is predominantly classified as Grade IV or V. The Nanshantaizi area also exhibits poor water quality, primarily due to high soil salinity in the vadose zone. Extensive irrigation leaches salts into the aquifer, exacerbating water quality issues. Groundwater in the western part of the study area, including Zhongwei City, Zhenluo Town, Mingsha Town, and Shikong Town in Zhongning County, generally meets Grade II or III standards.

The quality of groundwater in the Weining Plain for agricultural irrigation purposes was assessed using indicators such as TDS, SAR, RSC, %Na, and PI. The results showed that 192 of the 223 groundwater samples were rated as excellent, 17 as good, 9 as moderate, and 5 as poor based on TDS. According to SAR evaluation, 208 samples were excellent, 13 good, 1 moderate, and 1 poor. Based on RSC, 214 samples were suitable for irrigation, 3 generally suitable, and 6 unsuitable. Using %Na, 40.81% of samples were suitable, 43.05% generally suitable, and 16.14% unsuitable. Based on PI, 32 samples were suitable, 190 generally suitable, and 1 unsuitable. The Wilcox diagram indicated that 69.51% of the samples fell into Zones I and II (excellent and good), 11.66% into Zone III (acceptable), 11.66% into Zone IV (marginal), and 7.17% into Zone V (unsuitable). The USSL diagram showed that 190 samples were suitable for irrigation, while 33 may cause salinity or alkalinity hazards. Overall, the groundwater in the study area is generally suitable for irrigation. However, groundwater in Nanshantaizi is deemed unsuitable for irrigation due to its deep groundwater level depth and high extraction costs, despite meeting irrigation standards.

The Weining Plain faces several environmental challenges associated with groundwater, including industrial and agricultural pollution, secondary soil salinization, and groundwater level decline. The Meili Industrial Park and Shikong Industrial Park are major sources of industrial pollution. The Meili Paper Industry's wastewater irrigation has severely contaminated groundwater, with most areas classified as Grade IV water quality. The Shikong Industrial Park has caused significant heavy metal pollution in groundwater. Agricultural activities have led to extensive nitrate pollution in groundwater. Secondary soil salinization is particularly severe in the Nanshantaizi area, where improper irrigation practices have caused groundwater levels to rise, transporting poor-quality groundwater from Nanshantaizi to the alluvial plain and resulting in severe salinization downstream. Excessive groundwater extraction has led to continuous groundwater level declines in Zhongwei City. Outdated water resource management practices, coupled with a lack of coordinated allocation between groundwater and surface water, have exacerbated these groundwater environmental issues.

Human activities have significantly altered the groundwater flow system, recharge sources, and discharge pathways in the Weining Plain. The original groundwater flow system has shifted from being primarily rainfall-infiltration-driven to irrigation-infiltration-driven. Extensive groundwater extraction has caused water levels to decline in extraction centers, forming new groundwater flow systems. The irrigation of the paper industry's fast-growing forest and the water-lifting irrigation of Nanshantaizi have strengthened hydraulic connections with the Yellow River alluvial plain, creating a unified groundwater flow system. Groundwater recharge sources have expanded to include irrigation infiltration, canal leakage, wastewater infiltration, and more. Discharge pathways have shifted from evaporation and lateral outflow to a pattern dominated by drainage ditch discharge and artificial extraction, supplemented by evaporation and lateral outflow. Human activities have not only increased the number of recharge and discharge pathways but also altered the volumes of groundwater recharge and discharge, significantly impacting the groundwater system.

References

Aghazadeh N, Mogaddam AA (2010) Assessment of groundwater quality and its suitability for drinking and agricultural uses in the Oshnavieh Area, Northwest of Iran. J Environ Prot 1:30–40. https://doi.org/10.4236/jep.2010.11005

Almeida C, Quintar S, González P, Mallea M (2008) Assessment of irrigation water quality. A proposal of a quality profile. Environ Monit Assess 142:149–152. https://doi.org/10.1007/s10661-007-9916-7

Doneen LD (1964) Notes on water quality in Agriculture. Published as a water science and Engineering Paper 4001, Department of Water Sciences and Engineering. University of California

Gong M, Li P, Kong J (2025) Estimation of groundwater recharge and assessment of groundwater quality in the Weining Plain, China. Water 17(5):704. https://doi.org/10.3390/w17050704

Hakim MA, Juraimi AS, Begum M, Hasanuzzaman M, Uddin MK, Islam MM (2009) Suitability evaluation of groundwater for irrigation, drinking and industrial purposes. Am J Environ Sci 5(3):413–419

He S, Li P, Su F, Wang D, Ren X (2022) Identification and apportionment of shallow groundwater nitrate pollution in Weining Plain, northwest China, using hydrochemical indices, nitrate stable isotopes, and the new Bayesian stable isotope mixing model (MixSIAR). Environ Pollut 298:118852. https://doi.org/10.1016/j.envpol.2022.118852

Hwang C-L, Yoon K (1981) Multiple attribute decision making: Methods and applications a state-of-the-art survey. Springer-Verlag, Heidelberg. https://doi.org/10.1007/978-3-642-48318-9

Li P, Qian H, Howard KWF, Wu J, Lyu X (2014) Anthropogenic pollution and variability of manganese in alluvial sediments of the Yellow River, Ningxia, Northwest China. Environ Monit Assess 186:1385–1398. https://doi.org/10.1007/s10661-013-3461-3

Li PY, Qian H, Wu JH (2010a) Groundwater quality assessment based on improved water quality index in Pengyang County, Ningxia, Northwest China. E-J Chem 7(S1):S209–S216. https://doi.org/10.1155/2010/451304

Li PY, Qian H, Wu JH (2011a) Application of set pair analysis method based on entropy weight in groundwater quality assessment-a case study in Dongsheng City, Northwest China. E-J Chem 8(2):851–858. https://doi.org/10.1155/2011/879683

Li P, Qian H, Wu J (2011b) Hydrochemical formation mechanisms and quality assessment of groundwater with improved TOPSIS method in Pengyang county Northwest China. E-J Chem 8(3):1164–1173. https://doi.org/10.1155/2011/251918

Li PY, Qian H, Wu JH, Chen J (2013a) Sensitivity analysis of TOPSIS method in water quality assessment: I. Sensitivity to the parameter weights. Environ Monit Assess 185:2453–2461. https://doi.org/10.1007/s10661-012-2723-9

Li PY, Qian H, Wu JH, Zhang YQ, Zhang HB (2013c) Major ion chemistry of shallow groundwater in the Dongsheng Coalfield, Ordos Basin, China. Mine Water Environ 32(3):195–206. https://doi.org/10.1007/s10230-013-0234-8

Li PY, Wu JH, Qian H (2012) Groundwater quality assessment based on rough sets attribute reduction and TOPSIS method in a semi-arid area, China. Environ Monit Assess 184(8):4841–4854. https://doi.org/10.1007/s10661-011-2306-1

Li PY, Wu JH, Qian H, Chen J (2013b) Sensitivity analysis of TOPSIS method in water quality assessment II: Sensitivity to the index input data. Environ Monit Assess 185:2463–2474. https://doi.org/10.1007/s10661-012-2724-8

Li PY, Wu Q, Wu JH (2010b) Groundwater suitability for drinking and agricultural usage in Yinchuan area, China. Int J Environ Sci 1(6):1241–1249

Li Y, Li P, Liu L (2022) Source identification and potential ecological risk assessment of heavy metals in the Topsoil of the Weining Plain (Northwest China). Exposure and Health 14(2):281–294. https://doi.org/10.1007/s12403-021-00438-0

Liu L, Wu J, He S, Wang L (2022) Occurrence and distribution of groundwater fluoride and manganese in the Weining Plain (China) and their probabilistic health risk quantification. Exposure and Health 14(2):263–279. https://doi.org/10.1007/s12403-021-00434-4

Ragunath HM (1987) Groundwater. Wiley, New Delhi

Singh SK, Mahato MK, Neogi B, Singh KK (2010) Quality assessment of mine water in the Raniganj coalfield area, India. Mine Water Environ 29:248–262. https://doi.org/10.1007/s10230-010-0108-2

Singh V, Singh UC (2008) Assessment of groundwater quality of parts of Gwalior (India) for agricultural purposes. Indian J Sci Technol 1(4):1–5

Sivasankar V, Omine K, Msagati TAM, Senthil Kumar M, Chandramohan A (2014) Evaluation of groundwater quality in Madurai City, South India for drinking, irrigation and construction purposes. Arab J Geosci 7:3093–3107. https://doi.org/10.1007/s12517-013-0994-2

United States Salinity Laboratory (1954) Diagnosis and improvement of saline and alkali soils. US Dept of Agriculture (USDA), Agriculture Handbook 60, Washington

Wang SX (2013) Groundwater quality and its suitability for drinking and agricultural use in the Yanqi Basin of Xinjiang Province, Northwest China. Environ Monit Assess 185:7469–7484. https://doi.org/10.1007/s10661-013-3113-7

Wang Y (1983) Transformation of the Natural ecosystem in the Ningxia plain: A preliminary exploration of the impact of human activities on the ecological environment in the history of the Ningxia plain. Agric Hist China 4:10–22 (in Chinese)

Wang Y (2005) Evolution of Human-environment relationships in Ningxia. Ningxia People's Publishing House, Yinchuan (in Chinese)

Wilcox LV (1948) The quality of water for irrigation use. US Dept of Agriculture Tech Bull 1962, Washington

Wu JH, Li PY, Qian H, Chen J (2013a) Groundwater pollution in and around a paper wastewater-irrigated area, Northwest China. In: Proceeding of 2013 fourth international conference on digital manufacturing & automation. IEEE, pp 649–652. https://doi.org/10.1109/ICDMA.2013.154

Wu JH, Li PY, Qian H (2013b) Environmental chemistry of groundwater near an industrial area, Nortwest China. Asian J Chem 25(17):9795–9799. https://doi.org/10.14233/ajchem.2013.15355

Wu JH, Li PY, Qian H, Fang Y (2014) Assessment of soil salinization based on a low-cost method and its influencing factors in a semi-arid agricultural area, northwest China. Environ Earth Sci 71(8):3465–3475. https://doi.org/10.1007/s12665-013-2736-x

Xu F, Li P (2024) Biogeochemical mechanisms of iron (Fe) and manganese (Mn) in groundwater and soil profiles in the Zhongning section of the Weining Plain (northwest China). Sci Total Environ 939:173506. https://doi.org/10.1016/j.scitotenv.2024.173506

Yue Y (2014) State intervention in water management of the Ningxia plain during the Qing dynasty. Agric Archaeol 1:163–168 (in Chinese)

Zhang W (2006) On the relationship between the human activities and the historical evolution of the earth desertification in the Weining Plain Ningxia. Anc Mod Agric 3:44–48. (in Chinese)

Zhang J, Ren J (2011) The deficiencies and amendments of the calculation formulate of entropy and entropy weight in the theory of entropy. J Stat Inf 26(1):1–5

Zhang R, Liang X, Meng J et al (2011) Introduction to hydrogeology, 6th edn. Geological Publishing House, Beijing (in Chinese)

Zhang S (2010) A preliminary study on the environmental impact of water conservancy development in the Ningxia plain during the Qing dynasty. J Sanmenxia Polytech 9(4):88–92 (in Chinese)

Zheng X (2009) Groundwater pollution control. Huazhong University of Science and Technology Press, Wuhan (in Chinese)

Chapter 4
Health Risk Assessment of Groundwater in the Weining Plain

This chapter focuses on the health risk assessment of groundwater in the Weining Plain, building on the hydrogeochemical characteristics and groundwater quality issues identified in previous chapters. Groundwater is a critical resource for drinking and irrigation in the region, but rapid industrial and agricultural development has led to increasing contamination from pollutants such as nitrates, fluorides, and heavy metals, raising concerns about potential health impacts. To evaluate these risks, this study employs both deterministic and stochastic health risk assessment models, following the USEPA framework, to assess non-carcinogenic risks associated with key contaminants, including fluoride (F^-), nitrate (NO_3^-), nitrite (NO_2^-), and manganese (Mn). Groundwater samples were analyzed for various physicochemical parameters, and statistical methods were used to identify the exceedances of permissible limits. The deterministic approach quantifies the health risks through oral and dermal exposure pathways, while the stochastic approach, using Monte Carlo simulation, addresses uncertainties in exposure parameters. This chapter aims to provide a comprehensive understanding of the health risks posed by groundwater pollution in the Weining Plain, linking the findings from previous chapters to the potential impacts on human health.

4.1 Introduction

Groundwater is a vital natural resource for drinking and irrigation, and its quality significantly impacts human production and daily life (Moe and Rheingans 2006; Sheng et al. 2022). In recent years, rapid industrial and agricultural development has exacerbated groundwater pollution, making it a critical environmental issue that can no longer be ignored (Liu et al. 2025). Furthermore, severe groundwater pollution not only damages the ecological environment but also directly or indirectly affects human health through various pathways, with far-reaching consequences (Crayol

P. Li and V. Elumalai, *Groundwater Quality Under Agricultural Activities—Cases from China and South Africa*, Professional Practice in Earth Sciences,
https://doi.org/10.1007/978-3-031-98993-3_4

et al. 2024; Fida et al. 2024). The ingestion of water contaminated with high levels of pollutants, including nitrates, heavy metals, fluorides, and organic compounds, can lead to a range of health problems, from carcinogenic effects to non-carcinogenic diseases (Kavcar et al. 2009; Rajan et al. 2024; He et al. 2025).

Long-term exposure to high nitrate levels in water can cause various health issues, including methemoglobinemia, thyroid dysfunction, congenital disabilities, blue baby syndrome, and an increased risk of chronic diseases and cancer (Liu et al. 2023; Ward et al. 2018). Similarly, prolonged exposure to fluoride can lead to dental and skeletal fluorosis, as well as adverse effects on the thyroid, liver, kidneys, immune system, and nervous system (Kabir et al. 2020; Ghosh et al. 2013). The health risks associated with heavy metal contamination are equally concerning. Excessive intake of Cr^{6+} can cause nausea, vomiting, abdominal pain, and severe cases may lead to liver, kidney, and neurological damage, with long-term exposure increasing the risk of cancer (Chen et al. 2022). High exposure to lead can damage the nervous system, kidneys, and liver, induce genetic mutations, increase cancer risk, and, in severe cases, even lead to death (Abadin et al. 2019). While manganese is essential for neurological health and immune function, excessive intake can affect the nervous system, potentially leading to Parkinsonian symptoms and chronic manganese-induced pneumonia (Crossgrove and Zheng 2004). Therefore, understanding the health risks posed by groundwater pollution is crucial for ensuring public safety.

The Weining Plain, a typical Yellow River irrigation area, has a well-developed agricultural sector. However, the pursuit of high yields has led to excessive use of fertilizers and pesticides, resulting in severe nitrate pollution of groundwater, which poses a threat to human health (Li et al. 2014). Additionally, industrial activities in the Weining Plain contribute to water pollution, as wastewater discharge from some industrial enterprises has become a significant source of contamination, affecting both surface water and groundwater and jeopardizing the safety of drinking water for residents (Li et al. 2015). Therefore, conducting a health risk assessment in the Weining Plain is essential for comprehensively understanding the potential threats of groundwater pollution to human health. This assessment will provide a scientific basis for developing effective pollution control and health protection strategies, ultimately safeguarding the well-being of residents.

This chapter analyzes the groundwater composition of the Weining Plain and identifies key pollutants (F^-, NO_3^-, NO_2^-, and Mn) that may pose potential health risks to humans. Both deterministic and stochastic health risk assessments are conducted to evaluate the risks associated with these contaminants.

4.2 Health Risk Assessment Methods

The impact of various substances on human health can be quantitatively assessed through health risk evaluation. This study employs the health risk assessment model recommended by the United States Environmental Protection Agency (USEPA) to evaluate the health risks posed on adults and children. The assessment consists of four main components: hazard identification, dose–response assessment, exposure assessment, and risk characterization (USEPA 1991).

(1) Hazard Identification

Hazard identification involves recognizing the harmful effects of specific indicators. Its primary objective is to determine whether chemical, physical, or biological factors in the environment pose potential risks to human health (Silva and Beauvais, 2010). Based on the types of groundwater contaminants in the study area and the groundwater quality standards, this step identifies pollutants that may pose potential health risks to the population and determines their specific health hazards.

(2) Dose–Response Assessment

As the exposure dose of harmful substances increases, the probability and severity of adverse health effects in humans also increase. Dose–response assessment evaluates the adverse health effects of a specific chemical substance at given exposure levels and under specific exposure conditions (e.g., exposure pathways, duration, and frequency). Based on this assessment, substances in groundwater are classified as either carcinogenic or non-carcinogenic (Ruiz-Lara et al. 2025).

(3) Exposure Assessment

Exposure assessment focuses on determining the pathways of pollutant exposure, evaluation models, model parameters, and population exposure analysis. The exposure dose depends on the concentration of pollutants in the water and factors such as the exposure pathway, frequency, and duration. Exposure pathways primarily include ingestion (drinking), inhalation (breathing), and dermal contact (cleaning and washing) (Rashid et al. 2021).

(4) Risk Characterization

Risk characterization integrates the results of the previous three steps to provide a quantitative or qualitative description of the health risks posed by harmful substances (Panqing et al. 2023). It involves two key aspects: (1) Quantitative assessment of the health risk intensity posed by pollutants to the population, using specific data to measure the degree of risk. (2) In-depth analysis of potential risk levels, accompanied by clear and reasonable explanations of the results to help stakeholders understand the nature and impact of the risks. Risk characterization primarily includes carcinogenic risk assessment and non-carcinogenic risk assessment. The workflow of the health risk assessment is illustrated in Fig. 4.1.

4.3 Data Collection and Analysis

In this chapter, the study area focused at the alluvial plain of the Weining Plain, because the groundwater in the hilly desert and the Nanshantaizi is not used for drinking. A total of 145 groundwater samples were collected from the shallow aquifer

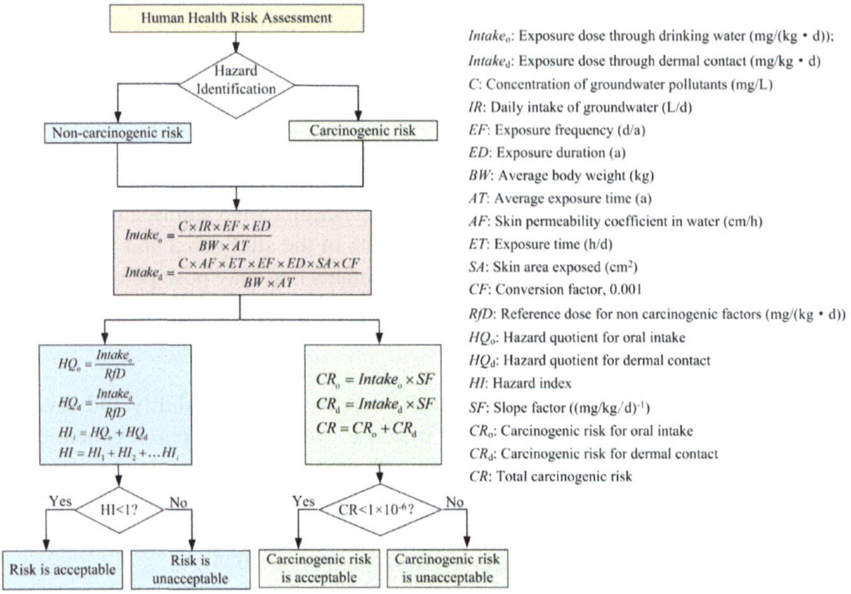

Fig. 4.1 Flow chart of human health risk assessment

in the alluvial plain of the Weining Plain in 2019. The locations of the sampling sites are shown in Fig. 4.2. The collection, storage, transportation, and processing of groundwater samples were strictly conducted in accordance with relevant standards (APHA 2005). During sampling, the pH of the water was measured on-site using a portable multi-parameter water quality meter. The water samples were then sent to the Soil and Water Testing Center of the Shaanxi Engineering Investigation and Research Institute for physicochemical analysis. The analyzed water quality parameters included K^+, Na^+, Ca^{2+}, Mg^{2+}, Cl^-, SO_4^{2-}, HCO_3^-, F^-, NH_4^+, NO_3^-, NO_2^-, Mn, Pb, Cr^{6+}, total dissolved solids (TDS), and total hardness (TH).

All samples were analyzed following the standard analytical methods recommended by the Ministry of Health of the People's Republic of China and the Standardization Administration of China (2006). Specifically, K^+, Na^+, Ca^{2+}, and Mg^{2+} were measured using atomic absorption spectrophotometry, Cl^-, SO_4^{2-}, HCO_3^-, and NH_4-N were determined using traditional titration methods, F^-, NO_3^-, and NO_2^- were analyzed using ion chromatography, Mn, Pb, and Cr^{6+} were measured using inductively coupled plasma atomic emission spectroscopy (ICP-AES), TDS was determined using the drying and weighing method, and TH was measured using EDTA titration.

The accuracy of the physicochemical analysis of groundwater samples was verified using the percentage of ion charge balance error (*%CBE*). The charge balance error for each sample was within the acceptable range of ± 5%, indicating that the physicochemical analysis results are reliable. The formula for calculating the charge balance error is as Eq. 4.1.

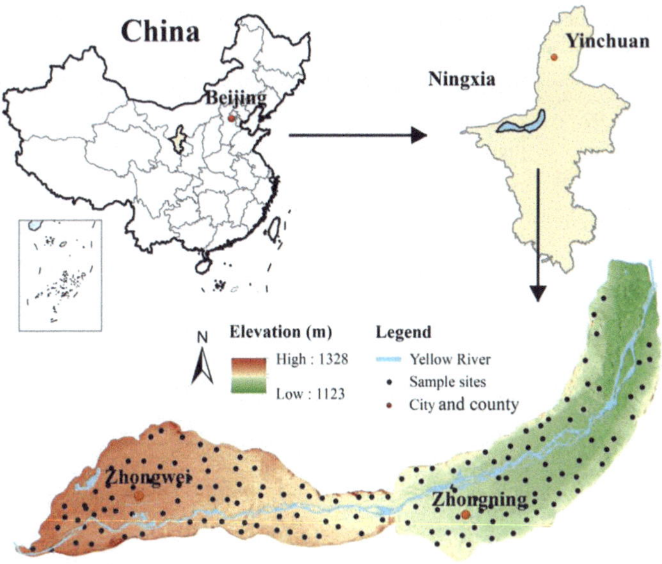

Fig. 4.2 Distribution of groundwater sampling sites

$$\%CBE = \frac{T_C - T_A}{T_C + T_A} \times 100\% \tag{4.1}$$

where, T_C and T_A represent the total concentrations of cations and anions, respectively, in milliequivalents per liter (meq/L).

4.4 Statistical Analysis of Major Water Quality Indicators

The residents of the Weining Plain rely entirely on groundwater for drinking water, making groundwater quality directly linked to their health and quality of life. As the primary source of drinking water, excessive concentrations of pollutants in groundwater can lead to various health issues, such as digestive system diseases, neurological damage, and cancer. Therefore, a systematic analysis of groundwater quality not only helps understand the basic pollution status in the study area but also provides a scientific basis for developing effective pollution control strategies.

This study conducted a statistical analysis of the major water quality indicators in the groundwater of the Weining Plain, and the results are presented in Table 4.1. The analyzed indicators include pH, TDS, TH, K^+, Na^+, Ca^{2+}, Mg^{2+}, Cl^-, SO_4^{2-}, HCO_3^-, F^-, NH_4^+, NO_3^-, NO_2^-, Mn, Pb, and Cr^{6+}. By comparing these values with the national drinking water standards (State Administration for Market Regulation of China and Standardization Administration of China 2017; WHO 2011), the exceedance rates of each indicator were evaluated.

Table 4.1 Statistics of major water quality parameters (Unit: mg/L except pH)

Statistics	Min	Max	Mean	Permissible limits	NSBL	RSBL (%)
pH	6.99	8.18	7.45	6.5–8.5	0	0.00
TDS	252.00	4280.00	1156.43	1000	61	42.07
TH	145.00	2302.00	604.40	450	104	71.72
K^+	2.09	48.30	10.84	12	38	26.21
Na^+	19.90	996.00	159.46	200	35	24.14
Ca^{2+}	26.10	511.00	116.86	200	11	7.59
Mg^{2+}	6.08	316.00	76.67	150	7	4.83
Cl^-	17.70	957.00	169.06	250	24	16.55
$SO_4{}^{2-}$	43.20	2113.00	362.45	250	75	51.72
$HCO_3{}^-$	97.60	726.00	397.70	600	7	4.83
F^-	0.17	5.10	0.81	1	30	20.69
NO_3–N	0.56	179.06	14.52	20	32	22.07
NO_2–N	0.00	1.89	0.05	0.02	53	36.55
NH_4–N	0.02	8.71	0.11	0.20	8	5.52
Mn	0.05	4.20	0.41	0.1	73	50.34
Pb	0.001	0.009	0.0016	0.05	0	0.00
Cr^{6+}	0.005	0.016	0.005	0.05	0	0.00

Note NSBL = number of samples beyond the permissible limit, RSBL = Ratio of samples beyond the permissible limit of national drinking water standards (State Administration for Market Regulation of China and Standardization Administration of China 2017; WHO 2011)

The statistical results reveal that groundwater in the study area exhibits significant exceedance of permissible limits. The main parameters exceeding the standards include TDS, TH, K^+, Na^+, Cl^-, $SO_4{}^{2-}$, F^-, NO_3-N, NO_2-N, and Mn. Among these, TH, $SO_4{}^{2-}$, and Mn have exceedance rates exceeding 50%, indicating that these pollutants are widespread and present at high concentrations in the groundwater. The exceedance rates for TDS and TH are 42.07% and 71.72%, respectively, reflecting the generally high mineralization and hardness of the groundwater, which may be related to the local geological background and agricultural activities. The high exceedance rate of $SO_4{}^{2-}$ (51.72%) may be attributed to industrial wastewater discharge and fertilizer use, in additional to some natural processes such as water–rock interactions and evaporation. The exceedance rates for F^-, NO_3-N, and NO_2-N are 20.69%, 22.07%, and 36.55%, respectively, indicating varying degrees of fluoride and nitrate pollution, likely due to dissolution of fluorite, and excessive use of agricultural fertilizers and pesticides. The exceedance rate for Mn is as high as 50.34%, suggesting that manganese pollution is widespread in the study area and may pose a potential threat to residents' health.

4.5 Health Risk Assessment

4.5.1 Deterministic Health Risk Assessment

In this study, the health hazards posed by F^-, NO_3^-, NO_2^-, and Mn were selected for the health risk assessment. They are limited to non-carcinogenic effects. Therefore, the health risk assessment model focuses solely on non-carcinogenic risks. This will help quantify the health hazards posed by these pollutants to residents and provide a scientific basis for targeted pollution control measures. NO_3^- and NO_2^- were converted to NO_3–N and NO_2–N, respectively, before the risk was assessed to match the reference dosage.

The average body weight, daily water intake, and exposure frequency, the exposure dose differ between adults and children. Thus, the population in the study area was divided into adults and children for separate analysis. Additionally, oral ingestion (drinking) and dermal contact (bathing and cleaning) were considered the major exposure pathways. The parameters for the health risk assessment model are listed in Table 4.2, and the reference doses for each pollutant are provided in Table 4.3.

The exposure dose parameters and pollutant reference doses were applied to the non-carcinogenic health risk assessment model shown in Fig. 4.1 to calculate the non-carcinogenic health risks for adults and children through drinking and dermal contact pathways. The results are presented in Table 4.4.

Table 4.2 Exposure dose parameters

Parameters	Units	Adults	Children
IR	L/d	1.5	0.7
EF_o	d/a	365	365
EF_d	d/a	200	200
ED	A	30	12
BW	kg	60	15
AT	d	$365 \times ED$	$365 \times ED$
AF	cm/h	0.001	0.001
ET	h/d	0.4	0.4
SA	cm^2	16,547	6597
CF	–	0.001	0.001

Table 4.3 Reference dose for each pollutant

Pollutant	RfD [mg/(kg·d)]	Reference
F^-	0.06	(Mahanta and Goswami, 2024)
NO_3–N	1.6	(Zeinali et al. 2024)
NO_2–N	0.1	(Zeinali et al. 2024)
Mn	0.14	(Machate et al. 2023)

Table 4.4 Deterministic non carcinogenic health risk assessment

Contaminants	Exposure pathways	Adults			Children		
		Min	Max	Mean	Min	Max	Mean
F^-	HQ_o	0.071	2.125	0.338	0.132	3.967	0.631
	HQ_d	1.71×10^{-4}	0.005	8.17×10^{-4}	2.73×10^{-4}	0.008	0.001
	HI_1	0.071	2.130	0.339	0.132	3.975	0.632
NO_3^-	HQ_o	0.009	2.798	0.227	0.016	5.223	0.424
	HQ_d	2.13×10^{-5}	0.007	5.49×10^{-4}	3.40×10^{-4}	0.011	8.75×10^{-4}
	HI_2	0.009	2.805	0.227	0.016	5.234	0.424
NO_2^-	HQ_o	2.28×10^{-4}	0.472	0.014	4.27×10^{-4}	0.880	0.026
	HQ_d	5.52×10^{-7}	0.001	3.32×10^{-5}	8.80×10^{-7}	0.002	5.30×10^{-5}
	HI_3	2.28×10^{-4}	0.473	0.014	4.27×10^{-4}	0.882	0.026
Mn	HQ_o	0.009	0.750	0.072	0.017	1.400	0.134
	HQ_d	2.16×10^{-5}	0.002	1.74×10^{-4}	3.44×10^{-5}	0.003	2.77×10^{-4}
	HI_4	0.009	0.752	0.072	0.017	1.403	0.134
Total	HQ_o	0.132	3.116	0.650	0.247	5.817	1.214
	HQ_d	3.20×10^{-4}	0.008	0.002	5.11×10^{-4}	0.012	0.003
	HI	0.133	3.124	0.652	0.248	5.829	1.217

As shown in Table 4.4, the health risk index (HI) for F^- in adults ranges from 0.071 to 2.130, with a mean of 0.339. Approximately 4.14% of the samples exceed the threshold (HI > 1), indicating that F^- in groundwater poses a non-carcinogenic health risk to adults. For NO_3^-, the HI ranges from 0.009 to 2.805, with a mean of 0.227 and an exceedance rate of 3.45%, which is lower than that of F^-. This suggests that both F^- and NO_3^- in groundwater have non-carcinogenic effects on adults. In contrast, the HI for NO_2^- and Mn ranges from 2.28×10^{-4} to 0.473 and 0.009 to 0.752, respectively, both below the threshold of 1. This indicates that the non-carcinogenic health risks posed by NO_2^- and Mn in groundwater are acceptable for adults.

The total non-carcinogenic health risk index for adults, calculated by summing the risk values of all pollutants, ranges from 0.133 to 3.124, with a mean of 0.652 and an exceedance rate of 15.17%. This suggests that adults in the study area face potential non-carcinogenic health risks, though not severe. The primary exposure pathway is oral ingestion, contributing 0.650 to the total HI, which accounts for 99.69% of the total health risk. The contribution of each pollutant to the total non-carcinogenic health risk for adults is ranked as follows: F^- (51.99%) > NO_3^- (34.82%) > Mn (11.04%) > NO_2^- (2.15%).

Children face more severe non-carcinogenic health risks compared to adults. The HI for F^- in children ranges from 0.132 to 3.975, with a mean of 0.632 and an exceedance rate of 11.72%, which is nearly three times higher than that for adults. This highlights the greater health hazards posed by F^- to children, which cannot be ignored. For NO_3^-, the HI ranges from 0.016 to 5.234, with a mean of 0.424 and an exceedance rate of 9.66%, lower than that of F^-. The HI for Mn ranges from 0.017 to 1.403, with a mean of 0.134, and only 3 groundwater samples exceed the threshold, indicating potential health risks for children. In contrast, the health risks posed by NO_2^- to children are lower than the risk posed by the other contaminants, with an HI ranging from 4.27×10^{-4} to 0.882.

Overall, F^-, NO_3^-, and Mn in groundwater pose varying degrees of health risks to children, with greater risk than that for adults. The total non-carcinogenic health risk index for children ranges from 0.248 to 5.829, with a mean of 1.217 and an exceedance rate of 46.90%. This means that nearly half of the children in the study area face unacceptable potential health risks, underscoring the need for urgent attention. Oral ingestion remains the primary exposure pathway, while health risks from dermal contact with groundwater are minimal. The contribution of each pollutant to the total health risk for children is ranked as follows: F^- (51.93%) > NO_3^- (34.84%) > Mn (11.01%) > NO_2^- (2.14%), which is consistent with the ranking for adults. F^- has the most significant potential health impact on both adults and children in the study area.

The spatial distribution of total non-carcinogenic health risks in the Weining Plain is shown in Fig. 4.3. The distribution patterns for adults and children are highly similar, with elevated risks primarily concentrated in the northwestern and western parts of Zhongwei City and the southern and northeastern parts of Zhongning County. Approximately 8.58% of the study area shows significant health risks to adults, while 62.96% of the area presents risks to children. This indicates that non-carcinogenic health risks are generally acceptable for adults in most areas, with relatively low hazard levels. In contrast, children face potential health risks in most regions.

4.5.2 *Stochastic Health Risk Assessment*

In groundwater pollution risk assessments, uncertainties in parameters and models are inevitable, leading to variability in the evaluation results. Therefore, stochastic analysis is an essential step in assessing the health risks posed by groundwater pollutants. Monte Carlo simulation, a widely used uncertainty analysis method, can effectively manage and evaluate the variability and uncertainty of factors affecting human health risk assessments (Eslami et al. 2022; Shalyari et al. 2019). Its fundamental principle is to define probability distributions for uncertain factors using randomness and statistical methods, followed by a random sampling to obtain simulated values at different confidence levels (Liu et al. 2022).

In this study, Oracle Crystal Ball software embedded in Excel was used to conduct the Monte Carlo simulations for groundwater health risks in the Weining Plain. The steps for uncertainty assessment are as follows:

Define Assumptions: Establish probability distributions for random parameters, including daily water intake (IR), exposure frequency (EF_o, EF_d), average body weight (BW), and pollutant concentrations. These are defined as assumption variables.

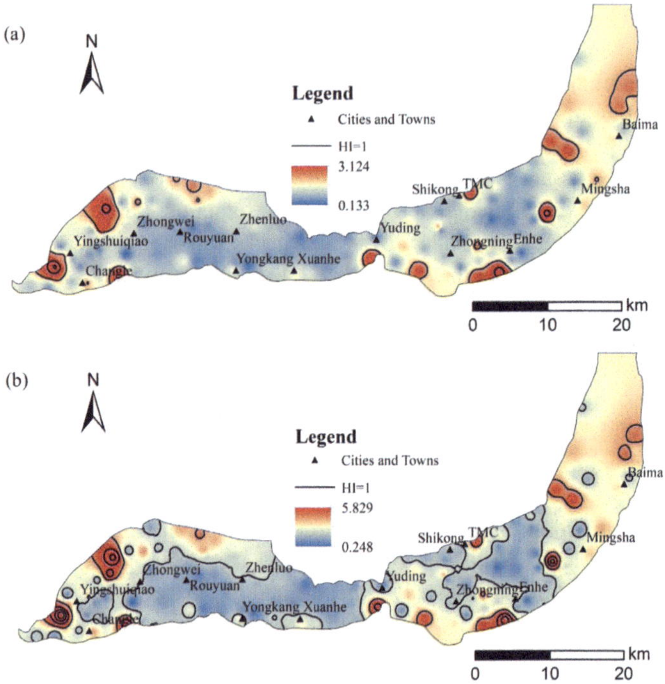

Fig. 4.3 Spatial distribution of total non-carcinogenic health risks for **a** adults; **b** children

Define Forecast Variables: Based on the health risk assessment models, define the non-carcinogenic health risks and total risks for adults and children as forecast variables.

Run Simulations: Set the number of trials to 10,000 and the confidence level to 95%. After simulation, statistical tables, frequency charts, and percentile tables can be obtained.

Sensitivity Analysis: Generate sensitivity charts to analyze the influence of different uncertain factors on the assessment results.

The concentrations of pollutants (F^-, NO_3^-, NO_2^-, Mn) were found to follow a lognormal distribution. The uncertainty factors, including daily water intake (*IR*, normal distribution), exposure frequency (EF_o, EF_d, triangular distribution), and average body weight (*BW*, lognormal distribution), were defined based on the investigation data (Smith 1994). The distribution parameters for pollutant concentrations and exposure factors are listed in Tables 4.5 and 4.6.

The Monte Carlo simulation results at the 95% confidence level are shown in Fig. 4.4. At a 95% confidence level, the average health risk index (HI) for adults in the Weining Plain is 0.52, following a lognormal distribution, with a 10.82% probability of exceeding the threshold of 1 (Fig. 4.4a). This indicates that adults in the study area face certain non-carcinogenic health risks, warranting attention. Among the four non-carcinogenic factors, the average HI of F^- is 0.28, with a 3.09% probability of

Table 4.5 Distribution of pollutant concentrations

Pollutants	Distribution
F^-	LN (0.811, 0.655)
NO_3–N	LN (14.520, 453.975)
NO_2–N	LN (0.055, 0.035)
Mn	LN (0.402, 0.402)

Table 4.6 Distribution of exposure parameters

Exposure parameters	Distribution	
	Adults	Children
IR	N (1.5, 0.15)	N (0.7, 0.07)
EF_o	TR (180, 345, 365)	TR (180, 345, 365)
EF_d	TR (180, 200, 220)	TR (180, 200, 220)
BW	LN (60, 6.0)	LN (15.0, 1.50)

Note LN—Lognormal, N—Normal, TR—Triangular

exceeding the threshold (Fig. 4.4c). The average HI of NO_3^- is 0.10, with a 2.10% probability of exceeding the threshold (Fig. 4.4e). The average HIs of NO_2^- and Mn are 0.01 and 0.06, respectively, with almost no exceedances (Fig. 4.4g) or only a 0.10% probability of exceeding the threshold (Fig. 4.4i), indicating acceptable health risks. Overall, adults in the study area face potential non-carcinogenic health risks, with F^- being the primary risk factor, followed by NO_3^-. The risks posed by Mn and NO_2^- are minimal.

For health risk of children, at the 95% confidence level, the average HI is 0.99, following a lognormal distribution, with a 35.62% probability of exceeding the threshold of 1 (Fig. 4.4b). This indicates that children in the study area face significant potential health risks, requiring urgent attention. Among the four non-carcinogenic factors, the average HI of F^- is 0.50, with a 12.56% probability of exceeding the threshold (Fig. 4.4d). The average HI of NO_3^- is 0.32, with a 6.39% probability of exceeding the threshold (Fig. 4.4f). NO_2^- and Mn pose low health risk, with the average HIs being 0.02 and 0.10, respectively, with almost no exceedances (Fig. 4.4h) or a 0.77% probability of exceeding the threshold (Fig. 4.4j). Nearly half of the study area poses unacceptable health risks to children, with F^- being the primary contributor.

The Monte Carlo simulation results are consistent with the deterministic model, showing that children face significantly higher health risks than adults, and F^- is the primary risk factor. The ranking of non-carcinogenic health risks for both adults and children is: $F^- > NO_3^- > Mn > NO_2^-$. The absolute value of sensitivity reflects the strength of the factor's influence on the results, while the sign (positive or negative) indicates the direction of influence (Gupta and Gupta 2023). The sensitivity results for total non-carcinogenic risks in adults and children are shown in Fig. 4.5.

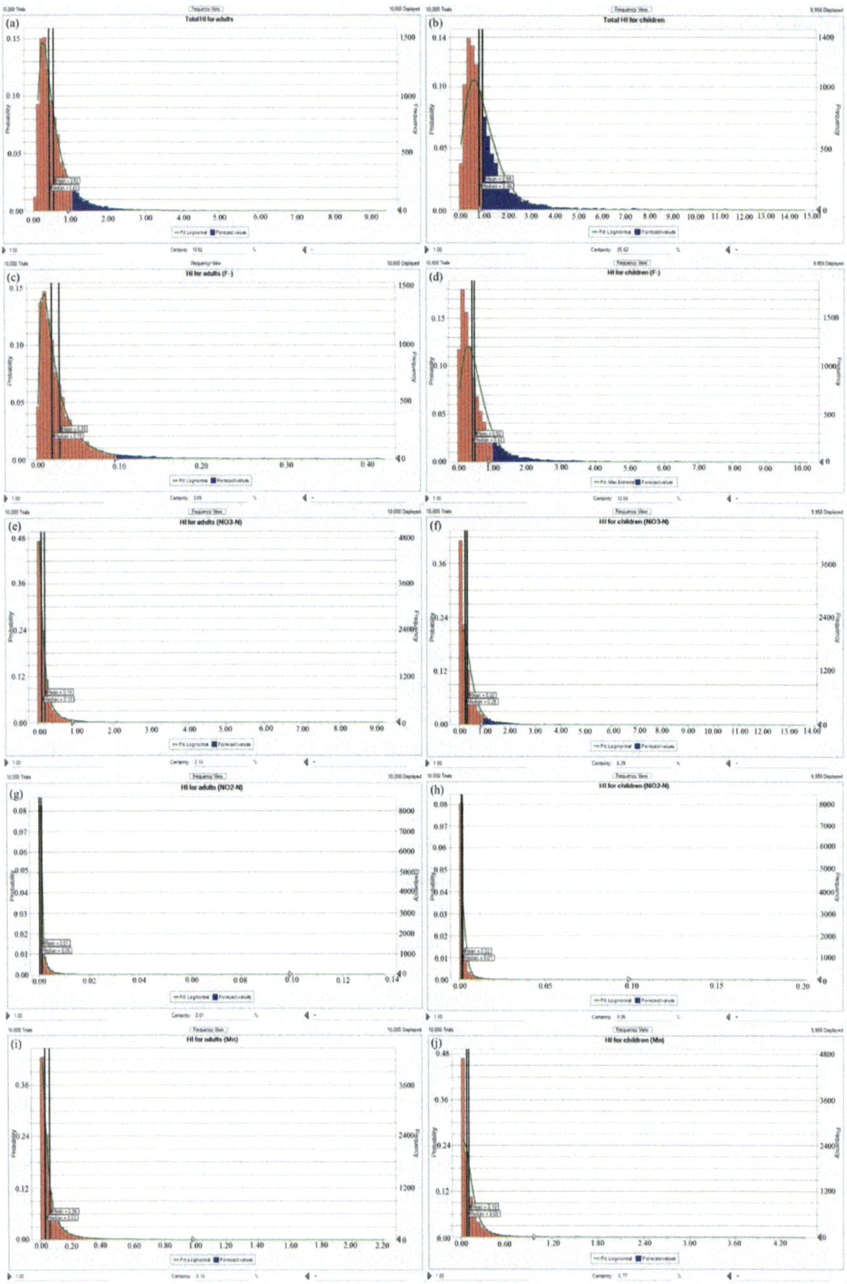

Fig. 4.4 Uncertainty simulation of non-carcinogenic risks. **a** Total HI for adults; **b** Total HI for children; **c** HI for adults regarding F^-; **d** HI for children regarding F^-; **e** HI for adults regarding NO_3-N; **f** HI for children regarding NO_3-N; **g** HI for adults regarding NO_2-N; **h** HI for children regarding NO_2-N; **i** HI for adults regarding Mn; **j** HI for children regarding Mn

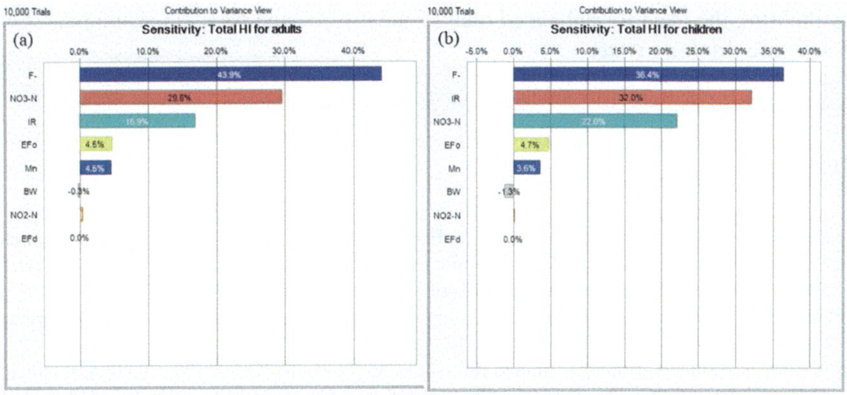

Fig. 4.5 Total non-carcinogeni risk sensitivity map for **a** adults; **b** children

According to the sensitivity analysis, the total non-carcinogenic risk is primarily influenced by pollutant concentrations in groundwater, daily water intake (*IR*), exposure frequency (EF_o, EF_d), and body weight (*BW*). F^- exhibits the highest sensitivity, contributing 43.9% and 36.4% to the total non-carcinogenic risks for adults and children, respectively. Sensitivity analysis shows that pollutant concentrations, daily water intake (*IR*), and exposure frequency (EF_o, EF_d) are positively correlated with the non-carcinogenic risks, while body weight (*BW*) is negatively correlated. This means that increased water intake raises health risks, and children with lighterweight than adults face higher risks. Daily water intake has the highest contribution among all exposure parameters, highlighting the need to control harmful substances in groundwater and strengthen the management of F^- and NO_3^- in the Weining Plain.

4.6 Conclusions

This study employed both deterministic and stochastic models to assess the non-carcinogenic health risks posed by F^-, NO_3^-, NO_2^-, and Mn in the Weining Plain for adults and children. The results indicate that F^- is the primary pollutant contributing to health risks, followed by NO_3^-, while the health hazards caused by NO_2^- and Mn are relatively minor and within acceptable limits.

The average health risk index (HI) of F^- for adults is 0.339, with an exceedance rate (HI > 1) of 4.14%; for children, the average HI is 0.632, with an exceedance rate of 11.72%. The average HI of NO_3-N for adults is 0.227, with an exceedance rate of 3.45%; for children, the average HI is 0.424, with an exceedance rate of 9.66%. The stochastic risk assessment indicates that the average total HI for adults is 0.52, with an exceedance probability of 10.82%; for children, the average total HI is 0.99, with an exceedance probability of 35.62%. Oral ingestion is the main exposure pathway, highlighting the need for enhanced management of F^- and NO_3^- in groundwater to ensure safe drinking water.

References

Abadin H, Taylor J, Buser MC, Scinicariello F, Przybyla J, Klotzbach JM, Diamond GL, Chappell LL, Mcllroy LA (2019) Toxicological profile for lead: draft for public comment. https://stacks.cdc.gov/view/cdc/78185/cdc_78185_DS1.pdf

APHA (2005) Standard methods for the examination of water and wastewater, 21st ed. American Public Health Association, Washington, DC, USA

Chen F, Ma J, Akhtar S, Khan ZI, Ahmad K, Ashfaq A, Nawaz H, Nadeem M (2022) Assessment of chromium toxicity and potential health implications of agriculturally diversely irrigated food crops in the semi-arid regions of South Asia. Agric Water Manag 272:107833. https://doi.org/10.1016/j.agwat.2022.107833

Crayol E, Huneau F, Garel E, Zuffianò LE, Limoni PP, Romanazzi A, Mattei A, Re V, Knoeller K, Polemio M (2024) Investigating pollution input to coastal groundwater-dependent ecosystems in dry Mediterranean agricultural regions. Sci Total Environ 954:176015. https://doi.org/10.1016/j.scitotenv.2024.176015

Crossgrove J, Zheng W (2004) Manganese toxicity upon overexposure. NMR in Biomedicine: an International Journal Devoted to the Development and Application of Magnetic Resonance in Vivo 17(8):544–553. https://doi.org/10.1002/nbm.931

Eslami H, Esmaeili A, Razaeian M, Salari M, Hosseini AZ, Mobini M, Barani A (2022) Potentially toxic metal concentration, spatial distribution, and health risk assessment in drinking groundwater resources of southeast Iran. Geosci Front 13(1):101276. https://doi.org/10.1016/j.gsf.2021.101276

Fida M, Li PY, Alam SMK, Wang YH, Nsabimana A, Sundar P (2024) Review of groundwater nitrate pollution from municipal landfill leachates: Implications for environmental and human health and leachate treatment technologies. Exposure and Health 16:1225–1249. https://doi.org/10.1007/s12403-023-00624-2

Ghosh A, Mukherjee K, Ghosh SK, Saha B (2013) Sources and toxicity of fluoride in the environment. Res Chem Intermed 39:2881–2915. https://doi.org/10.1007/s11164-012-0841-1

Gupta S, Gupta SK (2023) Application of Monte Carlo simulation for carcinogenic and non-carcinogenic risks assessment through multi-exposure pathways of heavy metals of river water and sediment, India. Environ Geochem Health 45:3465–3486. https://doi.org/10.1007/s10653-022-01421-7

He XD, Zheng ZW, Lu H (2025). Groundwater and public health: understanding the links and mitigating risks. In: Li PY, He XD, Wu JH, Elumalai V (eds) Sustainable groundwater and environment: Challenges and solutions. Springer Hydrogeology. Springer, Cham. https://doi.org/10.1007/978-3-031-82194-3_4

Kabir H, Gupta AK, Tripathy S (2020) Fluoride and human health: systematic appraisal of sources, exposures, metabolism, and toxicity. Crit Rev Environ Sci Technol 50(11):1116–1193. https://doi.org/10.1080/10643389.2019.1647028

Kavcar P, Sofuoglu A, Sofuoglu SC (2009) A health risk assessment for exposure to trace metals via drinking water ingestion pathway. Int J Hyg Environ Health 212:216–227. https://doi.org/10.1016/j.ijheh.2008.05.002

Li PY, Wu JH, Qian H, Lyu XS, Liu HW (2014) Origin and assessment of groundwater pollution and associated health risk: A case study in an industrial park, northwest China. Environ Geochem Health 36:693–712. https://doi.org/10.1007/s10653-013-9590-3

Li PY, Qian H, Howard KWF, Wu JH (2015) Heavy metal contamination of Yellow River alluvial sediments, northwest China. Environ Earth Sci 73:3403–3415. https://doi.org/10.1007/s12665-014-3628-4

Liu H, Hu XN, Zhu HH, Xing LT, Han Z, Hu K, Wang XZ, Huang LX (2025) Analysis of the hydrogeochemical characteristics of groundwater and identification of pollution sources in facility agriculture areas using self-organizing neural networks. Environ Earth Sci 84:161. https://doi.org/10.1007/s12665-025-12114-6

Liu LN, Wu JH, He S, Wang L (2022) Occurrence and distribution of groundwater fluoride and manganese in the Weining Plain (China) and their probabilistic health risk quantification. Expo Health 14:263–279. https://doi.org/10.1007/s12403-021-00434-4

Liu ZJ, Wang XH, Jia SQ, Mao BY (2023) Multi-methods to investigate spatiotemporal variations of nitrogen-nitrate and its risks to human health in China's largest fresh water lake (Poyang Lake). Sci Total Environ 863:160975. https://doi.org/10.1016/j.scitotenv.2022.160975

Machate DJ (2023) Anthropogenic hyperactivity for natural resources increases heavy metals concentrations in the environment: Toxicity of healthy food and cancer risks estimated. J Trace Elem Miner 4:100057. https://doi.org/10.1016/j.jtemin.2023.100057

Mahanta N, Goswami S (2024) Groundwater vulnerability to fluoride pollution and health risk assessment in the western part of Odisha, India. Environ Sci Pollut Res 31:35878–35896. https://doi.org/10.1007/s11356-024-33620-3

Ministry of Health of PRC, Administration S, of PRC (2006) Standard Examination Methods for Drinking Water; GB/T5750-2006. Standards Press of China, Beijing, China (In Chinese)

Moe CL, Rheingans RD (2006) Global challenges in water, sanitation and health. J Water Health 4:41–57. https://doi.org/10.2166/wh.2006.0043

Panqing Y, Abliz A, Xiaoli S, Aisaiduli H (2023) Human health-risk assessment of heavy metal–contaminated soil based on Monte Carlo simulation. Sci Rep 13:7033. https://doi.org/10.1038/s41598-023-33986-3

Rajan M, Karunanidhi D, Jaya J, Preethi B, Subramani T, Aravinthasamy P (2024) A comprehensive review on human health hazards due to groundwater contamination: A global perspective. Phys Chem Earth, Parts A/B/C 135:103637. https://doi.org/10.1016/j.pce.2024.103637

Rashid A, Ayub M, Javed A, Khan S, Gao XB, Li CC, Ullah Z, Sardar T, Muhammad J, Nazneen S (2021) Potentially harmful metals, and health risk evaluation in groundwater of Mardan, Pakistan: Application of geostatistical approach and geographic information system. Geosci Front 12(3):101128. https://doi.org/10.1016/j.gsf.2020.12.009

Ruiz-Lara K, Galar-Martínez M, García-Medina S et al (2025) Human health risk assessment by exposure to contaminants from an urban reservoir: a pilot study in the Madin Dam (México). Environ Monit Assess 197:193. https://doi.org/10.1007/s10661-025-13640-3

Shalyari N, Alinejad A, Hashemi AHG, RadFard M, Dehghani M (2019) Health risk assessment of nitrate in groundwater resources of Iranshahr using Monte Carlo simulation and geographic information system (GIS). MethodsX 6:1812–1821. https://doi.org/10.1016/j.mex.2019.07.024

Sheng DR, Meng XH, Wen XH, Wu J, Yu HJ, Wu M (2022) Contamination characteristics, source identification, and source-specific health risks of heavy metal(loid)s in groundwater of an arid oasis region in Northwest China. Sci Total Environ 841:156733. https://doi.org/10.1016/j.scitotenv.2022.156733

Silva MH, Beauvais SL (2010) Human health risk assessment of endosulfan. I: Toxicology and hazard identification. Regul Toxicol Pharmacol 56(1):4–17. https://doi.org/10.1016/j.yrtph.2009.08.013

Smith RL (1994) Use of Monte Carlo simulation for human exposure assessment at a superfund site. Risk Anal 14:433–439. https://doi.org/10.1111/j.1539-6924.1994.tb00261.x

State Administration for Market Regulation of China, Standardization Administration of China (2017) Standard for groundwater quality, GB/T 14848–2017. Available at https://openstd.samr.gov.cn/bzgk/std/newGbInfo?hcno=F745E3023BD5B10B9FB5314E0FFB5523. accessed March 24, 2025

USEPA (1991) Human health evaluation manual, supplemental guidance, standard default exposure factors. US environmental protection agency, Washington, USA, DC

Ward MH, Jones RR, Brender JD, De Kok TM, Weyer PJ, Nolan BT, Villanueva CM, Van Breda SG (2018) Drinking water nitrate and human health: An updated review. Int J Environ Res Public Health 15:1557. https://doi.org/10.3390/ijerph15071557

WHO (2011) Guidelines for drinking water quality. https://iris.who.int/handle/10665/44584 (Accessed on 24 Jan 2025)

Zeinali M, Heshmati A, Mohammadi Y, Nili-Ahmadabadi M, Nili-Ahmadabadi A (2024) Distribution of nitrate/nitrite and toxic metals in the soil-potato system and its health risk assessment in Iran. Environ Geochem Health 46:131. https://doi.org/10.1007/s10653-024-01897-5

Chapter 5
Land Use Land Cover Changes on Groundwater Quality in the Weining Plain

The quality of groundwater in the Weining Plain is increasingly threatened by pollution from agricultural and industrial activities, as well as changes in land use and land cover (LULC). Understanding the relationship between LULC patterns and groundwater quality is essential for sustainable water resource management and urban planning in this region. This chapter employs a comprehensive approach to analyze the impact of LULC changes on groundwater quality in the Weining Plain. LULC data from 2000 to 2020 are analyzed to quantify changes in land use, particularly the expansion of urban and rural settlements and the reduction of agricultural land. The Curved Streamline Searchlight Model (CS-SLM) is applied to determine the influence area of LULC types around monitoring wells. Spearman correlation analysis is used to evaluate the relationship between groundwater quality indicators and LULC patterns. The findings are expected to inform sustainable land use planning and groundwater management strategies, ensuring the protection of water resources in the face of ongoing urbanization and agricultural intensification.

5.1 Introduction

Water is the source of life, and the quantity, quality, spatiotemporal distribution, cycling characteristics, and availability of water resources often determine the population, society, economy, and ecological structure of a region (Li et al. 2016). In arid and semi-arid regions where surface water resources are limited, groundwater often serves as a critical, and sometimes the only, source of water for human activities and ecological balance (Li and Wu 2023). In recent years, under the dual pressures of climate change and human activities, groundwater environmental issues have become increasingly prominent, attracting growing attention from researchers (He and Wu 2019a).

P. Li and V. Elumalai, *Groundwater Quality Under Agricultural Activities—Cases from China and South Africa*, Professional Practice in Earth Sciences,
https://doi.org/10.1007/978-3-031-98993-3_5

Over the past few decades, rapid urban expansion has significantly altered the LULC patterns (Alqurashi and Kumar 2019). These changes play a crucial role in modifying the composition of groundwater solutes. New land cover patterns can alter groundwater circulation, leading to the dilution and transport of solutes (Barron et al. 2013). Additionally, urbanization introduces new pollution sources that may infiltrate aquifers and contaminate groundwater (Xu et al. 2022). For instance, Gu et al. (2012), through long-term research in the Greater Shanghai area, found that the increase in nitrate concentrations in groundwater was primarily driven by human activities, such as excessive use of nitrogen fertilizers and animal manure, discharge of domestic and industrial wastewater, and increased atmospheric nitrogen deposition. Jia et al. (2018) demonstrated that residential, commercial, industrial, and agricultural areas significantly influence the distribution of groundwater pollution. Li et al. (2021) analyzed the exposure of fluoride and arsenic under different land use types and found that human-induced land use changes can lead to variations in fluoride and arsenic concentrations in groundwater. Therefore, the impact of LULC patterns on groundwater pollution has become a critical issue in urban land use planning and water resource management and protection (Reddy et al. 2012).

However, previous studies investigating the correlation between groundwater quality and land use often employed circular contribution zones centered on monitoring wells (Worrall and Kolpin 2004; Singh et al. 2013). These studies overlooked the influence of regional and local groundwater flow on the movement and transformation of pollutants (Johnson and Belitz 2009). To address this limitation, Yan et al. (2016) proposed an improved Searchlight Model (SLM) that considers straight groundwater flow lines. However, since groundwater flow lines are typically curved, He et al. (2019) further developed the CS-SLM, which provides a more reasonable approach for studying the impact of LULC on groundwater quality.

The Weining Plain, with its millennia-long history of agricultural development, is one of the key development areas along the Yellow River Economic Belt in Ningxia. Long-term agricultural irrigation has resulted in relatively shallow groundwater levels in the region. Additionally, the aquifer in the Weining Plain consists of highly permeable alluvial gravel, making the groundwater environment susceptible to external pollution (Li et al. 2017). In recent decades, urbanization and agricultural development have significantly altered the LULC patterns in the Weining Plain. However, the impact of these changes on the groundwater environment remains unclear. Therefore, accurately describing the spatiotemporal distribution of groundwater pollution and analyzing the relationship between specific land use types and groundwater quality are essential for maintaining adequate urban water supply and achieving sustainable groundwater resource management (Janniche et al. 2012).

This chapter employs geostatistical models to analyze the spatial distribution of major exceedance indicators in the Weining Plain. The CS-SLM combined with Spearman correlation analysis is used to evaluate the relationship between groundwater indicators and LULC patterns. The objectives of this study are: (1) to determine the spatial distribution characteristics of major pollutants in the Weining Plain; (2) to quantify the changes in LULC patterns in the Weining Plain; and (3) to assess the relationship between LULC patterns and groundwater pollution.

5.2 Data Collection and Methods

5.2.1 Data Collection

The water quality data used in this chapter are the same as those in Chap. 4, collected from 145 sampling sites in the Weining Plain in 2019. Details regarding sample collection, preservation, transportation, and analytical methods are provided in Chap. 4. The analyzed parameters include pH, total dissolved solids (TDS), total hardness (TH), K^+, Na^+, Ca^{2+}, Mg^{2+}, Cl^-, SO_4^{2-}, HCO_3^-, F^-, NH_4^+, NO_3^-, NO_2^-, Mn, Pb, and Cr^{6+}. This chapter will focus on analyzing the impact of land use and land cover changes on groundwater quality based on these data.

Land use change data were obtained from the Resource and Environment Data Cloud Platform (http://www.resdc.cn) with a resolution of 1 km. Based on the Chinese Land Classification System, the land types were categorized into nine classes: Irrigated Land (IRR), Paddy Field (PAD), Forest (FOR), Grassland (GRA), Barren Land (BAR), River (RIV), Pond (PON), Urban Land (URB), and Rural Settlement (RUR). Geographic Information Systems (GIS) and geostatistical methods, including inverse distance weighting (IDW) interpolation, were employed to analyze the spatial distribution of pollutants.

5.2.2 Establishment of the CS-SLM

A critical step in studying the relationship between groundwater pollution and land use/land cover patterns is determining the potential influence area of land use types around monitoring wells (Jia et al. 2018). In this study, the CS-SLM was applied, which considers groundwater flow direction. The first step was to determine the optimal length and width of the contribution zone for this study. The contribution zone is defined as a domain within which the LULC types can significantly impact the groundwater quality in the domain (He and Wu 2019b). To achieve this, Spearman correlation analysis was conducted under different combinations of contribution zone lengths and widths. The analysis calculated the multiple linear regression correlation coefficients between groundwater quality indicators and the percentage of each land use/land cover type within the contribution zone. Given the high exceedance rate of nitrate (NO_3^-) in the Weining Plain and its strong correlation with land use patterns, nitrate was selected as the primary groundwater quality indicator for this analysis (Liu et al. 2022; Wang et al. 2024). The correlation coefficients between groundwater nitrate and quantified land use/land cover under different contribution zone scenarios are presented in Fig. 5.1.

The highest correlation coefficient was obtained for a contribution zone with a length of 2400 m and a width of 800 ms, making this configuration the most suitable for this study. Spearman correlation analysis was chosen because it does not require data to follow a normal distribution (Sharma et al. 2016). Based on this, SPSS 25

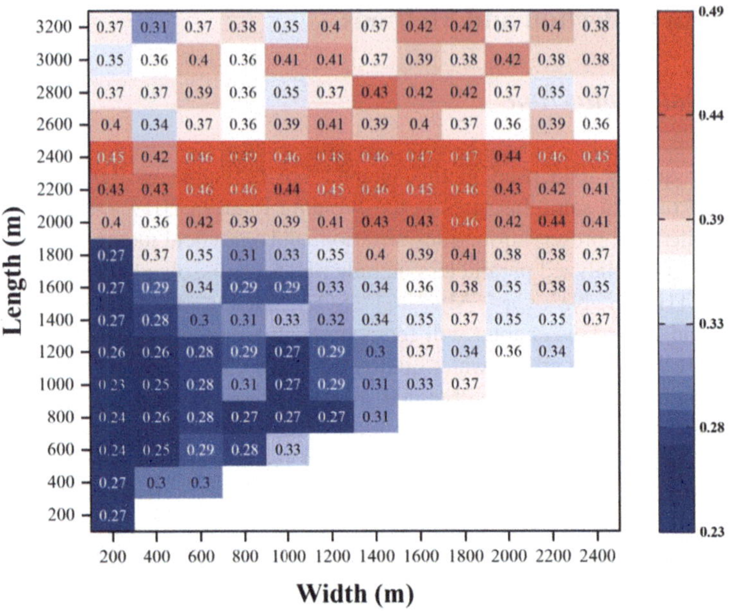

Fig. 5.1 Correlation coefficients between groundwater nitrate and quantified land use/land cover under different contribution zone scenarios

software was used to calculate the correlation coefficients between various water quality indicators and land use/land cover types, thereby determining the extent to which land use/land cover patterns influence groundwater quality indicators.

5.3 Spatial Distribution Characteristics of Major Pollutants

Statistical analysis of the groundwater samples in Chap. 4 revealed that the primary pollutants in the Weining Plain are Na^+, Cl^-, SO_4^{2-}, TDS, TH, Mn, F^-, and NO_3^-. The spatial distribution of these pollutants is illustrated in Fig. 5.2. Notably, Na^+, Cl^-, SO_4^{2-}, TDS, and TH exhibit similar spatial distribution patterns. Areas with high concentration of these pollutants are primarily located north of Dongyuan Town, south of Xuanhe Town and Zhongning County, the area between Shikong Town and Yuding Township, west of Mingsha Town, and west of Baima Town. Field investigations indicate that these areas are often associated with industrial land use and poor hydrogeological conditions, suggesting that the high concentrations of these indicators in groundwater may be attributed to intensive industrial activities and natural processes regulated by hydrogeological conditions (Jia et al. 2018).

Elevated Mn concentrations are predominantly found in the Yuding Town, southwest of Shikong Town and in the Changle Town and its eastern regions. The presence

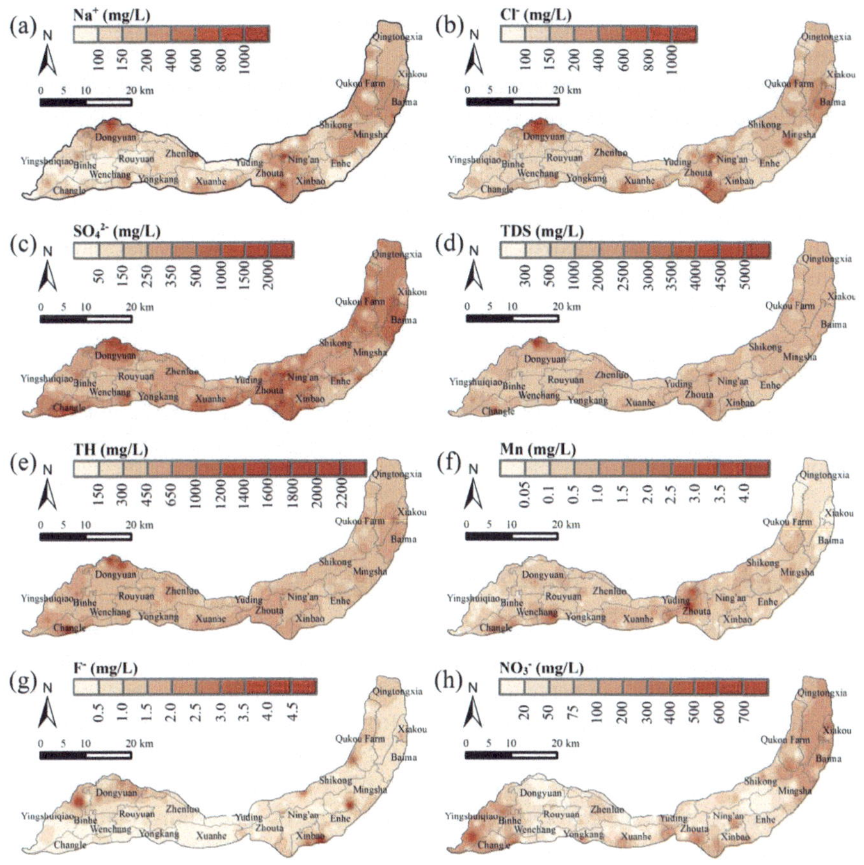

Fig. 5.2 Spatial distribution of major pollutants in the Weining plain

of manganese-rich minerals in the geological formations of the study area is a likely source of Mn in the groundwater (Sharma et al. 2021; Liu et al. 2022). Additionally, the Tianyuan Manganese Industry plant near Yuding Town and Shikong Town may contribute to the elevated Mn levels in groundwater (Li et al. 2014; Xu and Li 2024).

High F^- concentrations are mainly observed in the northwestern part of Zhongwei City and south of Zhongning County. The enrichment of F^- is likely due to the rapid decline in groundwater levels caused by excessive extraction, which facilitates the dissolution of fluoride-bearing minerals (Guo et al. 2018). Another significant source of high-fluoride groundwater is the recharge water from the Nanshantaizi area.

The Weining Plain has a long history of agricultural development, and NO_3^- is a common pollutant in agricultural regions. Groundwater nitrate concentrations are significantly influenced by land use patterns (Zhang et al. 2018). As shown in Fig. 5.3h, areas with high NO_3^- concentration are primarily distributed in Changle

Town, Xuanhe Town, southwest of Zhongning County, and northeast of the Mingsha-Baima area. Urban and agricultural lands are the main factors affecting nitrate concentrations (He et al. 2022a, 2022b). These regions are predominantly covered by agricultural land, where the widespread use of nitrogen fertilizers (e.g., urea, ammonium nitrate) leads to nitrogen leaching into groundwater. This suggests that groundwater quality in these areas is significantly impacted by intensive agricultural activities.

Furthermore, these regions are located downstream of the Nanshantaizi, where the introduction of Yellow River water for irrigation alters the groundwater flow regime. As irrigation water infiltrates through the soil and flows downstream, it recharges the groundwater in the alluvial plain, potentially leading to poorer water quality in downstream areas. Additionally, the rise in groundwater levels due to agricultural irrigation enhances the evaporation and concentration effects, further increasing nitrate levels (He et al. 2022b). This indicates that agricultural land use may contribute more significantly to elevated nitrate concentrations in the study area compared to urban land use.

5.4 Land Use and Land Cover Changes

The LULC patterns of the Weining Plain in 2000, 2010, and 2020 are illustrated in Fig. 5.3, while the area and percentage of each LULC type relative to the total study area are presented in Table 5.1. From 2000 to 2020, the Weining Plain exhibited consistent LULC patterns, with the nine LULC types ranked in descending order of area as follows: IRR, RIV, GRA, RUR, PAD, URB, FOR, BAR, and PON. Irrigated land, the dominant land cover type, is distributed across the entire plain. Paddy fields, forests, and grasslands are scattered throughout the region, while barren land is primarily located at the northern and southern edges of the plain, where it transitions into mountainous areas. Urban land is concentrated around Zhongning County and Zhongwei City, whereas rural settlements are dispersed within irrigated land.

Between 2000 and 2020, the area of irrigated land decreased from 631.80 km^2 to 599.06 km^2, accounting for 62.65% and 59.40% of the total plain area, respectively. In contrast, the area of paddy fields slightly increased from 51.34 to 54.64 km^2. Forest and grassland areas experienced an initial decline followed by an increase over the two decades, with forest area shrinking by 3.92 km^2 and grassland area expanding by 2.38 km^2. River areas decreased from 99.93 to 76.57 km^2, while pond areas steadily grew from 4.02 to 22.03 km^2. Urban land increased by 32.11 km^2, and rural settlements expanded by 8.76 km^2, reflecting ongoing urbanization. Barren land area continued to shrink, and decreased by 4.53 km^2 over the 20-year period, indicating intensified human land use activities.

To further understand the specific changes in LULC types from 2000 to 2020, a Sankey diagram of land use transitions was created (Fig. 5.4). Significant changes were observed in irrigated land, urban land, rural settlements, and rivers. A total of 12.55 km^2 of irrigated land was converted to urban land, 5.99 km^2 to rural settlements, and 12.46 km^2 to forests and grasslands. The conversion of irrigated land to forests

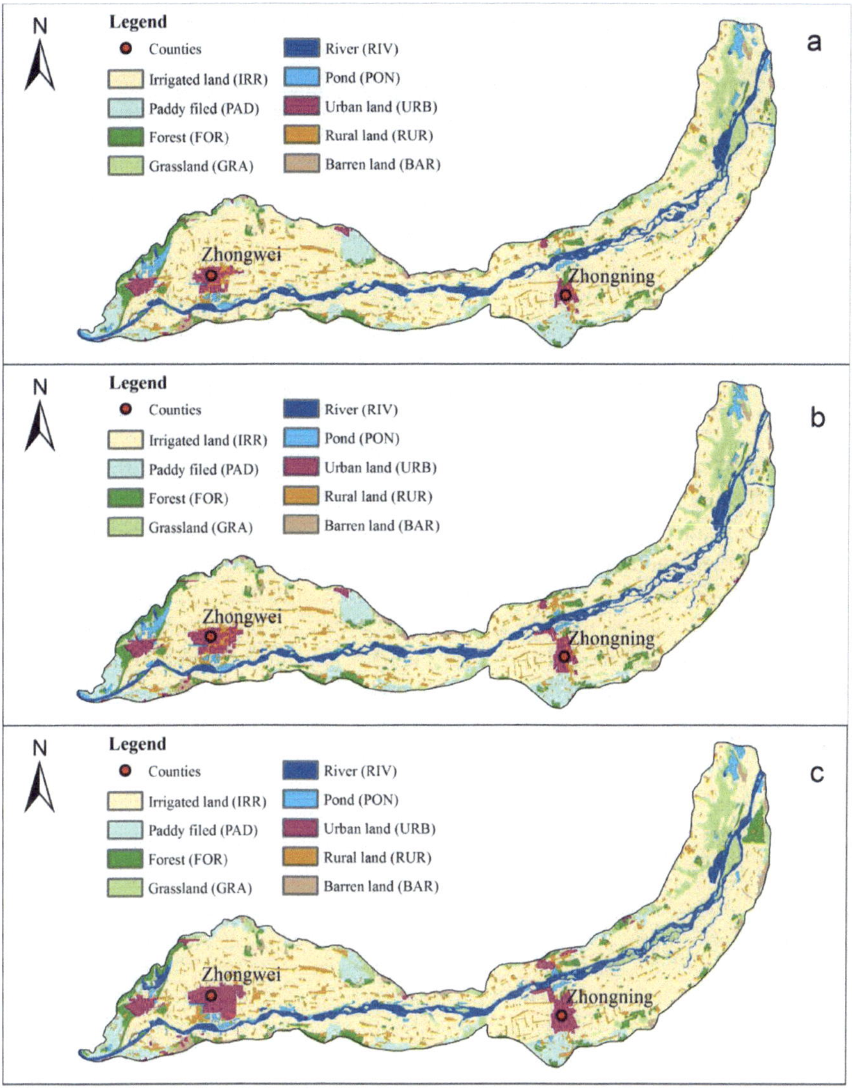

Fig. 5.3 Land use and land cover (LULC) patterns in different years in the Weining plain **a** 2000, **b** 2010, and **c** 2020

and grasslands was driven by the shift from crop cultivation to orchards and moderate reforestation efforts, while the conversion to rural and urban land was due to the expansion of human settlements. The increase in rural settlement area primarily resulted from the conversion of irrigated land (5.99 km^2) and paddy fields (1.25 km^2). Urban land expansion was mainly attributed to the conversion of irrigated land (12.55 km^2), rural settlements (4.79 km^2), forests (2.11 km^2), and grasslands (1.36

Table 5.1 Areas of LULC types in 2000, 2010 and 2020 (unit: km^2)

LULC types	2000	2010	2020
IRR	631.80 (62.65%)	629.21 (62.39%)	599.06 (59.40%)
PAD	51.34 (5.09%)	57.71 (5.72%)	54.64 (5.42%)
FOR	51.99 (5.16%)	46.11 (4.57%)	48.07 (4.77%)
GRA	72.70 (7.21%)	65.16 (6.46%)	75.08 (7.45%)
RIV	99.93 (9.91%)	79.70 (7.90%)	76.57 (7.59%)
PON	4.02 (0.40%)	19.31 (1.91%)	22.03 (2.18%)
URB	12.61 (1.25%)	24.90 (2.47%)	44.72 (4.43%)
RUR	55.61 (5.51%)	63.47 (6.29%)	64.37 (6.38%)
BAR	28.44 (2.82%)	22.86 (2.27%)	23.91 (2.37%)

km^2). Rivers were primarily from the conversion of irrigated land (7.02 km^2) and ponds (2.99 km^2). Overall, from 2000 to 2020, urban and rural settlements in the study area expanded significantly, with the increased area mainly originating from agricultural land, including irrigated land, paddy fields, and forests.

Before assessing the relationship between groundwater quality and LULC patterns, the percentage of each LULC type within the contribution domain for each well was calculated. As shown in Fig. 5.5, the mean percentages of LULC types within the contribution domain in descending order are: Irrigated Land > Rural Settlement > Forest > Barren Land > Urban Land > Grassland > Pond > Paddy Field > River. The results indicate that irrigated land, rural settlements, forests, urban land,

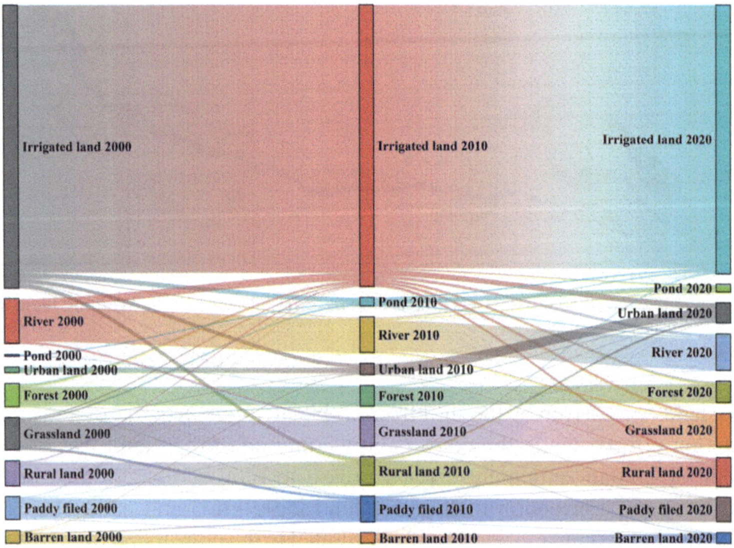

Fig. 5.4 Sankey diagram showing the land use transitions

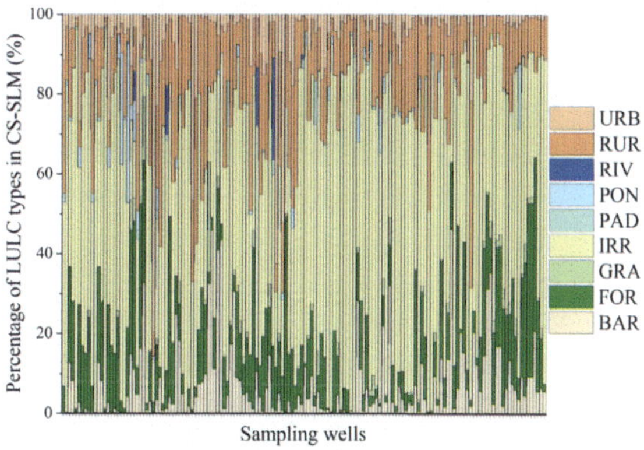

Fig. 5.5 Composition of LULC types in CS-SLM of the wells

and barren land are the dominant LULC types within the contribution zones of all sampling points, while other LULC types occupy smaller proportions. Irrigated land accounts for 10.81–95.08% in the contribution domain, with an average of 53.67%. For forests and grasslands, the maximum and average percentages within the contribution domain are 49.11%, 12.70% and 15.84%, 1.13%, respectively. Rural and urban settlements have average proportions of 16.46% and 5.80%, respectively.

5.5 Relationship Between Groundwater Quality Indicators and LULC Patterns

The number of groundwater samples, the maximum, minimum, and average concentrations of water quality indicators, and the percentage of groundwater parameters exceeding water quality standards across different LULC types in the Weining Plain are presented in Table 5.2. The results show that sampling points in barren land, rivers, grasslands, forests, and urban areas are limited, making it difficult to reflect the overall characteristics of these LULC types. Therefore, the discussion focuses on groundwater quality characteristics under irrigated land (IRR), paddy fields (PAD), and rural settlements (RUR).

The exceedance rates of Na^+ and Cl^- are highest in paddy fields, reaching 62.50%. This is likely due to slow water flow and salt accumulation in paddy fields, where excessive irrigation or poor drainage leads to the accumulation of Na^+ in groundwater (Che et al. 2021). This indicates that agricultural land use significantly influences the enrichment of Na^+ and Cl^- in groundwater. The exceedance rates of SO_4^{2-} are high across all three LULC types, suggesting strong evaporation-concentration effects and water–rock interactions on the concentration of SO_4^{2-} in the study area (Liu et al. 2022). The exceedance rates of F^- are highest in paddy fields, followed by irrigated land and rural settlements. This is attributed to the presence of fluoride in pesticides and fertilizers, which can infiltrate into groundwater, promoting fluoride enrichment in agricultural areas (Kumar et al. 2024), though the dissolution of fluoride bearing

Table 5.2 Statistical analysis of groundwater hydrochemical parameters at different LULCs (Unit: mg/L)

LULC types	Sample numbers	Na^+				Cl^-				SO_4^{2-}			
		Max	Min	Average	Percentage*	Max	Min	Average	Percentage*	Max	Min	Average	Percentage*
IRR	61	846.00	46.90	153.58	21.31	851.00	53.20	157.46	16.39	1441.00	134.00	315.28	49.18
PAD	8	1125.00	73.30	376.29	62.50	1808.00	95.70	468.18	50.00	1513.00	168.00	667.13	62.50
GRA	4	395.00	83.70	195.43	25.00	358.00	64.50	179.70	25.00	1501.00	159.00	625.25	50.00
URB	5	168.00	90.50	101.68	0.00	174.00	92.20	100.72	0.00	240.00	154.00	172.84	0.00
RUR	63	412.00	49.80	145.41	23.81	549.00	49.60	164.46	14.29	2113.00	110.00	366.89	57.14
LULC types	Sample numbers	Na^+				Cl^-				SO_4^{2-}			
		Max	Min	Average	Percentage*	Max	Min	Average	Percentage*	Max	Min	Average	Percentage*
IRR	61	302.00	0.00	41.51	55.74	5.10	0.17	0.81	21.31	3.38	0.03	0.41	54.10
PAD	8	126.00	9.63	82.35	75.00	4.90	0.24	1.52	37.50	0.05	0.03	0.03	0.00
GRA	4	57.93	18.90	36.26	75.00	3.02	0.91	2.04	75.00	0.11	0.03	0.05	25.00
URB	5	168.00	0.00	35.02	60.00	0.49	0.41	0.44	0.00	0.59	0.03	0.29	60.00
RUR	63	412.00	0.00	76.63	69.84	5.00	0.21	0.66	14.29	4.20	0.03	0.47	57.14

*indicates the percentage of groundwater hydrochemical parameters exceeding standards minerals can also induce high fluoride concentration in groundwater in the study area.

Due to the presence of manganese-rich minerals in the geological formations and manganese-related industries in the study area, the exceedance rates of Mn in irrigated land and rural settlements are high, exceeding 50% (Sharma et al. 2021). The exceedance rates of NO_3^- in all three LULC types exceed 55%, indicating severe nitrate pollution in the study area. As a typical agricultural region, the long-term excessive use of nitrogen fertilizers and pesticides in agricultural land is a major contributor to nitrate exceedances (Zhang et al. 2019). The accumulated nitrate in the vadose zonedue to excessive fertilizer application can be leached by improper irrigation practices, allowing nitrogen to migrate through the unsaturated zone into groundwater, thereby increasing groundwater nitrate concentrations (Niu et al. 2022; Xu et al. 2022). Additionally, sewage and livestock manure are potential sources of NO_3^- (Wang et al. 2024). In rural settlements, sampling points are primarily located in areas with sewage input. In rural areas, poorly designed or aging septic tanks and inadequate wastewater treatment facilities contribute to nitrogen infiltration into groundwater.

Further analysis of the correlation between groundwater quality indicators and LULC types is presented in Table 5.3. Most water quality parameters show negative correlations with barren land, forests, and grasslands. This suggests that the presence of barren land, forests, or grasslands within the contribution domain of wells can reduce the risk of groundwater quality deterioration, which is consistent with previous studies (Xu et al. 2019). Notably, forests show a significant negative correlation with NO_3^- at the 0.01 significance level. This is because forests and grasslands enhance soil water retention and reduce pollutant migration into groundwater through biochemical processes, such as root absorption (Laudon et al. 2016). Barren land, particularly undeveloped natural barren land, is associated with minimal agricultural activity, reducing the release of fertilizers, pesticides, and other pollutants into groundwater through surface runoff or infiltration (Wang et al. 2023). F^- shows a significant negative correlation with rivers. When more river water recharges groundwater, F^- concentration in groundwater will be further diluted, reducing the fluoride concentration in groundwater system (Shaji et al. 2024).

Irrigated land and paddy fields show positive correlations with most water quality indicators, particularly NO_3^- and Mn at the 0.01 significance level. Fertilizers, pesticides, insecticides, and manure used in agricultural activities are potential sources of nitrate pollution in groundwater (Su et al. 2024). The Weining Plain, with its over 2000-year history of agricultural development, has experienced rising groundwater levels due to long-term irrigation, resulting in shallow groundwater depths (Li et al. 2022). The evaporation-concentration effect further elevates nitrate levels (He et al. 2022b). Additionally, long-term fertilization has led to nitrate accumulation in the vadose zone. Rural settlements also show a positive correlation with NO_3^-. Field surveys reveal that rural settlements in the Weining Plain often lack proper sewage systems. Organic nitrogen in domestic wastewater undergoes mineralization and nitrification, and is converted to nitrate, contaminating groundwater (Liang et al. 2025). Furthermore, the disposal of human and animal waste on-site also contributes

Table 5.3 Correlations between groundwater quality parameters and LULC types

LULC types	Na^+	Cl^-	SO_4^{2-}	NO_3^-	F^-	Mn
BAR	−0.005	−0.014	0.019	0.188*	−0.045	−0.191*
FOR	−0.001	−0.065	0.020	−0.261**	0.022	−0.244**
GRA	−0.041	0.023	−0.023	−0.152	−0.009	−0.059
IRR	0.019	0.014	−0.039	0.252**	0.030	0.261**
PAD	0.060	0.150	0.067	0.256**	−0.095	0.260**
PON	0.024	0.026	0.037	0.013	−0.038	−0.044
RIV	0.020	−0.006	−0.009	−0.051	−0.197*	0.026
RUR	−0.002	0.085	0.066	0.134	−0.139	−0.049
URB	−0.009	0.065	0.015	−0.045	−0.132	−0.062

* indicates correlation is significant at the 0.05 level (2-tailed), and ** indicates correlation is significant at the 0.01 level (2-tailed)

to nitrate pollution. These sources of nitrate migrate through the vadose zone into groundwater, significantly increasing nitrate concentrations in groundwater. Areas with shallow groundwater depths, such as irrigated land and paddy fields, are more susceptible to surface pollutants (e.g., fertilizers, manure, and sewage), resulting in higher nitrate pollution levels.

The impact of urban and rural settlements on different water quality indicators is complex. In the study area, urban land shows negative correlations with Na^+, F^-, and Mn, but positive correlations with Cl^- and SO_4^{2-}. Notably, urban land shows a negative correlation with NO_3^-, which contradicts conventional understanding, because in conventional understanding urbanization usually leads to groundwater quality deterioration due to sewage leaks, domestic waste, and pollution from urban development (Zaryab et al. 2022). However, this contradiction requires consideration of the actual LULC patterns in the study area. As shown in Table 5.2, the average NO_3^- concentration in agricultural land is significantly higher than that in urban land. The increase in urban land area is often accompanied by a reduction in agricultural land, which may mitigate the negative impact of agricultural activities on groundwater quality (Li et al. 2020). That is to say, the transition from agricultural land to urban land can alleviate the groundwater nitrate pollution. Additionally, rivers show a negative correlation with groundwater nitrate. In the Weining Plain, groundwater discharges into the Yellow River in some sections, thus carrying nitrate away from the groundwater system. The percolation of river water into groundwater in other sections can dilute the groundwater nitrate, thus reducing the groundwater nitrate concentration. The correlation between Mn and LULC types are similar to that of NO_3^-, suggesting that high Mn concentrations may be related to soil solubility, acidity, and soil leaching. This implies that land use types play a significant role in determining pollutant sources and concentration patterns in groundwater.

5.6 Conclusions

This study identified the spatial distribution characteristics of major pollutants based on 145 collected groundwater samples, investigated the changes in LULC patterns in the Weining Plain, and evaluated the relationship between groundwater quality indicators and LULC patterns. The findings provide a basis for urban water source protection and sustainable groundwater resource management. The main conclusions are as follows:

- Na^+, Cl^-, SO_4^{2-}, TH, and TDS exhibit similar spatial distribution patterns, with exceedance areas consistent with industrial activity zones and poor hydrogeological conditions. Elevated Mn concentrations are mainly distributed in the Yuding Town and the eastern part of Changle Town, while high-fluoride groundwater is concentrated in the northwestern part of Zhongwei City and south of Zhongning County. NO_3^- is the primary pollutant in the Weining Plain, with exceedance areas overlapping with agricultural land.
- The LULC types in the Weining Plain, ranked by area, are irrigated land, rivers, grasslands, rural settlements, paddy fields, urban land, forests, barren land, and ponds. From 2000 to 2020, significant changes occurred in irrigated land, urban land, and rural settlements. Irrigated land decreased in area, primarily converting to urban land, rural settlements, forests, and grasslands. Rural and urban settlements expanded significantly, with the increased area mainly originating from agricultural land, including irrigated land, paddy fields, and forests. The optimal contribution zone for the CS-SLM was determined to be 2400 m in length and 800 m in width. The dominant LULC types within the contribution domain of of the sampling points are irrigated land, rural settlements, forests, urban land, and barren land.
- In groundwater samples, the exceedance rates of Na^+, Cl^-, and F^- are highest in paddy fields. SO_4^{2-} and NO_3^- also show high exceedance rates in paddy fields, irrigated land, and rural settlements. Mn exceedance rates are high in irrigated land and rural settlements. Correlation analysis indicates that irrigated land, paddy fields, and rural settlements are positively correlated with groundwater nitrate, suggesting that agricultural activities and rural pollution are potential sources of nitrate contamination. Forests, grasslands, urban land, and rivers show negative correlations with nitrate, contributing to the reduction of nitrate pollution in the Weining Plain.

References

Alqurashi AF, Kumar L (2019) An assessment of the impact of urbanization and land use changes in the fast-growing cities of Saudi Arabia. Geocarto Int 34(1):78–97. https://doi.org/10.1080/10106049.2017.1367423

Barron OV, Barr AD, Donn MJ (2013) Evolution of nutrient export under urban development in areas affected by shallow watertable. Sci Total Environ 443:491–504. https://doi.org/10.1016/j.scitotenv.2012.10.085

Che Z, Wang J, Li J (2021) Effects of water quality, irrigation amount and nitrogen applied on soil salinity and cotton production under mulched drip irrigation in arid Northwest China. Agric Water Manag 247:106738. https://doi.org/10.1016/j.agwat.2021.106738

Gu B, Dong X, Peng C, Luo W, Chang J, Ge Y (2012) The long-term impact of urbanization on nitrogen patterns and dynamics in Shanghai, China. Environ Pollut 171:30–37. https://doi.org/10.1016/j.envpol.2012.07.015

Guo X, Zuo R, Meng L, Wang J, Teng Y, Liu X, Chen M (2018) Seasonal and spatial variability of anthropogenic and natural factors influencing groundwater quality based on source apportionment. Int J Environ Res Public Health 15(2):279. https://doi.org/10.3390/ijerph15020279

He S, Wu J (2019a) Hydrogeochemical characteristics, groundwater quality, and health risks from hexavalent chromium and nitrate in groundwater of Huanhe formation in Wuqi county, northwest China. Expo Health 11:125–137. https://doi.org/10.1007/s12403-018-0289-7

He S, Wu J (2019b) Relationships of groundwater quality and associated health risks with land use/land cover patterns: a case study in a loess area, Northwest China. Hum Ecol Risk Assess 25(1–2):354–373. https://doi.org/10.1080/10807039.2019.1570463

He S, Li P, Su F, Wang D, Ren X (2022a) Identification and apportionment of shallow groundwater nitrate pollution in Weining plain, northwest China, using hydrochemical indices, nitrate stable isotopes, and the new Bayesian stable isotope mixing model (MixSIAR). Environ Pollut 298:118852. https://doi.org/10.1016/j.envpol.2022.118852

He S, Wu J, Wang D, He X (2022b) Predictive modeling of groundwater nitrate pollution and evaluating its main impact factors using random forest. Chemosphere 290:133388. https://doi.org/10.1016/j.chemosphere.2021.133388

He X, Wu J, He S (2019) Hydrochemical characteristics and quality evaluation of groundwater in terms of health risks in Luohe aquifer in Wuqi county of the Chinese loess plateau, northwest China. Hum Ecol Risk Assess 25(1–2):32–51. https://doi.org/10.1080/10807039.2018.1531693

Janniche GS, Spliid H, Albrechtsen H-J (2012) Microbial community-level physiological profiles (CLPP) and herbicide mineralization potential in groundwater affected by agricultural land use. J Contam Hydrol 140–141:45–55. https://doi.org/10.1016/j.jconhyd.2012.08.008

Jia Z, Bian J, Wang Y (2018) Impacts of urban land use on the spatial distribution of groundwater pollution, Harbin city, northeast China. J Contam Hydrol 215:29–38. https://doi.org/10.1016/j.jconhyd.2018.06.005

Johnson TD, Belitz K (2009) Assigning land use to supply wells for the statistical characterization of regional groundwater quality: correlating urban land use and VOC occurrence. J Hydrol 370(1–4):100–108. https://doi.org/10.1016/j.jhydrol.2009.02.056

Kumar P, Kumar M, Barnawi AB, Maurya P, Singh S, Shah D, Yadav VK, Kumar A, Kumar R, Yadav KK, Gacem A, Ahmad A, Patel A, Alreshidi MA, Singh V, Yaseen ZM, Cabral-Pinto MMS, Vinayak V, Wanale SG (2024) A review on fluoride contamination in groundwater and human health implications and its remediation: a sustainable approaches. Environ Toxicol Phar 106:104356. https://doi.org/10.1016/j.etap.2023.104356

Laudon H, Kuglerová L, Sponseller RA, Futter M, Nordin A, Bishop K, Lundmark T, Egnell G, Ågren AM (2016) The role of biogeochemical hotspots, landscape heterogeneity, and hydrological connectivity for minimizing forestry effects on water quality. Ambio 45:152–162. https://doi.org/10.1007/s13280-015-0751-8

Li P, Qian H, Howard KWF, Wu J, Lyu X (2014) Anthropogenic pollution and variability of manganese in alluvial sediments of the Yellow river, Ningxia, northwest China. Environ Monit Assess 186:1385–1398. https://doi.org/10.1007/s10661-013-3461-3

Li P, Wu J (2023) Water resources and sustainable development. Water 16(1):134. https://doi.org/10.3390/w16010134

Li P, Wu J, Qian H (2016) Hydrochemical appraisal of groundwater quality for drinking and irrigation purposes and the major influencing factors: a case study in and around Hua County, China. Arab J Geosci 9:15. https://doi.org/10.1007/s12517-015-2059-1

Li X, Gao Y, Qian H, Wu H (2017) Groundwater vulnerability and contamination risk assessment of the Weining plain, using a modified drastic model and quantized pollution loading method. Arab J Geosci 10:469. https://doi.org/10.1007/s12517-017-3255-y

Li Y, Bi Y, Mi W, Xie S, Ji L (2021) Land-use change caused by anthropogenic activities increase fluoride and arsenic pollution in groundwater and human health risk. J Hazard Mater 406:124337. https://doi.org/10.1016/j.jhazmat.2020.124337

Li Y, Li P, Liu L (2022) Source identification and potential ecological risk assessment of heavy metals in the topsoil of the Weining plain (northwest China). Expo Health 14:281–294. https://doi.org/10.1007/s12403-021-00438-0

Li Y, Wang F, Feng J, Lv J, Liu Q, Nan F, Liu X, Xu L, Xie S (2020) Increased health threats from land use change caused by anthropogenic activity in an endemic fluorosis and arsenicosis area. Environ Pollut 261(3):114130. https://doi.org/10.1016/j.envpol.2020.114130

Liang H, Li P, Elumalai V, Tian Y, Kou X (2025) Hydrochemical characteristics, groundwater nitrate sources and potential health risks in a typical alluvial plain of northwest China. Phys Chem Earth 139:103903. https://doi.org/10.1016/j.pce.2025.103903

Liu L, Wu J, He S, Wang L (2022) Occurrence and distribution of groundwater fluoride and manganese in the Weining plain (China) and their probabilistic health risk quantification. Expo Health 14:263–279. https://doi.org/10.1007/s12403-021-00434-4

Niu X, Jia X, Yang X, Wang J, Wei X, Wu L, Shao M (2022) Tracing the sources and fate of NO_3^- in the vadose zone-groundwater system of a thousand-year-cultivated region. Environ Sci Technol 56(13):9335–9345. https://doi.org/10.1021/acs.est.1c06289

Reddy AGS, Saibaba B, Sudarshan G (2012) Hydrogeochemical characterization of contaminated groundwater in Patancheru industrial area, southern India. Environ Monit Assess 184:3557–3576. https://doi.org/10.1007/s10661-011-2208-2

Shaji E, Sarath KV, Santosh M, Krishnaprasad PK, Arya BK, Babu MS (2024) Fluoride contamination in groundwater: a global review of the status, processes, challenges, and remedial measures. Geosci Front 15(2):101734. https://doi.org/10.1016/j.gsf.2023.101734

Sharma GK, Jena RK, Ray P, Yadav KK, Moharana PC, Cabral-Pinto MMS, Bordoloi G (2021) Evaluating the geochemistry of groundwater contamination with iron and manganese and probabilistic human health risk assessment in endemic areas of the world's largest River Island India. Environ Toxicol Phar 87:103690. https://doi.org/10.1016/j.etap.2021.103690

Sharma S, Roy A, Agrawal M (2016) Spatial variations in water quality of river Ganga with respect to land uses in Varanasi. Environ Sci Pollut Res 23:21872–21882. https://doi.org/10.1007/s11356-016-7411-9

Singh SK, Srivastava PK, Pandey AC, Gautam SK (2013) Integrated assessment of groundwater influenced by a confluence river system: concurrence with remote sensing and geochemical modelling. Water Resour Manage 27:4291–4313. https://doi.org/10.1007/s11269-013-0408-y

Su F, He S, Zhou X, Yu F, Qiang S, Ma H, Guan Z, Zhang T (2024) Assessing impacts of land use/land cover patterns to shallow groundwater nitrate pollution in an agricultural-dominant area in northwest China using random forest. Environ Earth Sci 83:584. https://doi.org/10.1007/s12665-024-11856-z

Wang M, Jiang T, Mao Y, Wang F, Yu J, Zhu C (2023) Current situation of agricultural non-point source pollution and its control. Water Air Soil Pollut 234:471. https://doi.org/10.1007/s11270-023-06462-x

Wang D, Wu J, Li P, Li L, Yang J, Zhang P, He S, Kou X, Wang Y, (2024) Seasonal nitrate variations, risks, and sources in groundwater under different land use types in a thousand-year-cultivated region, Northwestern China. Environ Res 251(2):118699. https://doi.org/10.1016/j.envres.2024.118699

Worrall F, Kolpin DW (2004) Aquifer vulnerability to pesticide pollution—combining soil, land-use and aquifer properties with molecular descriptors. J Hydrol 293(1–4):191–204. https://doi.org/10.1016/j.jhydrol.2004.01.013

Xu F, Li P (2024) Biogeochemical mechanisms of iron (Fe) and manganese (Mn) in groundwater and soil profiles in the Zhongning section of the Weining Plain (northwest China). Sci Total Environ 939:173506. https://doi.org/10.1016/j.scitotenv.2024.173506

Xu F, Li P, Chen W, He S, Li F, Mu D, Elumalai V (2022) Impacts of land use/land cover patterns on groundwater quality in the Guanzhong Basin of northwest China. Geocarto Int 37(27):16769–16785. https://doi.org/10.1080/10106049.2022.2115153

Xu G, Li P, Lu K, Tantai Z, Zhang J, Ren Z, Wang X, Yu K, Shi P, Cheng Y (2019) Seasonal changes in water quality and its main influencing factors in the Dan river basin. CATENA 173(2):131–140. https://doi.org/10.1016/j.catena.2018.10.014

Yan B, Xiao C, Liang X, Fang Z (2016) Impacts of urban land use on nitrate contamination in groundwater, Jilin city, northeast China. Arab J Geosci 9(2):105. https://doi.org/10.1007/s12517-015-2052-8

Zaryab A, Nassery HR, Alijani F (2022) The effects of urbanization on the groundwater system of the Kabul shallow aquifers, Afghanistan. Hydrogeol J 30:429–443. https://doi.org/10.1007/s10040-021-02445-6

Zhang Q, Xu P, Qian H (2019) Assessment of groundwater quality and human health risk (HHR) evaluation of nitrate in the central-western Guanzhong Basin, China. Int J Environ Res Public Health 16(21):4246. https://doi.org/10.3390/ijerph16214246

Zhang Y, Wu J, Xu B (2018) Human health risk assessment of groundwater nitrogen pollution in Jinghui canal irrigation area of the loess region, northwest China. Environ Earth Sci 77:273. https://doi.org/10.1007/s12665-018-7456-9

Chapter 6
Hydrogeochemical Assessment of Groundwater in KwaZulu-Natal Coastal Aquifers

In order to assess water quality, it is necessary to define the basic hydrogeochemical properties and governing elements of groundwater chemistry, which are crucial to knowing the whole groundwater ecosystem. Extensive agricultural activities significantly impact groundwater chemistry, influencing chemical composition and spatial distribution, revealing mechanisms of contamination and transformation through agricultural environments. This chapter explores the hydrogeochemical characteristics of the groundwater of the three catchments (i.e., Maputaland or upper, uMhlathuze or middle, and Tugela or lower) situated in the eastern coastal region of South Africa. The hydrogeochemical properties of groundwater in three separate catchments demonstrated differences in water type influenced by natural processes and human activities. According to major ion chemistry, Piper, and Chadha plot interpretation, most of the water facies were determined as Cl-Na facies, suggesting gradual saltwater contamination and ion exchange. Different levels of seawater intrusion are confirmed by the Seawater Mixing Index (SMI); the coastal catchment exhibits the greatest SMI values, which align with higher concentrations of Cl^- and Na^+. Statistical investigation utilizing SPSS, encompassing principal component analysis (PCA), determined primary pollutant sources, differentiating between runoff from agriculture (elevated NO_3^-, K^+), industrial effluents, and seawater intrusion.

6.1 Introduction

The significance of ensuring a consistent supply of high-quality water has drawn much focus in recent years. A comprehensive understanding of water availability, different uses of water in hydrological processes, and efficient regional management methods are essential (Hanasaki et al. 2008; Schlosser et al. 2014). Groundwater is a necessary supply of freshwater because of its extensive availability, relatively low cost, and superior quality (Hu et al. 2017; Liao et al. 2025). Since groundwater

P. Li and V. Elumalai, *Groundwater Quality Under Agricultural Activities—Cases from China and South Africa*, Professional Practice in Earth Sciences,
https://doi.org/10.1007/978-3-031-98993-3_6

irrigates 38% of the world's cultivable land (301 million hectares) and accounts for 43% of all irrigation water usage, it is essential for supplying water demands in homes, businesses, and agriculture (Shah et al. 2007; Zhang et al. 2019; Singh et al. 2025).

The susceptibility of groundwater hydrochemical makeup to external influences leads to a dynamic process termed hydrochemical evolution (Dong and Gao 2022). As groundwater moves through subterranean systems and interacts with nearby rocks, minerals, and environmental materials, its hydrochemical components which are produced by atmospheric rainfall, weathering of rocks, and evaporation crystallisation, go through dynamic changes. These modifications led to variations in hydrochemical classifications and an elevation in total dissolved solids (Wang et al. 2022). Climate and hydrogeological characteristics govern the variability in groundwater recharge, discharge, hydrochemistry, and groundwater dynamics (Haaf et al. 2020; Kabir et al. 2023). As worldwide population growth, urbanisation, industrialisation, and fertiliser application increase, human activities significantly affect groundwater systems, frequently resulting in ecosystem degradation and elevated health hazards for humans (Abotalib et al. 2019; Li et al. 2020). Anthropogenic activities significantly impact groundwater quality, enabling the examination of hydrochemical components essential to groundwater hydrology investigation. In addition to revealing the genesis and development of groundwater hydrochemistry, examining changes in its constituent hydrochemicals provides information on the routes and traits of groundwater cycling.

Coastal agriculture's sustainability and adaptability are being severely impacted by seawater intrusion (SWI), a phenomenon where saltwater moves inland and into freshwater aquifers and surface water systems (Kirwan and Gedan 2019; Mondal et al. 2023; Tackley et al. 2024). The worldwide effect of SWI on coastal cultivation is still not well known, although there is increased awareness and research activity. SWI impact extends hundreds of meters across impacted farms, making minimal output losses hard to determine (Tivianton et al. 2021). It is crucial to know how SWI affects agricultural land to choose appropriate mitigation and adaptation strategies for each phase of this change. The current study reveals the saline water intrusion in the three catchments (these are Maputaland, uMhlathuze, and Tugela) of the eastern coastal regions of South Africa. Additionally, the hydrogeochemical characteristics of the regions, the controlling mechanisms of these hydrochemical dynamics, and the sources of contamination have been investigated.

6.2 Study Area

The current study focuses on three catchments of the eastern coastal regions of South Africa. These catchments are Maputaland, uMhlathuze, and Tugela, which are named in this study as upper, middle, and lower catchments, respectively, based on their location from the northern to southern part of the region. The upper catchment is located between 26°89′ to 27°86′ S latitudes and 32°12′ to 32°88′ E longitudes, and within the Maputaland coastal plain in northern KwaZulu-Natal, South Africa. The coastal region of Maputaland (upper) in northern KwaZulu Natal extends from Mtunzini in the south to the northern border with Mozambique (Fig. 6.1). This research location of the upper catchment, which is over 4,400 km^2, is under the administration of the uMkhanyakude District municipality. The primary source of water for household, agricultural, and consumption is groundwater in this region. Additionally, the middle (uMhlathuze) catchment is situated within this study's other two included catchments, which cover an approximate area of 792 km^2. The Babanango Hills are the beginning of the catchment. The two main rivers are the Mhlathuze in the south, the Enseleni in the east, and the Indian Ocean border along the region's eastern part. In this catchment, two coastal towns named Richards Bay and the neighbouring town of Empangeni are situated roughly 180 km northeast of Durban, which is the provincial capital of KwaZulu-Natal, South Africa. The study area's lower catchment and southernmost region encompass over 840 km^2 from the Bushmans River's convergence to the river discharge at the Indian Ocean, passing through Mandini town and the Isithebe industrial sector (Fig. 6.1). One of the longest rivers in the country, the Tugela River, flows nearly 405 km long, mostly flowing through the KwaZulu-Natal province and discharged into the Indian Ocean (NRIO 1986).

The current study focuses on three catchments of the eastern coastal regions of South Africa. These catchments are Maputaland, uMhlathuze, and Tugela, which are named in this study as upper, middle, and lower catchments, respectively, based on their location from the northern to southern part of the region. The upper catchment is located between 26°89′ to 27°86′ S latitudes and 32°12′ to 32°88′ E longitudes, and within the Maputaland coastal plain in northern KwaZulu-Natal, South Africa. The

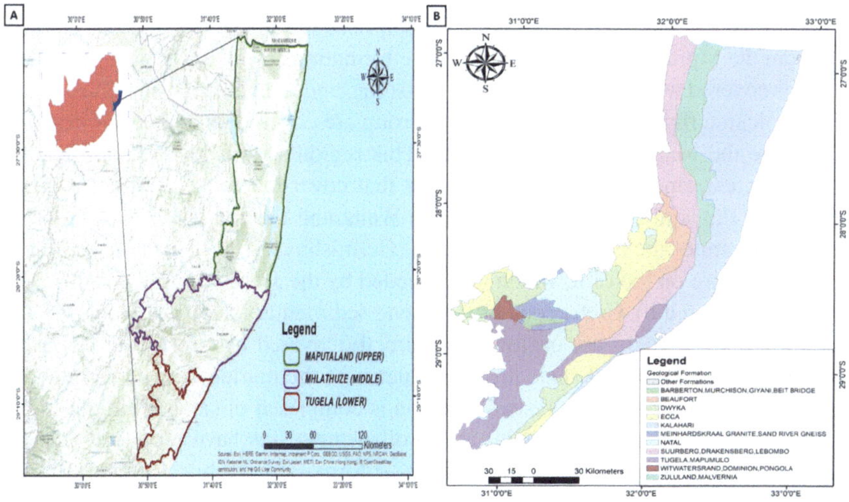

Fig. 6.1 Maps illustrating the study area (**a**) and the geological formations (**b**)

coastal region of Maputaland (upper) in northern KwaZulu-Natal extends from Mtunzini in the south to the northern border with Mozambique (Fig. 6.1a). This research location of the upper catchment, which is over 4,400 km^2, is under the administration of the uMkhanyakude District municipality. The primary source of water for household, agricultural, and consumption is groundwater in this region. Additionally, the middle (uMhlathuze) catchment is situated within this study's other two included catchments, which cover an approximate area of 792 km^2. The Babanango Hills are the beginning of the catchment. The two main rivers are the Mhlathuze in the south, the Enseleni in the east, and the Indian Ocean border along the region's eastern part. In this catchment, two coastal towns named Richards Bay and the neighbouring town of Empangeni are situated roughly 180 km northeast of Durban, which is the provincial capital of KwaZulu-Natal, South Africa. The study area's lower catchment and southernmost region encompass over 840 km^2 from the Bushmans River's convergence to the river discharge at the Indian Ocean, passing through Mandini town and the Isithebe industrial sector (Fig. 6.1). One of the longest rivers in the country, the Tugela River, flows nearly 405 km long, mostly flowing through the KwaZulu-Natal province and discharged into the Indian Ocean (NRIO 1986).

6.2.1 Geology of the Study Area

The geology of the study areas shows a diversity in nature due to their different formations (Fig. 6.1b). In terms of geology, the coastal plain of the upper catchment is constituted by unconsolidated to semi-consolidated sediments spanning from the Cretaceous to the Quaternary of the Maputaland group. Maputaland group is underlain by the Zululand group, which comprises basaltic rocks and conglomerates from the Mpilo, Msunduze, and Movene formations (Watkeys et al. 1993). This region is covered with arenite, sandstone, siltstone, and alluvial deposits. The consolidated volcanic rhyolites of the Lebombo Group are covered by a series of deposits that define the Maputaland coastal plain. This region's wetlands are significantly affected by extremely permeable sediments that cover the area, resulting in quick recharge to the aquifers (Mkhwanazi 2010; Weitz and Demlie 2014). Precambrian granite constitutes the basement in this area (Germishuyse 1999). Triassic sedimentary deposits are on top of it, which is succeeded by the Jurassic basaltic rocks. The basalts are covered by Cretaceous to Holocene sedimentary deposits. However, the granitic basement rocks are occasionally directly covered by Cretaceous deposits. Sandstone, mudstone, and shale are constituents of the marine-derived Cretaceous deposits. The majority of the research region is comprised of sands from the Pleistocene to Holocene epochs. Fluvial and aeolian processes have produced silt and clay in these fine-grained sands. The outermost formation exhibits adequate permeability, allowing for the penetration of rainwater (Elumalai et al. 2017). In addition to the two overlaying catchments, the lower catchment's geological settings are completely diverse. Basaltic lava from the Drakensberg, Stormberg, and Beaufort layers, as well as ancient granites and gneisses, sandstone, and rocks from the

Dwyka and Ecca series, are all present in the catchment areas. Despite the overall complexity of the Tugela Catchment's geology, the coastal aquifer is supported by complex sedimentary deposits of the Karoo super group and Natal Group Sandstone as well as hard rocks of the basement complex. The aquifer system in the studied area consists of several rocks exhibiting significant porosity resulting from fracture and weathering processes over time. Fractured and intergranular aquifers make up every lithological unit found in the research region. Groundwater is present in intergranular, fractured karst aquifers and in certain quaternary floodplain layers. The normal range of groundwater level depths is 10 to 30 m below ground level in this region (Ntanganedzeni et al. 2018).

6.2.2 Climatic Conditions

The climatic conditions of the three consecutive catchments of the study areas are nearly similar to each other. The upper catchment's climate is characterised as humid subtropical, featuring warm summers with an average temperature of 21 °C and an average annual precipitation of 600 mm inland to nearly 1000 mm along the coastline (Ramsay 1996; Porat and Botha 2008). Both the tropical and middle-latitude weather patterns contribute to this region's precipitation (Weitz and Demlie 2014). Likewise, the middle catchment shows throughout January and March, the average daily temperature ranges from 21 to 29 °C, whereas from June to August, it ranges from 12 to 23 °C. In contrast to other parts of Africa, this region has a humid and moist climate (DWAF 2000). This catchment, one of South Africa's greatest rainfall zones, receives annual precipitation between 850 mm in the west and nearly 1200 mm along the coastal strip (Elumalai et al. 2023). The climate in coastal regions doesn't shift with time; instead, it is humid and hot along the coast. Overall, the areas surrounding the Drakensberg Mountains have a wet and cold climate, while the regions extending from Colenso down towards the coast have a dry to warm climate. In the highlands and surrounding areas of this catchment, the average annual rainfall is roughly 1500 mm, whereas in the central areas, it is roughly 650 mm (Garland and Broderick 1992). Rainfall is slightly higher around the coast, accounting for 30% of total rainfall, with the wet summer season (October to March) receiving the majority of rainfall (>80%) (Schulze 1979).

6.3 Sample Collection and Analysis

6.3.1 Sampling

A few preliminary surveys were carried out for each of the samplings before the final sampling due to the identification of the borewells wells, the ease of accessibility, and the proximity to the anthropogenic activities in the study areas. A total of 128 groundwater samples were collected from the three catchments. The sampling locations and their respective sample IDs are demonstrated in Table 6.1. Out of these, for the upper catchment, 53 borewells were selected throughout the northern KwaZulu Natal coastal regions, spanning from St. Lucia to Manguzi (Fig. 6.2). Similarly, a total of 39 and 36 groundwater samples were collected from the borewells of the middle and lower catchments, respectively. Two sets of samples were collected from each location of the study area. In the field, all samples were filtered through a 0.45 μm Millipore membrane filter after being collected at each sampling point of each catchment's site. After that, samples were kept for major ions and nutrient analysis in high-density polyethylene bottles. Out of these, one set of filtered samples was acidified with 0.5 mL of concentrated HNO_3 to prevent the precipitation of cations. All samples were immediately stored in a cooler at 4 °C to preserve their integrity during transportation to the laboratory. The temperature, pH, total dissolved solids (TDS), electrical conductivity (EC), and oxidation–reduction potential (ORP) were measured at the site using an Aqua Probe A-700 m. Where possible during the sample period, a water level meter was used to determine the water level in a few chosen monitoring wells.

6.3.2 Data Analysis

The collected samples were analysed based on the earlier proposed methodology. Anions (Cl^- and SO_4^{2-}) and nutrients (NO_3^-) were analysed via Ion Chromatography (Metrohm 930 IC Compact Flex), whereas HCO_3^- was evaluated using normal titration techniques as recommended by the American Public Health Association (APHA 2007). The cations (Na^+, K^+, Ca^{2+}, and Mg^{2+}) were analysed using ICP-OES (NexION 2000 ICP-OES). Standards and blanks were executed periodically to verify the precision of the analysis. The Ion balance error was ultimately assessed as within the permissible limit (±5). The study area, sampling locations, and spatial distribution maps were generated by using ArcGIS 10.8. Piper and Chadha plots, which were also produced from the data through geochemical methods, were used to classify the different types of hydrochemical facies. The Piper diagram depicts the primary cations and anions as percentages in milligram equivalents per litre (meq/L). The diamond-shaped sections are divided into nine zones, each displaying unique hydrochemical characteristics, facilitating the examination of the formation conditions of different aquifers (Piper 1944). The Chadha Plot delineates the categorisation

Table 6.1 Details of the sampling locations

Catchments	Sample ID	Latitude	Longitude	Catchments	Sample ID	Latitude	Longitude
Maputaland (Upper)	MP-1	−27.481	32.5704	uMhlathuze (Middle)	MH-12	−28.7273	31.93537
	MP-2	−27.476	32.5051		MH-13	−28.8213	31.86604
	MP-3	−27.475	32.5219		MH-14	−28.6975	31.96515
	MP-4	−27.565	32.489		MH-15	−28.7566	31.97493
	MP-5	−27.655	32.4033		MH-16	−28.7898	31.93872
	MP-6	−28.382	32.4082		MH-17	−28.8013	31.95576
	MP-7	−28.373	32.4115		MH-18	−28.8179	31.94553
	MP-8	−28.379	32.3769		MH-19	−28.754	31.91997
	MP-9	−28.375	32.3783		MH-20	−28.7889	31.83775
	MP-10	−28.369	32.3749		MH-21	−28.8034	31.89186
	MP-11	−28.366	32.3736		MH-22	−28.7625	31.86161
	MP-12	−28.353	32.3595		MH-23	−28.8535	31.88894
	MP-13	−28.373	32.3637		MH-24	−28.8469	31.86715
	MP-14	−28.385	32.3501		MH-25	−28.8256	31.90123
	MP-15	−27.512	32.5867		MH-26	−28.7272	31.87396
	MP-16	−27.518	32.5897		MH-27	−28.7659	31.92892
	MP-17	−27.568	32.4849		MH-28	−28.7668	31.91742
	MP-18	−27.575	32.4711		MH-29	−28.7753	31.92253
	MP-19	−27.561	32.4112		MH-30	−28.7591	32.02434
	MP-20	−27.572	32.446		MH-31	−28.7659	32.01284
	MP-21	−27.486	32.5871		MH-32	−28.7549	32.01199
	MP-22	−27.483	32.5631		MH-33	−28.7689	32.02179
	MP-23	−27.475	32.6631		MH-34	−28.7723	32.03371
	MP-24	−27.359	32.6506		MH-35	−28.7779	32.02349
	MP-25	−27.516	32.6431		MH-36	−28.7664	31.99878
	MP-26	−27.505	32.5953		MH-37	−28.774	32.00773
	MP-27	−27.54	32.5796		MH-38	−28.7864	32.03371
	MP-28	−27.544	32.5829		MH-39	−28.7319	32.03755
	MP-29	−27.556	32.5307	Tugela (Lower)	TU-1	−29.2197	31.47416
	MP-30	−27.565	32.5243		TU-2	−29.2189	31.50583
	MP-31	−27.542	32.5481		TU-3	−29.1918	31.48128
	MP-32	−27.509	32.6469		TU-4	−29.189	31.43122
	MP-33	−27.494	32.5973		TU-5	−29.2189	31.42669
	MP-34	−27.403	32.5554		TU-6	−29.2078	31.42585
	MP-35	−27.413	32.5552		TU-7	−29.2053	31.46675
	MP-36	−27.414	32.5546		TU-8	−29.2042	31.43664

(continued)

Table 6.1 (continued)

Catchments	Sample ID	Latitude	Longitude	Catchments	Sample ID	Latitude	Longitude
	MP-37	−27.352	32.5281		TU-9	−29.1968	31.44233
	MP-38	−27.331	32.5489		TU-10	−29.0983	31.35928
	MP-39	−27.164	32.7143		TU-11	−29.131	31.44336
	MP-40	−27.146	32.7009		TU-12	−29.1865	31.37085
	MP-41	−27.14	32.6964		TU-13	−29.2055	31.35894
	MP-42	−27.126	32.6935		TU-14	−29.1812	31.35719
	MP-43	−27.118	32.6973		TU-15	−29.1899	31.35039
	MP-44	−27.1	32.6989		TU-16	−29.1948	31.2955
	MP-45	−27.056	32.6875		TU-17	−29.1909	31.49186
	MP-46	−27.049	32.6877		TU-18	−29.1997	31.48972
	MP-47	−26.904	32.7834		TU-19	−29.1814	31.46217
	MP-48	−26.891	32.8035		TU-20	−29.1736	31.47806
	MP-49	−26.891	32.8034		TU-21	−29.1598	31.4615
	MP-50	−26.963	32.7671		TU-22	−29.1796	31.48542
	MP-51	−26.974	32.7581		TU-23	−29.2603	31.42131
	MP-52	−26.99	32.7411		TU-24	−29.2495	31.4339
	MP-53	−26.998	32.7353		TU-25	−29.2478	31.41628
uMhlathuze (Middle)	MH-1	−28.7668	31.81469		TU-26	−29.2436	31.43742
	MH-2	−28.7001	31.84858		TU-27	−29.2283	31.46836
	MH-3	−28.6965	31.93794		TU-28	−29.2314	31.48842
	MH-4	−28.6713	31.9554		TU-29	−29.2304	31.48808
	MH-5	−28.6549	31.95437		TU-30	−29.2816	31.42281
	MH-6	−28.6729	31.93126		TU-31	−29.2504	31.39772
	MH-7	−28.6996	31.89994		TU-32	−29.2481	31.37556
	MH-8	−28.6518	31.91586		TU-33	−29.2372	31.37394
	MH-9	−28.6724	31.89326		TU-34	−29.2011	31.38158
	MH-10	−28.6508	31.93434		TU-35	−29.2107	31.37268
	MH-11	−28.716	31.98723		TU-36	−29.2046	31.37417

of water types based on ion concentrations across all three catchments' groundwater, providing a comprehensive analysis of hydrogeochemical facies (Chadha 1999). The hydrochemical composition of groundwater and its formation process were both clearly depicted by the Gibbs diagram (Gibbs 1970). The seawater mixing index (*SMI*) was introduced by Park et al. (2005). *SMI* is extensively utilised for evaluating seawater mixing through the analysis of four principal ions: Mg^{2+}, Na^+, Cl^-, and SO_4^{2-} (Sajil Kumar 2016; Kanagaraj et al. 2018; Goswami and Rai 2024). It is calculated using the following Eq. (6.1) Park et al. (2005):

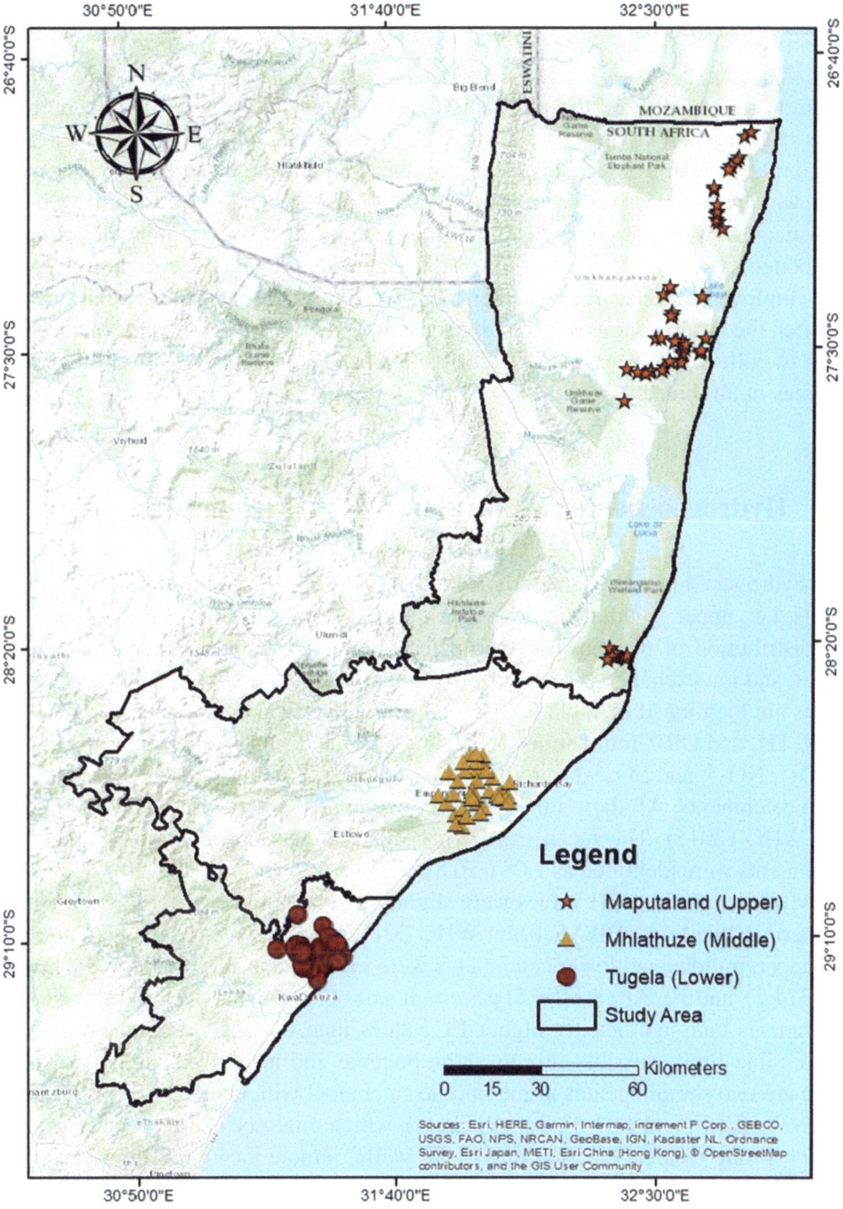

Fig. 6.2 Sampling locations map of three catchments

$$\mathrm{SMI} = \mathrm{a} \times \frac{C_{\mathrm{Na}}}{T_{\mathrm{Na}}} + \mathrm{b} \times \frac{C_{\mathrm{Mg}}}{T_{\mathrm{Mg}}} + \mathrm{c} \times \frac{C_{\mathrm{Cl}}}{T_{\mathrm{Cl}}} + \mathrm{d} \times \frac{C_{\mathrm{SO_4}}}{T_{\mathrm{SO_4}}} \quad (6.1)$$

where a, b, c, and d are the constants that represent the respective concentration proportions of Na^+, Mg^{2+},Cl^-, and $SO_4{}^{2-}$ in seawater. C denotes the measured concentration in mg/L, whereas T signifies the local threshold values of the analysed ions, which can be derived from the interpretation of cumulative probability curves. Statistical analysis was conducted by SPSS Version 27.0.1 to determine the principal components (PCs) and factors based on the Eigenvalues (>1). The correlation matrix was employed to determine whether there were positive or negative relationships between the observed ions of groundwater samples. PCA was performed on the variables to collectively determine the correlations and sources of ions in groundwater samples of all three catchments.

6.4 Hydrogeochemical Characteristics of Groundwater

The hydrogeochemical characteristics of groundwater throughout the three catchments, i.e., upper, middle, and lower, exhibit the geochemical signatures regarding the distribution of the physical parameters, major ions, nutrients, hydrochemical facies, and ion exchange. The descriptive statistics of the findings of all three catchments are mentioned in Table 6.2. A total of five physical parameters, i.e., pH, EC, TDS, TH, and ORP, four cations (Na^+, K^+, Ca^{2+}, and Mg^{2+}), and four anions (Cl^-, $HCO_3{}^-$, $NO_3{}^-$, and $SO_4{}^{2-}$) were measured for all 128 number of samples from the three catchments. All these outcomes were compared with those of DWAF (1996) and WHO (2011). The mean (±SD) pH values of the upper, middle, and lower catchments were identified as 7.08 (±0.87), 7.39 (±1.49), and 7.13 (±0.32), respectively. Although these pH values were slightly alkaline in nature, these were within the permissible limit (DWAF 1996; WHO 2011). The average (±SD) EC values of the upper, middle, and lower catchments were measured as 330.1 (±227.6), 1806.3 (±4324.2), and 1928.6 (±1106.5) μS/cm. It was observed that the middle and lower catchments had significantly higher EC values than the upper catchment and the permissible limit, possibly due to anthropogenic and natural activities. However, there are two sampling sites in the middle catchment with unusual EC, i.e., MH-28 (26,260 μS/cm) and MH-33 (9844 μS/cm), which might be caused by evaporation and/or a long residence period due to the wells' remote location. The average TH values of the upper, middle, and lower catchments were determined as follows: 99.34, 189.83, and 232.59 mg/L. Based on the TH values, it was observed that the groundwater of the upper and middle catchments belongs to the moderate class, while the lower catchment shows a hard water type (Sawyer and McCarty 1978). The mean TDS values of the upper, middle, and lower catchments were determined as 211.26, 1156, and 1667.5 mg/L, which show that only the upper catchment's groundwater belong to fresh water while the saline water or ionic concentrations might influence the other types. The average ORP of the upper, middle, and lower catchments were measured as 74.97, 0.05, and −3.02. The findings of the ORP demonstrate that the upper and middle catchments are in transition zones, while the lower catchment show mildly reducing conditions, i.e., reduction of iron or sulfate production. In addition to the physical parameters of the three catchments, the four major cations and four major anions were measured, and these are illustrated in terms of their concentration with the help of a pie chart in Fig. 6.3. According to the mean concentration of the major ions measured in groundwater of the three catchments, the dominant order of anions and cations were $Cl^- > HCO_3^- > SO_4^{2-} > NO_3^-$ and Na +> $Ca^{2+} > Mg^{2+} > K^+$, respectively. The upper and lower catchments' ion concentrations were determined in a similar order to the above. However, the groundwater of the middle catchment showed the major cations in the following order: $Na^+ > K^+ > Ca^{2+} > Mg^{2+}$. It was also noted that Cl^- concentrations of the middle and lower catchments were beyond the permissible limit, while most of the other ion concentrations were within the permissible limit (DWAF 1996; WHO 2011). The hydrochemical properties of groundwater are typically impacted by the presence of abnormal anion concentrations close to areas of industrial and human activity (Rao et al. 2016; Feng et al. 2020). Determining the hydrogeochemistry of their groundwater exposes both anthropogenic and natural influences.

The hydrochemical facies of the groundwater of the three catchments were properly interpreted using the Trilinear or Piper diagram, which classifies water types based on the relative concentrations of major cations (Na^+, K^+, Ca^{2+}, Mg^{2+}) and anions (HCO_3^-, Cl^-, SO_4^{2-}, F^-). These facies can be distinguished using the Piper diagram, which also clarifies the geochemical development and potential pollutants of diverse water sources (Piper 1944). Figure 6.4 illustrates that nearly 55% of the upper catchment samples belong to the Na-Cl type, followed by approximately 40% occurring within the Ca-Mg.Cl type. That indicates that most of the groundwater samples were influenced by seawater due to the near-coastal regions, and a few were recharged with fresh water. Similarly, the samples of the middle catchment in the Piper plot demarcated Na-Cl type(nearly 65%),followed by Ca-Mg·Cl followed by type (approximately 30%). The remaining samples of the upper and middle catchments belong to Ca-Mg·HCO_3 type facies, which generally indicate a mineral weathering process. However, all the samples of the lower catchment showed Na-Cl type facies, which means all of its water samples were influenced by saline water.

In order to explain the hydrochemical genesis and water–rock interaction processes, groundwater kinds are categorised using the Chadha plot, a modified version of the Piper trilinear diagram, which depends on the relative concentrations of the primary cations and anions (Chadha 1999). The four primary hydrochemical facies represented in this diagram are high acidic anions surpassing weaker acidic anions, alkali metals surpassing alkaline earth, and vice versa. The indication of the types of facies in the Chadha plot is nearly identical to the above-mentioned Piper diagram (Fig. 6.4). It illustrated that among the samples, almost 50%, 65%, and 100% of the upper, middle and lower catchments belong to Na-Cl type facies, which clearly stated that the seawater interaction in these regions is due to the near approach of the coastline. However, the remaining majority of the samples of the upper and middle catchments were identified as Ca-Mg·Cl type, except for four samples of

Table 6.2 Representing the statistical findings of the physical parameters and detected major ions

Catchment		Maputaland				Mhlathuze				Tugela				Permissible limits	
Statistical parameters		Min	Max	Mean	S.D	Min	Max	Mean	S.D	Min	Max	Mean	S.D	DWAF (1996)	WHO (2011)
Physical Parameters	ORP (mV)	−4.10	231.00	74.97	107.19	−21.60	8.23	0.05	6.16	−8.30	−0.20	−3.02	1.57	–	–
	pH	5.28	8.34	7.08	0.87	0.20	9.50	7.39	1.49	6.46	7.89	7.13	0.32	6.0–9.0	6.5–8.5
	EC (μS/cm)	115.00	1194.00	330.09	227.65	140.00	26,260.00	1806.26	4324.16	408.00	6849.00	1928.63	1106.46	<700	1500.00
	TH	23.10	463.55	99.34	88.20	31.40	1296.90	189.83	233.50	65.30	761.66	232.59	155.23	50–100	100–500
	TDS	73.60	764.16	211.26	145.69	89.60	16,806.40	1156.00	2767.46	256.00	19,230.00	1667.53	3065.36	450–1000	500–1000
Cations	Na^+	14.00	126.00	40.07	26.71	15.00	5121.00	233.28	823.86	48.00	6971.00	490.69	1126.49	<100	200.00
	K^+	0.47	17.66	3.54	2.65	4.00	634.00	46.41	140.76	1.94	10.00	4.93	2.42	<50	12.00
	Ca^{2+}	4.00	154.00	25.05	25.55	4.00	217.00	34.51	45.99	13.00	147.23	49.39	28.40	<32	200.00
	Mg^{2+}	1.79	56.00	8.96	8.63	3.00	184.00	25.26	31.46	8.00	112.00	26.61	21.88	<30	50.00
Anions	Cl^-	27.00	289.00	77.90	47.64	40.00	7657.00	370.19	1223.29	95.00	9926.00	753.04	1613.01	<100	250.00
	HCO_3^-	14.00	312.00	69.59	67.86	12.00	1903.00	179.55	306.06	48.80	311.10	159.00	62.95	–	–
	SO_4^{2-}	1.75	53.00	12.22	11.00	2.00	374.00	23.57	59.18	4.83	1330.00	88.90	219.79	<200	250.00
	NO_3^-	0.00	112.00	6.69	17.30	1.00	29.00	11.82	6.72	0.00	16.32	8.24	4.88	<6	45.00

**Note* Units of all parameters are in mg/L, except ORP (mV) and EC (μS/cm)

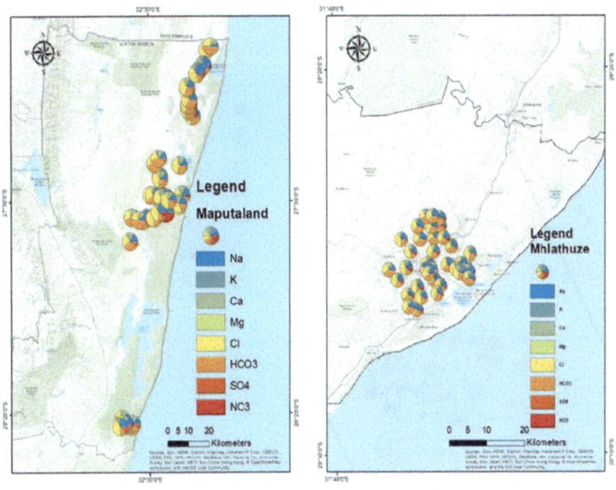

Fig. 6.3 Pie chart showing the spatial distribution of major ions in the water samples of the three catchments

upper and two samples of middle catchments, were detected as Ca-Mg.HCO_3 type facies. Moreover, the presence of Na-Cl and Ca-Mg.Cl facies suggests the impact of ion exchange mechanisms, evaporation, and extended storage/residence periods (Rajmohan et al. 2021; Elumalai et al. 2023).

6.5 Impact of Seawater Intrusion

The seawater mixing index (SMI) is valuable for quantitatively evaluating the extent of relative saline water mixing in freshwater (Kanagaraj et al. 2018). Seawater intrusion (SWI) can be determined through both SMI and ionic ratios, which correspond to Cl- fluctuation in freshwater. Both human activity and natural processes contribute to soil salinity in farming areas, which changes the biochemistry of the soil and has a negative impact on crop health and production. Evapotranspiration, which carries away dissolved salts, naturally raises the salinity of the soil, particularly in places with inadequate drainage. Increasing levels of salinity are also caused by atmospheric salt deposition and weathering of salt-bearing bedrock. Anthropogenic activities, particularly the widespread application of fertilisers, add to salinity in the soil (Su et al. 2025).

6.5.1 Evaluation of Seawater Mixing Index (SMI)

The SMI values were calculated based on the Eq. 6.1. Figure 6.5a–d demonstrates the steps of the derivation of the regional threshold values (T). Based on the intersection

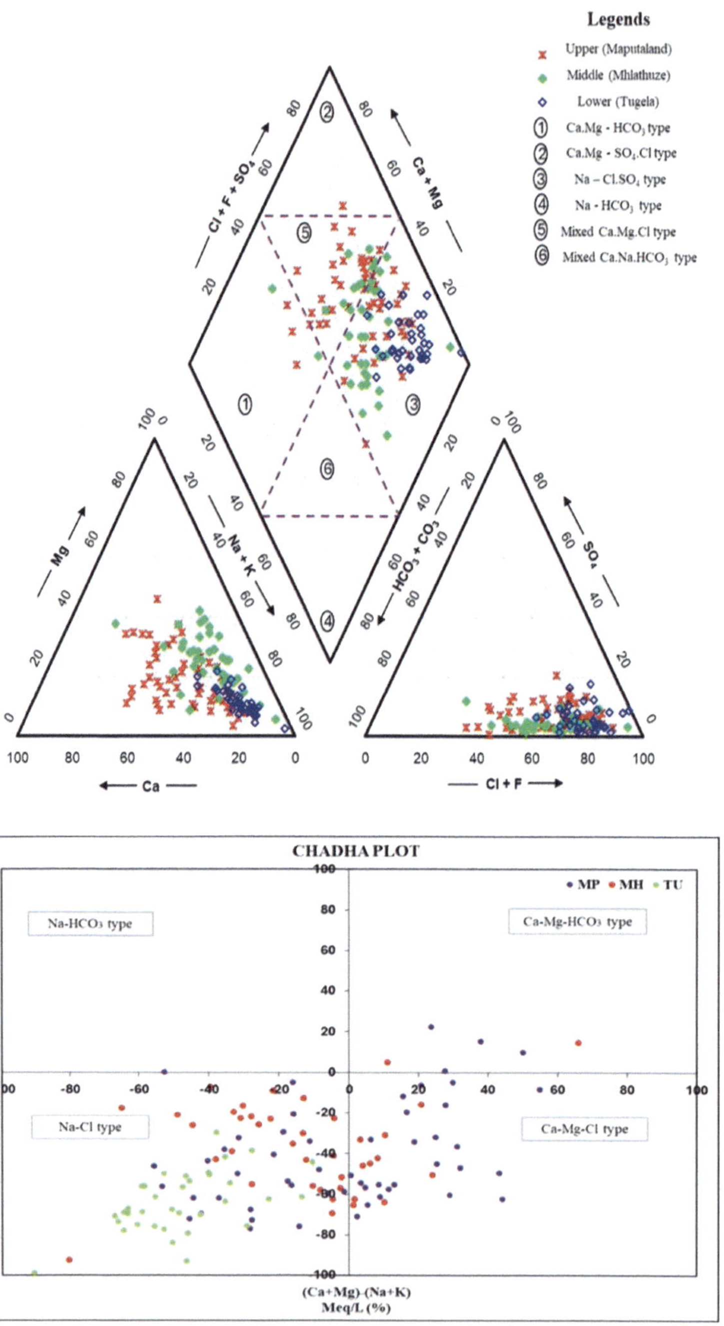

Fig. 6.4 Analysis of water chemistry through Piper diagram and Chadha plot showing the hydrogeochemical facies of groundwater

point (or IP: Inflection Point) of the plots for the corresponding ion, the T value(s) were measured (Park et al. 2005). The determined T values for the corresponding ions Na^+ (17 and 56 mg/L), Mg^{2+} (21 and 15 mg/L), Cl^- (231 mg/L), and SO_4^{2-} (18 mh/L). However, as two IPs were determined for Na and Mg, the highest concentrations were chosen for the calculation of *SMI* in this study. According to Park et al. (2005), if the *SMI* value is greater than 1, then there might be possibilities of mixing seawater either through natural or anthropogenic activities. Moreover, it was determined that the *SMI* values of these regions vary from 0.20 to 69.17, while the average (±SD) *SMI* value was calculated as 2.33 (±7.50). Figure 6.5e illustrates that nearly 1% of each ionic concentration is strongly related to saline water influence based on the *SMI* values. Figure 6.5 (f) demonstrates the *SMI* values with respect to the sample IDs. These findings show that nearly 9.5%, 28.2%, and 77% of the upper, middle, and lower catchments were affected by saline water intrusion. Additionally, the evaluation of *SMI* values helped to classify them in terms of pollution (Goswami and Rai 2024) and has been represented in Table 6.3. It was observed that 64.84% of samples ($n = 83$) belong to fresh water, i.e., $SMI < 1$, while 13.28% ($n = 17$) samples were slightly polluted by saline water intrusion. The moderately polluted, highly polluted, and dangerously polluted classes of *SMI* were identified by 17.97%, 0.78%, and 3.13% of samples, respectively. The overall results indicate that groundwater chemistry is highly influenced by the seawater intrusion.

6.5.2 *Identification of Seawater Intrusion Through Ionic Ratios*

Apart from the SMI evaluation, seawater intrusion can be derived based on the associated ionic ratios. The ionic ratios such as Cl^-/HCO_3^-, Na^+/Cl^-, SO_4^{2-}/Cl^-, and EC were plotted against the Cl^- concentration of the groundwater to determine the SWI in these catchments (shown in Fig. 6.6). It was noticed that three samples of the lower and 1 sample of the middle catchments were strongly affected by SWI, as shown in Cl^-/HCO_3^- versus Cl^- plot (Fig. 6.6a), while nearly 40% of the lower catchment and three samples of the middle catchment are slightly affected by SWI. The Na^+/Cl^- versus Cl^- plot (Fig. 6.6b) illustrates that nearly 70% of the lower, four samples of the middle, and one sample of the upper catchments were affected by SWI. Likewise, the SO_4^{2-}/Cl^- versus Cl^- plot (Fig. 6.6c) reveals that nearly 50% of all samples belong to the seawater interaction zone. EC versus Cl^- plot (Fig. 6.6d) demonstrates that only two samples from the middle and lower samples were affected through SWI.

The influence of SWI on agricultural soil results in three principal chemical transformations: higher ionic strength, sulfidation, and alkalinisation (Tully et al. 2019). SWI increases the ionic strength of the soil, which improves ion exchange activities and has a major impact on the availability of nutrients like phosphorus (P) and nitrogen (N) (Weston et al. 2010). According to earlier research, SWI increases

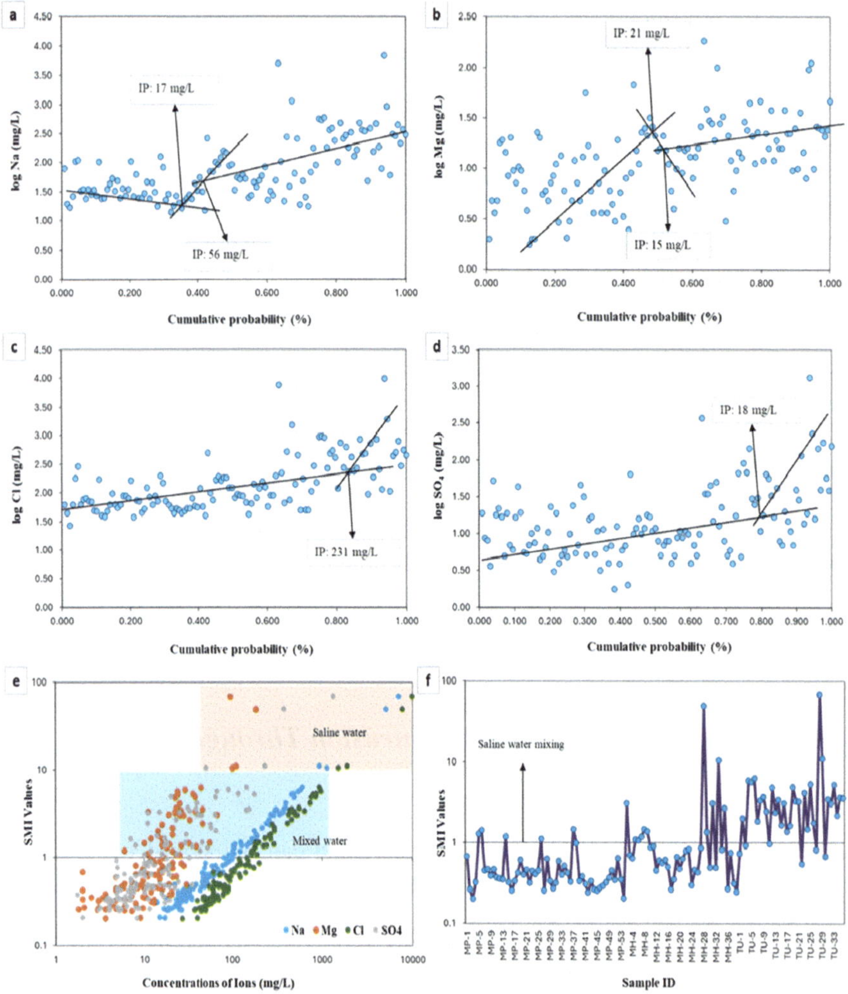

Fig. 6.5 Demonstration of the regional threshold values of Na^{+} (**a**), Mg^{2+} (**b**), Cl^{-} (**c**), and SO_4^{2-} (**d**); illustration of SMI values with respect to the ions and sampling points. (*IP: Inflection Point)

Table 6.3 Seawater mixing index (SMI) classification

SMI range	Classes	Number of samples	% of Samples
<1.0	Freshwater	83	64.84
1.0–2.0	Slightly polluted	17	13.28
2.0–6.0	Moderately polluted	23	17.97
6.0–10.0	Highly polluted	1	0.78
>10.0	Dangerously polluted	4	3.13

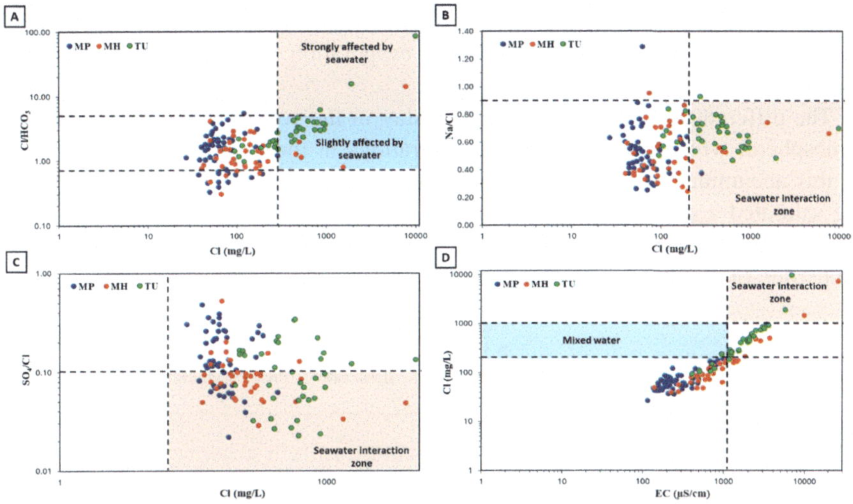

Fig. 6.6 An illustration of the seawater intrusion through ionic ratios: **a** Cl^- versus Cl^-/HCO_3^-, **b** Cl^- versus Na^+/Cl^-, **c** Cl^- versus SO_4^{2-}/Cl^-, **d** EC versus Cl^-

competition for ion exchange sites with cations including Na^+, Ca^{2+}, and Mg^{2+} that are common in marine environments, which in turn enhances the removal of ammonium from soil particles (Weston et al. 2010; Steinmuller and Chambers 2018). By removing ammonium, this desorption process contributes to plant stress and confuses nutrient dynamics, resulting in nutrition deficiencies and increased leaching threats (Su et al. 2025). In this study area, lithological characteristics and excessive groundwater use are indicators of natural and anthropogenic factors that impact SWI and groundwater evolution in these coastal catchments, especially in the lower one (Tugela).

6.6 Major Controlling Mechanisms of Groundwater Chemistry

Gibbs determined that the dissolved salt level of groundwater was primarily controlled by three mechanisms: ambient rainfall, water–rock interactions, and evaporative crystallisation, which were represented through the Gibbs plot (Gibbs 1970). It includes two distinct scatter plots demonstrating anions (i.e., TDS versus $Cl^-/Cl^- + HCO_3^-$) and cations (i.e., TDS versus $Na^+ + K^+/Na^+ + K^+ + Ca^{2+}$). Figure 6.7 (A: TDS versus $Na^+ + K^+/Na^+ + K^+ + Ca^{2+}$) demonstrated that nearly 50% of the samples of three catchments were shown by rock weathering, and the remaining samples of upper and middle catchments belonged to no significant interaction process. All samples of the middle catchment were located within the weathering zone, suggesting a predominance of rock-water interactions in Fig. 6.7 (B: TDS versus $Cl^-/Cl^- + HCO_3^-$). Whereas, most samples of the lower catchment and the samples of the upper catchment belonged to evaporation and precipitation processes, respectively.

The different ions in the groundwater stay in the soil as it evaporates into the atmosphere, which raises the ion concentrations in the groundwater. As a result of this accumulation, some less soluble minerals precipitate, ultimately invading the saturated zone and raising the salinity of the groundwater. The groundwater body and geology were identified as the primary determinants of the hydrochemical properties of shallow groundwater in the research area (Wang et al. 2017; Feng

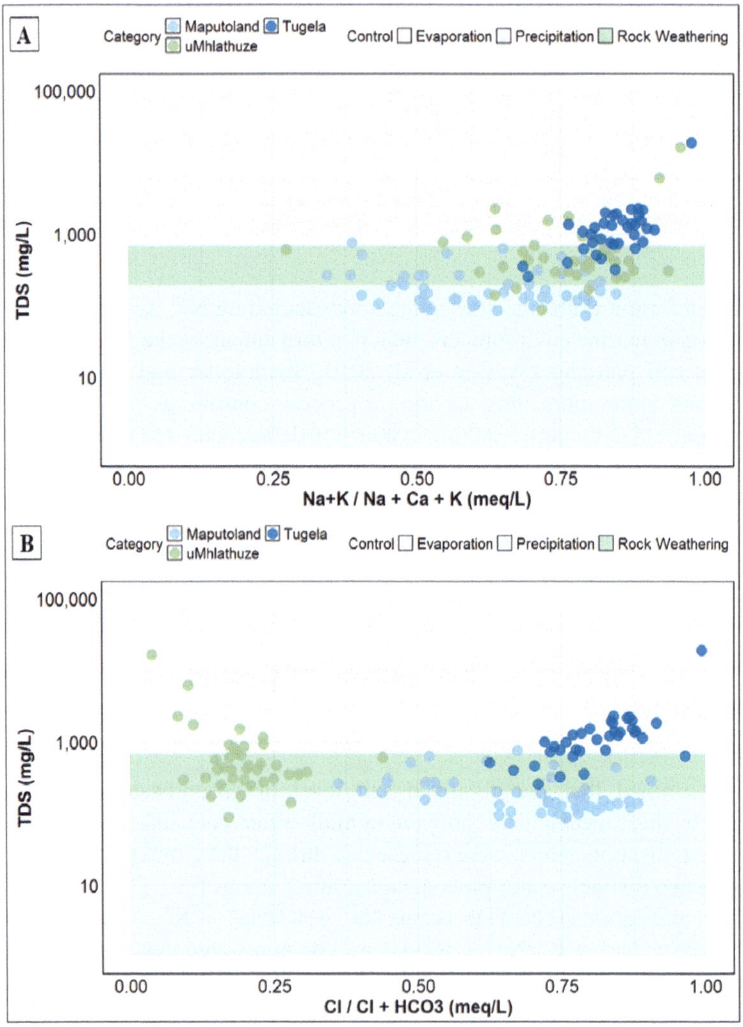

Fig. 6.7 Gibbs plot demonstrating the controlling mechanisms of water chemistry

et al. 2020). Groundwater chemistry is significantly influenced by processes such as carbonate disintegration, weathering of silicates, ion exchange, precipitation, and evaporation (Ansari et al. 2019). The HCO_3^- versus Ca^{2+} scatter plot (Fig. 6.8a) demonstrated that nearly all of the samples approached the equiline (1:1), which indicated that although the dissolution of calcite and dolomite is a source of Ca^{2+}, additional processes also influence the Ca^{2+} content in groundwater (Vikas et al. 2014). Similarly, ($HCO_3^- + SO_4^{2-}$) versus ($Ca^{2+} + Mg^{2+}$) scatter plot (Fig. 6.8b) of the study area revealed mineral weathering by more than 90% of the samples, as these samples were near the equiline. The reduction of $Ca^{2+} + Mg^{2+}$ relative to $HCO_3^- + SO_4^{2-}$ suggests that $Ca^{2+} + Mg^{2+}$ is extracted from groundwater by cation exchange reactions or the weathering of Na^+/K^+ silicates, which should have increased HCO_3^- levels (Rajmohan et al. 2021). Likewise, the Cl^- versus Na (Fig. 6.8c) plot determined the halite dissolution as most of the samples were attached to the equiline (1:1), and the previous hydrogeochemical facies determination already clarified that these catchments are highly influenced by saline water intrusion. The SO_4^{2-} against Ca^{2+} graph (Fig. 6.8d) unveiled that most of the water samples fall above the 1:1 line. That implies the reverse ion exchange process is predominant, resulting in the ingress of Ca^{2+} ions into the groundwater by dissolution, whereas Na^+ ions are mineralised on the rock surface (Gupta et al. 2023). The (Na–Cl) versus {(Ca + Mg)–(HCO_3 + SO_4)} plot demonstrated that there was no significant ion exchange as most of the samples were attached to the centre of the graph (Fig. 6.8e). Since calcium-bicarbonate comprised most of the groundwater, carbonate dissolution was primarily responsible for controlling rock-water interactions.

An understanding of the ion exchange reactions that take place with the groundwater and the surrounding environment during residence or travel can be gained by analysing the chloroalkaline indices (CAI) (Schoeller 1965, 1967). The following relations are used to evaluate chloroalkaline indices 1 and 2 (i.e., CAI1 and CAI2) for each type of water sample of the study area. The following formulas are used to derive CAI1 (Eq. 6.2) and CAI2 (Eq. 6.3):

$$\mathrm{CAI1} = \frac{\mathrm{Cl^-} - \left(\mathrm{Na^+} + \mathrm{K^+}\right)}{\mathrm{Cl^-}} \tag{6.2}$$

$$\mathrm{CAI2} = \frac{\mathrm{Cl^-} - \left(\mathrm{Na^+} + \mathrm{K^+}\right)}{\left(\mathrm{SO_4^{2-}} + \mathrm{HCO_3^-} + \mathrm{CO_3^{2-}} + \mathrm{NO_3^-}\right)} \tag{6.3}$$

Cation exchange and reverse ion exchange reactions were explained by CAI1 and CAI2's negative and positive values, respectively. Figure 6.8f demonstrates that the groundwater chemistry in these three catchments is governed by both cation exchange and reverse ion exchange reactions. The CAI1 observed between −1.1 and 0.6 of the upper catchment samples, simultaneously with the CAI2 of the upper catchment samples varying from −0.4 to 0.4, with an R^2 of 0.94. Additionally, CAI1 of the middle and lower catchments' samples occur within –0.5 to 0.5, and −0.42 to 0.35, respectively. At the same time, the CAI2 of the middle catchment varies between

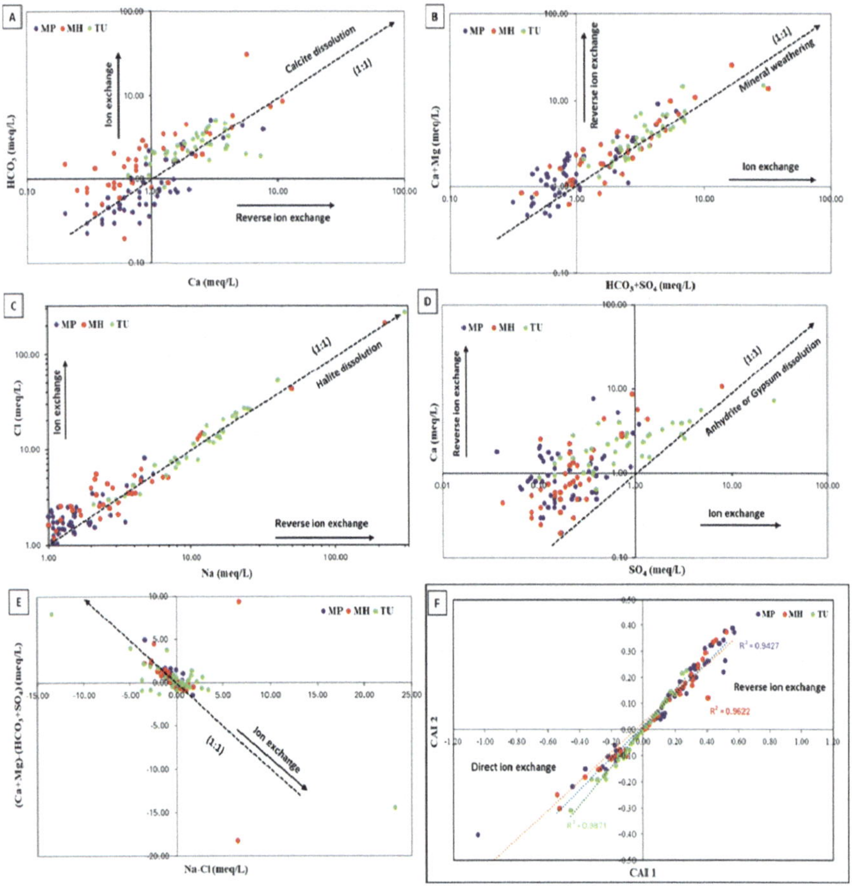

Fig. 6.8 An illustration of the relationship between major ions of each type of water: **a** HCO_3^- versus Ca, **b** ($HCO_3^- + SO_4^{2-}$) versus ($Ca^{2+} + Mg^{2+}$), **c** Na^+ versus Cl^-, **d** SO_4^{2-} versus Ca^{2+}, **e** ($Ca^{2+} + Mg^{2+}$)–($HCO_3^- + SO_4^{2-}$) versus Na^+-Cl^-, and **f** CAI1 versus CAI2

−0.3 and 0.4, with an R^2 of 0.96. Apart from that, CAI2 of the lower catchment occurred within −0.30 to 0.23, with an R^2 of 0.98.

6.7 Influencing Factors and Sources of Contaminants

The source of the pollutants in these three catchments' groundwater was identified using principal component analysis (PCA) or factor analysis (FA). The statistical analysis can elucidate the links across variables to accurately represent actual issues. Two such statistical techniques that are frequently applied in groundwater hydrogeochemical studies are the correlation matrix and factor analysis (Khanoranga 2019).

The Varimax rotation method was employed using Kaiser normalisation. Groundwater ions have a particular interaction with one another; hence, a correlation study of water chemistry data is important (Li et al. 2019). Table 6.4 demonstrates the heat map of the correlation matrix of all three catchments of their physical parameters and major ions. The heat map clearly indicates that the green cells represent correlated and linked parameters or ions, respectively, while the red colour boxes signify an effective negative correlation between the parameters or ions. The intensity of the cells explained the nature of the correlation. The correlation coefficients of EC with TH, TDS, Na^+, K^+, Ca^{2+}, Mg^{2+}, and Cl^- were strongly correlated (>0.700). Ion exchange and SWI appear to be the main mechanisms governing the groundwater chemistry in this region, based on the high positive correlation of EC with the other major ions, resulting in the increase of groundwater salinity. Similarly, TH was strongly correlated with Ca^{2+}, Mg^{2+}, Cl^-, Na^+, and TDS. A shared reason resulting from excessive evaporation, the disintegration of halite rock, and human activity is indicated by the strong correlation between Na^+ and Cl^-. The strong correlation between Ca^{2+} and Mg^{2+} results from the breakdown of calcium-magnesium-bearing minerals, including aragonite, calcite, dolomite, and clays (Papazotos et al. 2019). TDS indicated strong correlation matrices between Na^+, Cl^-, and Mg^{2+}, while K showed a strong correlation matrix with HCO_3^-.

Table 6.4 Heat map showing the correlation matrix of the physical parameters and major ions detected from all three catchments

Parameters	ORP	pH	EC	TH	TDS	Na	K	Ca	Mg	Cl	HCO_3	SO_4	NO_3
ORP	1.000												
pH	0.332	1.000											
EC	−0.156	−0.008	1.000										
TH	−0.265	−0.043	0.867	1.000									
TDS	−0.131	−0.009	0.817	0.771	1.000								
Na	−0.103	−0.004	0.750	0.718	0.982	1.000							
K	−0.069	−0.033	0.824	0.652	0.593	0.476	1.000						
Ca	−0.241	−0.069	0.743	0.939	0.680	0.620	0.537	1.000					
Mg	−0.258	−0.014	0.888	0.948	0.773	0.731	0.688	0.781	1.000				
Cl	−0.113	−0.014	0.767	0.743	0.979	0.998	0.479	0.644	0.755	1.000			
HCO_3	−0.189	0.005	0.554	0.608	0.399	0.266	0.778	0.565	0.582	0.263	1.000		
SO_4	−0.096	0.004	0.476	0.572	0.859	0.916	0.178	0.515	0.563	0.910	0.114	1.000	
NO_3	0.255	0.171	0.168	0.129	0.132	0.117	0.118	0.083	0.158	0.118	0.066	0.083	1.000

The molar relationship between different factors governing hydrogeochemical processes in groundwater can be explained by factor analysis. This technique is effective for comprehending the hydrogeochemical progression of groundwater (Minhas et al. 2019). The Kaiser–Meyer–Olkin (KMO = 0.78) sample adequacy was measured, and Bartlett's test of sphericity ($p < 0.001$) indicated that the sample data set collected for this study was suitable for PCA. The detected principal components (PCs) are shown in Table 6.5, along with their component loadings, Eigen values (>1), percentage of variance, and cumulative percentages. The scree plot (Fig. 6.9) was determined based on the Varimax with Kaiser normalisation rotation method, which was used to summarise the predominant factors. Three PCs (PC1, PC2, and PC3) were kept, as illustrated in Table 6.5. From the three PCs, two distinct causes, mostly geogenic and anthropogenic, were identified as important governing factors for every kind of water sample in the study area. According to the Kaiser criterion selection calculation, the initial five PCs with a large variance input were found to have eigenvalues of 7.236, 1.704, and 1.496 (Fig. 6.9) and respective contribution variance rates of 37.110%, 31.395%, and 11.770% (Table 6.5), with a cumulative contribution variance percentage of 80.28% (Table 6.5). The rotated space matrix of the initial three PCs can account for most of the facts derived from the water samples, and the corresponding three factors were also plotted (shown in Fig. 6.9). It has been observed that the highest number of parameters were loaded with PC1, and the parameters are EC, TDS, Na^+, Mg^{2+}, SO_4^{2-}, Cl^-, and TH. The significant SO_4^{2-} the sulphate fertiliser or oxidation could cause loadings in PC1. Similarly, PC2 was loaded with EC, TH, K^+, Ca^{2+}, Mg^{2+}, and HCO_3^-. The higher concentration of NO_3^- in PC3 may have originated due to anthropogenic factors. PC1 significantly influences the ionic characteristics of all three catchments and demonstrates the natural geochemical progression arising from rock-water interactions. The PCA indicates that the chemistry of each water type is affected by multiple processes, such as ion exchange, mineral disintegration, and anthropogenic influences.

6.8 Conclusions

In this current study, three catchments named upper (Maputaland), middle (uMhlathuze), and lower (Tugela) were considered as the study areas, situated in the eastern coastal regions of South Africa. A total of 128 groundwater samples were considered to analyse the hydrogeochemical characteristics, controlling mechanisms, and impact of seawater intrusion in the regions. It was observed that throughout the three catchments, the average concentrations of the major cations and anions were determined in the following order: $Na^+ > Ca^{2+} > Mg^{2+} > K^+$ and $Cl^- > HCO_3^- > SO_4^{2-} > NO_3^-$, respectively. Significant regional changes were observed in certain geochemical variables according to hydrogeochemical characteristics. The trilinear Piper diagram and Chadha plot revealed the major hydrochemical facies as Na-Cl type and Ca-Mg·Cl type across the three catchments. SMI evaluation demonstrated that 64.84% of the water samples belong to the freshwater category, while 13.28% of

Table 6.5 Rotated component matrix of the physical parameters and major ions of all water samples

Parameters	Communalities	Components		
		1	2	3
ORP	0.645	−0.081	−0.208	0.771
pH	0.493	−0.010	−0.012	0.702
EC	0.882	0.559	0.754	0.042
TH	0.905	0.586	0.744	−0.086
TDS	0.957	0.885	0.416	0.025
Na	0.977	0.948	0.279	0.029
K	0.817	0.187	0.881	0.075
Ca	0.739	0.517	0.675	−0.123
Mg	0.875	0.587	0.727	-0.041
Cl	0.986	0.948	0.295	0.018
HCO_3	0.793	−0.006	0.890	−0.024
SO_4	0.927	0.962	0.020	−0.006
NO_3	0.441	0.086	0.156	0.639
Percentage variance explained		37.110	31.395	11.770
Total variance explained		80.28%		
Eigen value		Greater than 1		

the samples were slightly polluted by SWI. At the same time, SMI values unveiled that nearly 17.97%, 0.78%, and 3.13% of samples were of moderately polluted, highly polluted, and dangerously polluted classes, respectively, which demarcated the strong influence of saline water in these catchments. Additionally, the ionic ratios between Cl^- against Cl^-/HCO_3^-, Na^+/Cl^-, SO_4^{2-}/Cl^-, and EC illustrated the SWI in these catchments. Influences from rock-water interaction, silicate weathering, reverse ion exchange, seawater intrusion, and anthropogenic activities were the hydrogeochemical processes that governed the chemistry of groundwater in the study areas. The reverse ion exchange process is responsible for the increased concentration of some major ions when the salinity rises. PCA and correlation matrices revealed that the leading cause of cation and anion enrichment in the groundwater of these catchments is the interaction between water and rocks. The results of this study will help administrators and lawmakers who are in charge of managing these three catchments of water resources in the regions sustainably.

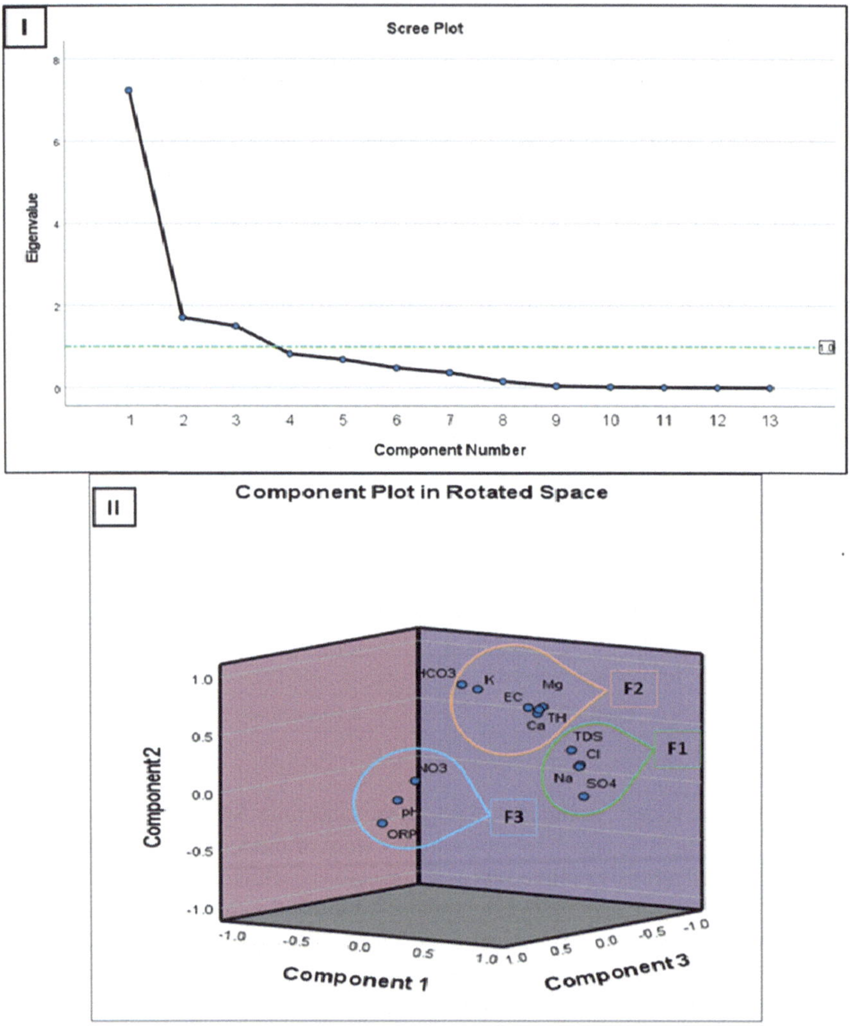

Fig. 6.9 Statistical findings from PCA: (I) Scree plot, (II) Component plot showing the factors

References

Abotalib AZ, Heggy E, Scabbia G, Mazzoni A (2019) Groundwater dynamics in fossil fractured carbonate aquifers in eastern Arabian peninsula: a preliminary investigation. J Hydrol 571:460–470. https://doi.org/10.1016/J.JHYDROL.2019.02.013

Ansari MA, Deodhar A, Kumar US (2019) Modeling of geochemical processes and multivariate statistical analysis for hydrochemical assessment of spring water of the outer Himalaya, India. Environ Earth Sci 78(24):1–17. https://doi.org/10.1007/S12665-019-8682-5/FIGURES/13

Chadha DK (1999) A proposed new diagram for geochemical classification of natural waters and interpretation of chemical data. Hydrogeol J 7(5):431–439. https://doi.org/10.1007/S100400050216/METRICS

Dong H, Gao Z (2022) Theoretical progress of groundwater chemical evolution based on Tóthian theory: a review. Front Mar Sci 9:972426

DWAF (1996) Water quality guidelines for domestic use. Department of Water Affairs and Forestry

DWAF (2000) Strategic environmental assessment for water use Mhlathuze catchment—KZN. Report No SEA-01/2000

Elumalai V, Brindha K, Sithole B, Lakshmanan E (2017) Spatial interpolation methods and geostatistics for mapping groundwater contamination in a coastal area. Environ Sci Pollut Res 24(12):11601–11617. https://doi.org/10.1007/s11356-017-8681-6

Elumalai V, Rajmohan N, Sithole B, Li P, Uthandi S, van Tol J (2023) Geochemical evolution and the processes controlling groundwater chemistry using ionic ratios, geochemical modeling and chemometric analysis in uMhlathuze catchment, KwaZulu-Natal, South Africa. Chemosphere 312:137179. https://doi.org/10.1016/j.chemosphere.2022.137179

Feng W, Qian H, Xu P, Hou K (2020) Hydrochemical characteristic of groundwater and its impact on crop yields in the Baojixia irrigation area, China. Water (Switzerland) 12(5):1443. https://doi.org/10.3390/w12051443

Garland G, Broderick D (1992) Changes in the extent of erosion in the Tugela catchment, 1944–1981. S Afr Geogr J 74(2):45–48. https://doi.org/10.1080/03736245.1992.10586395

Germishuyse T (1999) A geohydrological study of the Richards bay area. University of Zululand

Gibbs RJ (1970) Mechanisms controlling world water chemistry. Science (1979) 170(3962):1088–1090. https://doi.org/10.1126/SCIENCE.170.3962.1088

Goswami S, Rai AK (2024) Identifying intrusion of seawater in coastal aquifers by modified GALDIT (M-GALDIT) index. Groundw Sustain Dev 25:101173. https://doi.org/10.1016/j.gsd.2024.101173

Gupta D, Chaudhary S, Singh A, Shukla R, Mishra VK (2023) Hydrochemical assessment of groundwater quality in the Narmada river basin (central India). Water Supply 23(2):459–481. https://doi.org/10.2166/ws.2022.409

Haaf E, Giese M, Heudorfer B, Stahl K, Barthel R (2020) Physiographic and climatic controls on regional groundwater dynamics. Water Resour Res 56(10):e2019WR026545. https://doi.org/10.1029/2019WR026545

Hanasaki N, Kanae S, Oki T, Masuda K, Motoya K, Shirakawa N, Shen Y, Tanaka K (2008) An integrated model for the assessment of global water resources—part 2: applications and assessments. Hydrol Earth Syst Sci 12(4):1027–1037. https://doi.org/10.5194/HESS-12-1027-2008

Hu S, Shi L, Zha Y, Williams M, Lin L (2017) Simultaneous state-parameter estimation supports the evaluation of data assimilation performance and measurement design for soil-water-atmosphere-plant system. J Hydrol (Amst) 555:812–831. https://doi.org/10.1016/J.JHYDROL.2017.10.061

Kabir T, Pokhrel Y, Felfelani F (2023) Climatic and anthropogenic controls on groundwater dynamics in the Mekong river Basin. J Hydrol (Amst) 622:129761. https://doi.org/10.1016/J.JHYDROL.2023.129761

Kanagaraj G, Elango L, Sridhar SGD, Gowrisankar G (2018) Hydrogeochemical processes and influence of seawater intrusion in coastal aquifers south of Chennai, Tamil Nadu, India. Environ Sci Pollut Res 25(9):8989–9011. https://doi.org/10.1007/s11356-017-0910-5

Khanoranga KS (2019) An assessment of groundwater quality for irrigation and drinking purposes around brick kilns in three districts of Balochistan province, Pakistan, through water quality index and multivariate statistical approaches. J Geochem Explor 197:14–26. https://doi.org/10.1016/J.GEXPLO.2018.11.007

Kirwan ML, Gedan KB (2019) Sea-level driven land conversion and the formation of ghost forests. Nat Clim Chang 9(6):450–457. https://doi.org/10.1038/s41558-019-0488-7

Li P, Tian R, Liu R (2019) Solute geochemistry and multivariate analysis of water quality in the Guohua phosphorite mine, Guizhou province, China. Expo Health 11(2):81–94. https://doi.org/10.1007/s12403-018-0277-y

Li X, Tang C, Cao Y, Li D (2020) A multiple isotope (H, O, N, C and S) approach to elucidate the hydrochemical evolution of shallow groundwater in a rapidly urbanised area of the pearl river delta, China. Sci Total Environ 724:137930. https://doi.org/10.1016/J.SCITOTENV.2020.137930

Liao F, Fu K, Zhang W, Song H, Kong Y, Wang Z, Tang J (2025) Stabilisation mechanism and remediation effectiveness of Pb and cd in agricultural soil using nonmetallic minerals. Sci Rep 15(1):1–13. https://doi.org/10.1038/s41598-025-96970-z

Minhas PS, Qadir M, Yadav RK (2019) Groundwater irrigation induced soil sodification and response options. Agric Water Manag 215:74–85. https://doi.org/10.1016/J.AGWAT.2018.12.030

Mkhwanazi MN (2010) Establishment of the relationship between the sediments mineral composition and groundwater quality of the primary aquifers in the Maputaland coastal plain. Master thesis of University of Zululand. https://hdl.handle.net/10530/614. Accessed 23 Apr 2025

Mondal P, Walter M, Miller J, Epanchin-Niell R, Gedan K, Yawatkar V, Nguyen E, Tully KL (2023) The spread and cost of saltwater intrusion in the US mid-atlantic. Nat Sustain 6(11):1352–1362. https://doi.org/10.1038/S41893-023-01186-6

NRIO (1986) Basic physical geography/hydro data for Natal "estuaries". South African Estuary Information System, Stellonbosch

Ntanganedzeni B, Elumalai V, Rajmohan N (2018) Coastal aquifer contamination and geochemical processes evaluation in Tugela catchment, South Africa-geochemical and statistical approaches. Water 10(6):687. https://doi.org/10.3390/w10060687

Papazotos P, Koumantakis I, Vasileiou E (2019) Hydrogeochemical assessment and suitability of groundwater in a typical Mediterranean coastal area: a case study of the Marathon basin, NE Attica, Greece. HydroResearch 2:49–59. https://doi.org/10.1016/J.HYDRES.2019.11.002

Park SC, Yun ST, Chae GT, Yoo IS, Shin KS, Heo CH, Lee SK (2005) Regional hydrochemical study on salinisation of coastal aquifers, western coastal area of South Korea. J Hydrol (Amst) 313(3–4):182–194. https://doi.org/10.1016/j.jhydrol.2005.03.001

Piper AM (1944) A graphic procedure in the geochemical interpretation of water-analyses. EOS Trans Am Geophys Union 25(6):914–928. https://doi.org/10.1029/TR025i006p00914

Porat N, Botha G (2008) The luminescence chronology of dune development on the Maputaland coastal plain, southeast Africa. Quat Sci Rev 27:1024–1046. https://doi.org/10.1016/j.quascirev.2008.01.017

Rajmohan N, Masoud MHZ, Niyazi BAM (2021) Impact of evaporation on groundwater salinity in the arid coastal aquifer, western Saudi Arabia. Catena (Amst) 196:104864. https://doi.org/10.1016/j.catena.2020.104864

Ramsay PJ (1996) 9000 Years of sea-level change along the southern African coastline. Quat Int 31:71–75. https://doi.org/10.1016/1040-6182(95)00040-P

Rao PN, Prasad KM, Madhusudhan BJ, Krishna VSR, Anand AVSS, Madhnure P (2016) Impact of urbanisation on groundwater quality in Vijayawada urban agglomeration, the new capital region of Andhra Pradesh, India–a baseline study. J Geol Soc India 87:539–552. https://doi.org/10.1007/s12594-016-0428-3

Sajil Kumar PJ (2016) Deciphering the groundwater–saline water interaction in a complex coastal aquifer in South India using statistical and hydrochemical mixing models. Model Earth Syst Environ 2(4):1–11. https://doi.org/10.1007/s40808-016-0251-2

Sawyer CN, McCarty PL (1978) Chemistry for environmental engineering, 3rd edn. McGraw Hill Book Company, New York

Schlosser CA, Strzepek K, Gao X, Fant C, Blanc É, Paltsev S, Jacoby H, Reilly J, Gueneau A (2014) The future of global water stress: an integrated assessment. Earths Future 2(8):341–361. https://doi.org/10.1002/2014EF000238

Schoeller H (1965) Qualitative evaluation of groundwater resources. In: Methods and techniques of groundwater investigations and development. UNESCO

Schoeller H (1967) Geochemistry of groundwater: an international guide for research and practice. UNESCO

Schulze RE (1979) Hydrology and water resources of the Drakensberg, 2nd edn. Natal Town and Regional Planning Commission

Shah T, Bruke J, Vullholth K (2007) Groundwater: a global assessment of scale and significance. In: Molden D (eds) Water for food water for life. Routledge, London

Singh A, Cutting NG, Kaur S, Biwalkar N (2025) Assessing groundwater quality and suitability for agricultural use in Punjab, India: a spatial and temporal analysis. Environ Monit Assess 197(2):132. https://doi.org/10.1007/s10661-024-13508-y

Steinmuller HE, Chambers LG (2018) Can saltwater intrusion accelerate nutrient export from freshwater wetland soils? an experimental approach. Soil Sci Soc Am J 82(1):283–292. https://doi.org/10.2136/sssaj2017.05.0162

Su Q, Kambale RD, Tzeng JH, Amy GL, Ladner DA, Karthikeyan R (2025) The growing trend of saltwater intrusion and its impact on coastal agriculture: challenges and opportunities. Sci Total Environ 966:178710

Tackley HA, Kurylyk BL, Lake CB (2024) Impacts of groundwater dynamics around a macro-tidal river on agricultural soil salinity. Sci Total Environ 927:172344. https://doi.org/10.1016/J.SCITOTENV.2024.172344

Tivianton TA, Barus B, Purwanto MYJ, Anwar S, Widiatmaka, Laudiansyah R(2021) Temporal NDVI analysis to detect the effects of seawater intrusion on rice growth in coastal areas. IOP Conf Ser Earth Environ Sci 662(1):012021. https://doi.org/10.1088/1755-1315/662/1/012021

Tully KL, Weissman D, Wyner WJ, Miller J, Jordan T (2019) Soils in transition: saltwater intrusion alters soil chemistry in agricultural fields. Biogeochemistry 142(3):339–356. https://doi.org/10.1007/s10533-019-00538-9

Vikas C, Kushwaha R, Ahmad W, Prasannakumar V, Dhanya PV, Reghunath R, Kushwaha R, Prasannakumar V, Reghunath R, Dhanya PV (2014) Hydrochemical appraisal and geochemical evolution of groundwater with special reference to nitrate contamination in aquifers of a semi-arid terrain of NW India. Water Qual Exposure Health 7(3):347–361. https://doi.org/10.1007/S12403-014-0155-1

Wang X, Zhou X, Zheng Y, Song C, Long M, Chen T, Ren Z, Yang M, Li X, Guo J (2017) Hydrochemical characteristics and mixing behavior of thermal springs along the Bijiang river in the Lanping basin of China. Environ Earth Sci 76(14):1–19. https://doi.org/10.1007/S12665-017-6821-4/FIGURES/12

Wang Z, Guo X, Kuang Y, Chen Q, Luo M, Zhou H (2022) Recharge sources and hydrogeochemical evolution of groundwater in a heterogeneous karst water system in Hubei province, central China. Appl Geochem 136:105165. https://doi.org/10.1016/J.APGEOCHEM.2021.105165

Watkeys MK, Mason TR, Goodman PS (1993) The rôle of geology in the development of Maputaland, South Africa. J Afr Earth Sci 16:205–221. https://doi.org/10.1016/0899-5362(93)90168-P

Weitz J, Demlie M (2014) Conceptual modelling of groundwater-surface water interactions in the Lake Sibayi catchment, eastern South Africa. J Afr Earth Sc 99:613–624. https://doi.org/10.1016/j.jafrearsci.2013.11.018

Weston NB, Giblin AE, Banta GT, Hopkinson CS, Tucker J (2010) The effects of varying salinity on ammonium exchange in estuarine sediments of the Parker river, Massachusetts. Estuaries Coasts 33:985–1003. https://doi.org/10.1007/s12237-010-9282-5

WHO (2011) Guidelines for drinking-water quality. https://iris.who.int/handle/10665/44584. Accessed 14 Mar 2025

Zhang K, Xie X, Zhu B, Meng S, Yao Y (2019) Unexpected groundwater recovery with decreasing agricultural irrigation in the yellow river basin. Agric Water Manag 213:858–867. https://doi.org/10.1016/J.AGWAT.2018.12.009

Chapter 7
Status of Groundwater Quality in KwaZulu-Natal Coastal Aquifers

This chapter focussed on assessing the geochemical characteristics and quality of groundwater in the eastern coastal aquifers of South Africa such as Maputoland, uMhlathuze, and Tugela to determine its suitability for domestic, agricultural, and industrial uses. Given the agricultural dominance in the region, understanding groundwater quality is crucial for sustaining soil productivity and public health. These physico-chemical, ionic data were used to compute various hydrochemical indices such as Water Quality Index (WQI), Sodium Percentage (%Na), Kelly's Ratio, Magnesium Hazard, Sodium Adsorption Ratio (SAR), Permeability Index, and Corrosivity Ratio, following standard procedures and using WHO and SANS guidelines. Spatial variation was mapped to identify zones of concern. The results indicated spatial differences in groundwater chemistry, with elevated salinity and contamination levels in the central and southern aquifers. The chapter provided a comprehensive evaluation groundwater suitability and highlighted the importance of regular monitoring to inform sustainable water management in coastal agricultural regions.

7.1 Introduction

Groundwater quality coastal aquifers have been a long-standing concern due to the unique hydrogeological conditions they pose along with the growing demand due to increasing population coastal regions. It serves 44% of the world population by fulfilling their drinking water needs along with irrigational needs to meet global growing food demands (Senarathne et al. 2021). These resources are closer to the ocean making them vulnerable towards seawater intrusion due to over extraction and imbalance in hydraulic gradient between freshwater and seawater contact zones (Pitchaimani et al. 2024; Panthi et al. 2022). The intrusion of saline water makes groundwater unsuitable for both drinking and agricultural activity, vastly limiting

P. Li and V. Elumalai, *Groundwater Quality Under Agricultural Activities—Cases from China and South Africa*, Professional Practice in Earth Sciences,
https://doi.org/10.1007/978-3-031-98993-3_7

their use directly influencing public health and impacting food security (Irfeey et al. 2023; Kaushal et al. 2021; Miller et al. 2021). The elevated levels of ions such as sodium, chloride, calcium, magnesium along with sulfates, which might lead to degraded soil quality and reduced crop yield in case of long-term usage (Salcedo et al. 2022). On the other hand, intense agricultural activities have increased the nitrate levels in soil and groundwater causing eutrophication and serious health concerns in humans (Akinnawo 2023; Craswell 2021; Verma et al. 2023; Ramalingam et al. 2022).

The explosion of population, intense agriculture, rapid urbanization and industrialization in coastal regions has been a concerning factor for decades leading to over extraction (Senarathne et al. 2021; Kundzewicz and Doell 2009). Human activities such as pumping, application of manure, fertilizers and wastewater infusions are affecting the quarter quality directly. Especially in arid and semi-arid regions, where the availability of groundwater is erratic due to changing precipitation patterns, groundwater suitability for agriculture is highly dependent controlled salinity, alkalinity, and soil properties (Al-Mashreki et al. 2023; Abukhadra et al. 2022; Yin et al. 2022). Concentration of Na^+, Ca^{2+}, Mg^{2+}, K^+, HCO_3^-, SO_4^{2-}, and Cl^- in groundwater define their saline property and suitability for agriculture along with define the status on soil (Salam et al. 2021; Rajmohan et al. 2017; Elumalai et al. 2017). Both total amount of salt in soil and types of salt ions carried by water could be responsible for crop failure by causing coil degradation. So use of untreated saline water in agricultural activities exacerbates these underlying issues causing long term damage by soil pore blockage and reduced water filtration (Hakami et al. 2024; Salcedo et al. 2022; Shaki and Adeloye 2006). Additionally, precipitation of various minerals from geological sources with volatile geochemical environment could cause reduced porosity and permeability hampering local or global food supply and demand (Gastmans et al. 2010).

Monitoring the groundwater quality in coastal aquifers is essential for assessing the levels of contamination and evolution of sustainable resource management and practices. This requires various integrated approaches that combine hydrogeological contamination assessment. In such scenario quality parameters such as total dissolved solids (TDS), electrical conductivity (EC), pH, major ions (Na^+, K^+, Ca^{2+}, Mg^{2+}, Cl^-, SO_4^{2-}, HCO_3^-), nutrients (NO_3^-) play a vital role in understanding the extent of contamination in various water resources. Their integration with tools like water quality index (WQI), and various irrigational quality index along with spatial distribution provides a comprehensive over of water quality status and identification of hotspots for prioritized intervention and mitigation strategies. So this chapter deals with application of WQI and irrigational indices such as SAR, %Na, Magnesium hazard, Permeability index, Corrosivity ratio to assess the level of impact in the eastern coastal aquifers of KwaZulu-Natal, South Africa.

7.2 Materials and Methods

To assessment the water quality in South Africa's eastern coastal aquifers, 128 groundwater samples were collected from various industrial, agricultural and rural and urban areas along 3 major aquifers such as Maputoland, uMhlathuze and Tugela. The boreholes were flushed for at least 10 min before collecting the samples to avoid rainwater and represent surrounding lithological and anthropogenic influence accurately. The physicochemical parameters such as pH, EC and TDS were measured in the field with the help of YSI portable multiprobe. The samples were filtered through 0.45 μm membranes with the help of a portable hand pump and sealed in high-density polyethylene (HDPE) bottles to avoid any atmospheric contact. The samples were carried to the International research centre for water science and technology (IRC-WST) laboratories situated in University of Zululand and stored at 4 °C for further analysis. Anions such as CO_3^{2-} and HCO_3^- were analysed using titrimetric method using APHA (2005) protocols. The cations such as Ca^{2+}, Mg^{2+}, Na^+, and K^+, anions such as Cl^- and SO_4^{2-} along with nutrients NO_3^- were analyzed with the help of Metrohm Ion chromatograph.

7.2.1 *Water Quality Index (WQI)*

The Water Quality Index (WQI) is a useful tool for assessing the overall quality of water for human consumption by combining various physico-chemical parameters into a single number that reflects the general status of the water. In this study, the WQI was calculated using the WHO (2011) drinking water quality standards as a reference framework. The process involves several sequential steps, starting with the selection of appropriate water quality parameters that influence human health. These typically include pH, Total Dissolved Solids (TDS), nitrate (NO_3^-), fluoride (F^-), chloride (Cl^-), sulfate (SO_4^{2-}), calcium (Ca^{2+}), magnesium (Mg^{2+}), bicarbonate (HCO_3^-), carbonates (CO_3^{2-}), and electrical conductivity (EC), among others.

After collecting and analysing water samples from the selected sampling points, the quality rating scale (Q_i) for each parameter is calculated using the formula as per Eq. 7.1.

$$Q_i = \left(\frac{C_i}{S_i}\right) \times 100 \tag{7.1}$$

where, Q_i is the quality rating for the i-th parameter, C_i is the measured concentration of that parameter, and S_i is the standard permissible limit set by WHO (2011). For parameters such as pH, which are defined by an acceptable range, the ideal value or midpoint can be used, or deviations from the optimal range can be factored into Q_i.

To account for the relative importance of each parameter in influencing water quality, a unit weight (W_i) is assigned. This weight is calculated inversely proportional to the standard value using the Eq. 7.2.

$$W_i = \frac{1/S_i}{\sum(1/S_i)} \tag{7.2}$$

This ensures that parameters with lower permissible limits (and higher health risks if exceeded) receive greater weight in the final index.

The overall WQI is then calculated using the weighted arithmetic mean method as per Eq. 7.3.

$$\text{WQI} = \frac{\sum Q_i/W_i}{\sum W_i} \tag{7.3}$$

This formula combines the quality ratings and their corresponding weights, producing a single index value.

Based on the final WQI value, water quality is classified into one of several categories: excellent (WQI 0–50), good (50–100), poor (100–200), very poor (200–300), or unsuitable for drinking (>300). A lower WQI indicates better water quality and higher suitability for consumption, whereas a higher WQI indicates degraded quality and potential health risks. The interpretation of WQI allows decision-makers and water resource managers to easily communicate water quality information and prioritize areas requiring treatment or intervention.

7.2.2 *Irrigational and Industrial Suitability*

The Percent Sodium (%Na) indicates the proportion of sodium ions relative to the total concentration of major cations in water given by Wilcox (1955). Excess sodium in irrigation water can displace calcium and magnesium in soil, leading to poor soil structure and reduced permeability. The formula for %Na is as per Eq. 7.4.

$$\text{Na\%} = \frac{\text{Na}^+ + \text{K}^+}{\text{Ca}^{2+} + \text{Mg}^{2+} + \text{Na}^+ + \text{K}^+} \times 100 \tag{7.4}$$

All ionic concentrations should be expressed in milliequivalents per liter (meq/L). A %Na value above 60 typically indicates a risk of sodium accumulation and poor infiltration, which can be harmful to soil and crop health.

The Kelly's Ratio (KR) is another parameter used to assess sodium hazards in irrigation water (Kelly, 1940, 1963). It is the ratio of sodium to the sum of calcium and magnesium (Eq. 7.5):

$$KR = \frac{Na^+}{\left(Ca^{2+} + Mg^{2+}\right)} \tag{7.5}$$

If KR is less than 1, the water is considered suitable for irrigation. A KR greater than 1 indicates an excessive sodium level that may lead to soil degradation over time.

The Magnesium Hazard (MH) evaluates the effect of magnesium in water (Szabolcs and Darab 1964). While both calcium and magnesium are essential for plants, excess magnesium can adversely affect soil quality by increasing alkalinity and reducing soil aggregation. It is calculated as per Eq. 7.6.

$$MH = \frac{Mg^{2+}}{Ca^{2+} + Mg^{2+}} \times 100 \tag{7.6}$$

Water with MH values above 50% is considered harmful to crops and soil, as high magnesium content can lead to reduced soil permeability and aeration.

The Sodium Absorption Ratio (SAR) is a widely used indicator that reflects the potential for sodium to be absorbed by soil in exchange for calcium and magnesium, thus affecting soil structure given by Richard (1954). The formula is shown in Eq. 7.7.

$$SAR = \frac{Na^+}{\sqrt{\frac{Ca^{2+} + Mg^{2+}}{2}}} \tag{7.7}$$

SAR values below 10 are generally considered safe for irrigation, while higher values may lead to sodium-induced soil compaction and reduced permeability.

The Permeability Index (PI) is used to assess the long-term effect of irrigation water on soil permeability. It accounts for the effect of sodium, calcium, magnesium, and bicarbonate ions, and is given by Doneen (1964) and calculated using Eq. 7.8.

$$PI = \frac{Na^+ + \sqrt{HCO_3^-}}{Ca^{2+} + Mg^{2+} + Na^+} \times 100 \tag{7.8}$$

where the square root of bicarbonate is taken to reflect its buffering effect. PI values are classified into three classes: Class I (>75%), Class II (25–75%), and Class III (<25%). Water with a higher PI is generally more suitable for irrigation, as it poses less risk of reducing soil permeability.

All ion concentrations used in these indices must be converted to meq/L using the conversion.

The Corrosivity Ratio (CR) is a hydrochemical index used to evaluate the likelihood of water corroding metallic infrastructure. The ion-based method considers the balance between aggressive anions (Cl^- and SO_4^{2-}) and the buffering ions (HCO_3^- and CO_3^{2-}), which help neutralize the corrosive effect (Raman, 1985). The formula used for this calculation is as per Eq. 7.9.

$$\text{Corrosivity Ratio (CR)} = \frac{Cl^- + 2(SO_4^{2-})}{2[HCO_3^- + CO_3^{2-}/100]} \quad (7.9)$$

In this method, all ionic concentrations must be expressed in milliequivalents per liter (meq/L), since the equation is based on the ionic charge balance rather than mass. After converting all relevant ion concentrations into meq/L, the values of Cl^- and SO_4^{2-} are summed to represent the corrosive force, while the sum of HCO_3^- and CO_3^{2-} represents the neutralizing or buffering capacity. The resulting ratio gives a direct indication of the water's corrosivity. A CR value less than 1 indicates that the water is non-corrosive, a value equal to 1 suggests marginal corrosion potential, and a CR greater than 1 signifies that the water is corrosive. This method is particularly applicable in environments where chloride and sulfate levels are elevated due to either natural geochemical processes or anthropogenic contamination, such as from industrial discharge or agricultural runoff. Meanwhile, higher concentrations of bicarbonate and carbonate ions provide a buffering effect that mitigates corrosion. The CR therefore serves as a practical tool to assess infrastructure risk and helps inform mitigation strategies in groundwater management and water treatment systems.

7.3 Spatial Distribution and Descriptive Statistics of Water Quality Parameters

The descriptive statistics of groundwater quality parameters show significant variability and deviations from normal distribution (Table 7.1). The parameters reflect diverse hydrogeochemical and anthropogenic influences across the study area. Sodium (Na^+) concentrations (14.00–6971.00 mg/L, mean = 225.68 mg/L) frequently exceed both WHO and SANS limits (200 mg/L), with extreme skewness and kurtosis indicating irregular but severe enrichment (Fig. 7.1Na). Potassium (K^+), with a mean of 16.99 mg/L, also surpasses the WHO guideline of 12 mg/L and exhibits similar statistical extremes in the coastal zone central region in uMhlathuze catchment hinting at anthropogenic variability in the region. Ca^{2+} and Mg^{2+} mostly fall within permissible ranges but show irregular distributions (skewness > 2), suggesting lithological variability and comparatively higher weathering in southern region of the study area (Fig. 7.1Ca & Mg). Cl^- levels, averaging 356.84 mg/L, exceed standards and show a highly skewed and peaked distribution, indicating occasional high contamination events, possibly from seawater influence in the coastal parts of central region. HCO_3^- and SO_4^{2-} also show broad concentration ranges and high kurtosis, reflecting irregular geochemical contributions. Their higher concentration in southern catchment suggests a highly weathering environment probably from intense agriculture in this part (Ntanganedzeni et al. 2018). NO_3^- presents as a potential pollutant with a mean of 8.69 mg/L and extreme variability. The widespread agriculture in central to southern region might be responsible for such variability

(Fig. 7.1). pH values range widely (0.20–9.50), though most samples are neutral to slightly alkaline, probably due to the influence of seawater intrusion in the coastal region (Fig. 7.1pH). This was evident in studies conducted by Mthembu et al. (2020) and Elumalai et al. (2017). Electrical conductivity (EC) and total dissolved solids (TDS) are particularly elevated (means of 1464.77 μS/cm and 908.69 mg/L respectively), often exceeding drinking water limits and suggesting high salinity in parts of the aquifer. The he pronounced skewness and kurtosis values across nearly all parameters indicate that the groundwater quality is highly heterogeneous. The Maputoland aquifer (northern region) major ions including Na^+, K^+, Cl^-, and NO_3^-, were observed in low concentration compared to central and southern regions (Fig. 7.1). Calcium and magnesium levels are moderate, likely influenced by natural weathering of silicate and carbonate minerals (Elumalai et al. 2017; Ntanganedzeni et al. 2018) in southern aquifers. The low NO_3^- and SO_4^{2-} levels further stipulate towards minimal agricultural or sewage-related pollution in this region (Torres-Martínez et al. 2021). The uMhlathuze aquifer, located centrally, exhibits more pronounced spatial variability in ion concentrations. Elevated Na^+ and K^+ levels are observed in localized zones, particularly near the industrial and urban areas, implying possible anthropogenic inputs such as wastewater discharge or agricultural runoff. Similarly, Cl^- and SO_4^{2-} concentrations are significantly higher in certain pockets, especially in the southern portion of this aquifer, which may be linked to fertilizer use, industrial effluents, or leaching from landfill sites. The spatial patterns of Ca^{2+} and Mg^{2+} also reflect variation in lithology and groundwater-rock interactions (Elumalai et al. 2017; Ntanganedzeni et al. 2018). In the Tugela aquifer (southern region), several ions, notably sodium, calcium, chloride, sulfate, and nitrate, exhibit elevated concentrations, especially near the coastal and urbanized fringe. The extremely high sodium levels in the southwestern and coastal parts could be attributed to seawater intrusion, domestic wastewater percolation, or saline water encroachment. Overall, northern region retains a good geochemical status with probable good condition of water quality with less anthropogenic impact whereas central and southern zone stipulate towards moderate to high contamination, probably from higher human activities due to industrial zones and intense agricultural activities (Fig. 7.1).

7.4 Water Quality Index

The spatial distribution map of the Water Quality Index (WQI) reveals distinct regional patterns in groundwater quality across the Maputoland, uMhlathuze, and Tugela aquifers (Fig. 7.2). The northern region, corresponding to the Maputoland and uMlathuze aquifer, is predominantly characterized by high-quality water, with majority of the area falling within the excellent WQI range. This suggests limited anthropogenic impact and relatively pristine hydrogeological conditions, likely due to protected recharge zones and lower levels of intensive land use (Ntanganedzeni et al. 2018). The central region, representing the uMhlathuze aquifer, also displays generally good water quality; however, there are localized zones of concern. Around

Table 7.1 Descriptive statistics of water quality parameters and their guidelines

Parameters	Unit	Min	Max	Mean	Std. Dev	Skewness		Kurtosis		WHO (2011)	SANS (2006)
						Statistic	Std. Error	Statistic	Std. Error		
pH	/	0.20	9.50	7.19	1.01	-2.68	0.21	17.08	0.42	65.-8.5	5–9.5
EC	μS/cm	115.00	32,050.00	1464.77	3733.89	6.83	0.21	50.38	0.42	1500.00	1000.00
TDS	mg/L	73.60	19,230.00	908.69	2296.71	6.81	0.21	49.57	0.42	500.00	/
Na^+	mg/L	14.00	6971.00	225.68	766.42	7.55	0.21	60.22	0.42	200.00	200.00
K^+	mg/L	0.47	634.00	16.99	79.47	7.49	0.21	56.59	0.42	12.00	50.00
Ca^{2+}	mg/L	4.00	217.00	34.78	34.97	2.53	0.21	8.08	0.42	200.00	150.00
Mg^{2+}	mg/L	1.79	184.00	18.89	23.00	4.24	0.21	23.80	0.42	150.00	70.00
Cl^-	mg/L	27.00	9926.00	356.84	1114.78	7.33	0.21	56.97	0.42	250.00	200.00
HCO_3^-	mg/L	12.00	1903.00	128.24	183.08	7.43	0.21	70.21	0.42	/	/
SO_4^{2-}	mg/L	1.75	1330.00	37.24	124.44	9.18	0.21	93.60	0.42	500.00	400.00
NO_3^-	mg/L	0.00	112.00	8.69	12.14	5.61	0.21	43.40	0.42	45.00	45.00

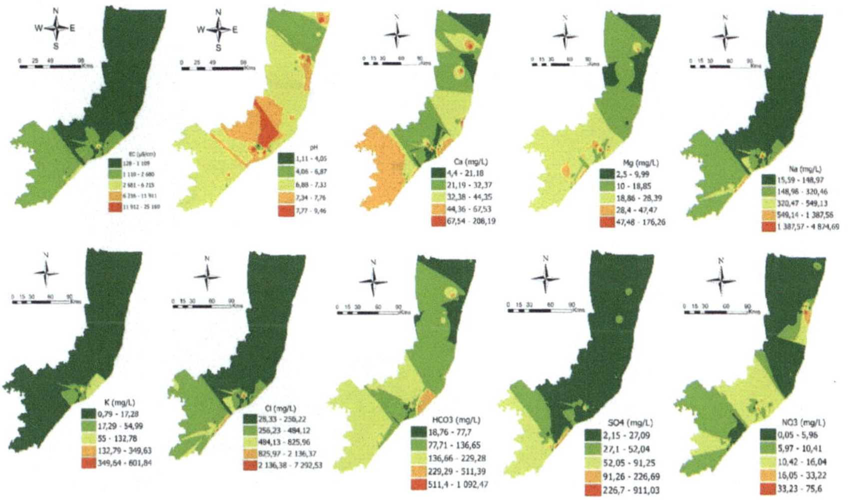

Fig. 7.1 Spatial distribution of considered water quality parameters

the central-southern portion of the catchment, where urban settlements and industrial activities situated as part of Empangeni and Richards Bay industrial areas degradation of water quality is evident, particularly due nitrate and salinity issues (Fig. 7.2) (Elumalai et al. 2023; Ponnusamy and Elumalai 2022). In the southern region, which encompasses the Tugela aquifer, the water quality remains mostly good but with slightly more widespread areas showing signs of moderate degradation. These areas may be influenced by land use practices such as intensive agriculture, livestock farming, and settlements lacking adequate wastewater management. A few isolated hotspots with very poor water quality are visible, primarily near densely populated or industrialized zones, indicating the need for targeted monitoring and remediation. The average WQI values suggest a south ward contamination trend with an order of contamination in considered aquifers as follows: Tugela aquifer > uMhlathuze > Maputoland The majority of the aquifer systems maintain good to excellent water quality, where critical zones in the uMhlathuze and Tugela regions show significant concern regarding gradual degradation of water quality (Table 7.2).

7.5 Groundwater Suitability for Agricultural Purpose

Based on Fig. 7.3, the suitability of water from the Maputoland (MP), uMhlathuze (MH), and Tugela (TU) aquifers for irrigation is categorized by Permeability Index, USSL diagram and Wilcox classification. The majority of the MP samples were categorized in Class I and Class II, suggesting moderate to high permeability and minimal risk. Class II and Class III were found predominantly categorized with uMhlthuze

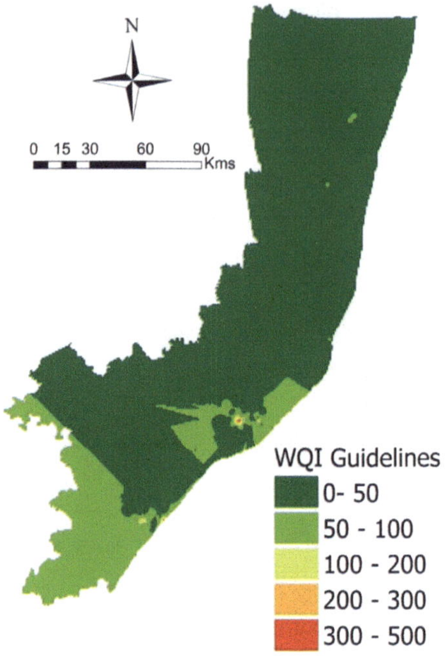

Fig. 7.2 Spatial distributing of acquired WQI values in the study area

Table 7.2 Descriptive statistics of acquired WQI values in considered aquifers

Aquifer	Minimum WQI	Maximum WQI	Average WQI	Standard deviation
Maputoland (northern aquifer)	12.89	69.75	23.42	11.57
uMhlathuze (central aquifer)	11.67	673.49	57.48	112.08
Tugela (southern aquifer)	19.17	430.34	59.9	74.74
Overall	11.67	673.49	46.93	82.81

and Tugale aquifer samples indicating a higher unsuitability of groundwater for irrigational activities. This is an indication of higher probability of causing salinity problems or reducing soil permeability in central and southern region. The USSL diagram further classifies this irrigation suitability based on salinity and sodium hazards. Maputoland (MP) generally falls in the C1-S1 category, indicating low salinity and sodium hazards, and excellent irrigation water quality. uMhlathuze (MH) shows a wider range, mostly in C1-S1 and C2-S1, suggesting low to medium salinity hazards and low sodium hazard, indicating good to fair irrigation water quality. Tugela (TU) samples primarily fall in the C3-S2, C3-S3, and even C4-S4 categories, indicating high to very high salinity and medium to very high sodium hazards, and poor irrigation water quality. Similarly, Wilcox diagram classifies irrigation water based on electrical conductivity (EC) and sodium percentage (%Na). As with the USSL diagram, EC indicates salinity hazard, with higher values indicating greater hazard, while high %Na water can damage soil structure. Similar observations were noticed in this classification as well, that placed MP samples in excellent to good category whereas MH and TU samples were found to be doubtful to unsuitable category. The majority of sample in southern region with many samples having high salinity and sodium percentage, making it generally unsuitable for irrigation without careful management and potentially limiting its use to salt-tolerant crops (Hakami et al. 2024).

Similarly, Fig. 7.4 represents several indices used to assess the suitability of water for agriculture, primarily irrigation, highlighting potential long-term problems (Richards 1954), where %Na and Kelly's Ratio indicate sodium hazards that pose a significant threat to soil structure. These observations stipulate towards clay particle dispersion, reduced permeability, poor drainage, and soil hardening (AL-Areeq et al. 2023), with a substantial portion of the area showing high %Na values > 60 and Kelly's Ratio > 1 (Fig. 7.4). On the other hand, spatial distribution of Mg Hazard indicates potential magnesium-related issues especially in central region

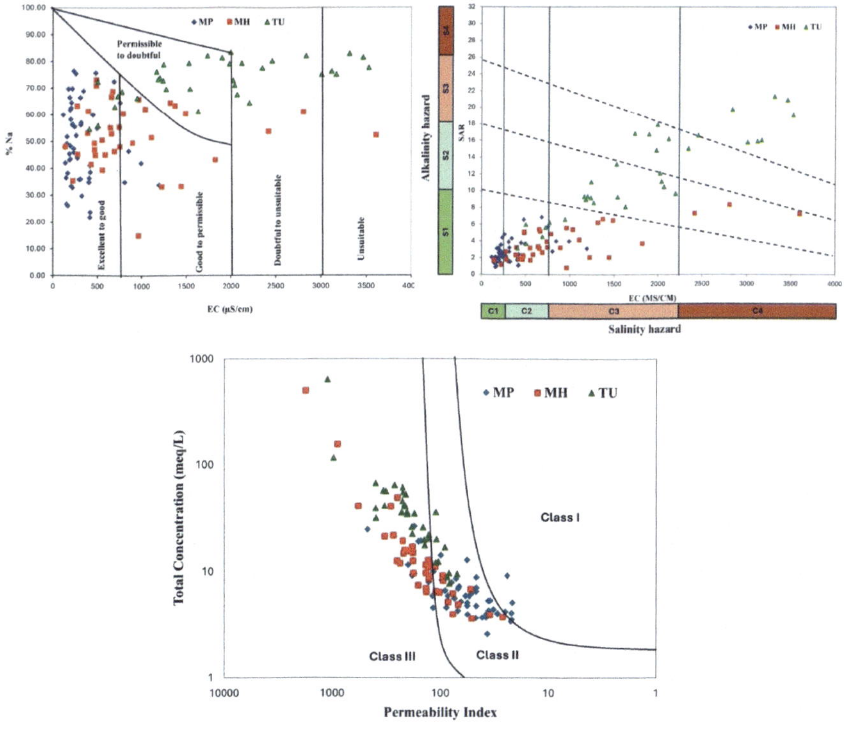

Fig. 7.3 USSL, Wilcox and Doneen's plot showing irrigational suitability of groundwater

(Fig. 7.4) with majority of samples from uMhlathuze catchment crossing the guidelines values (>50), which may lead to soil compaction and reduced water infiltration (Singh et al. 2018). The SAR values found in good to doubtful category in all catchments except the coastal region of uMhlathuze catchment where seawater intrusion and industrial activities are dominant. This stipulates towards the limited rate of sodicity of the soil in the considered region (Qadir et al. 2021).

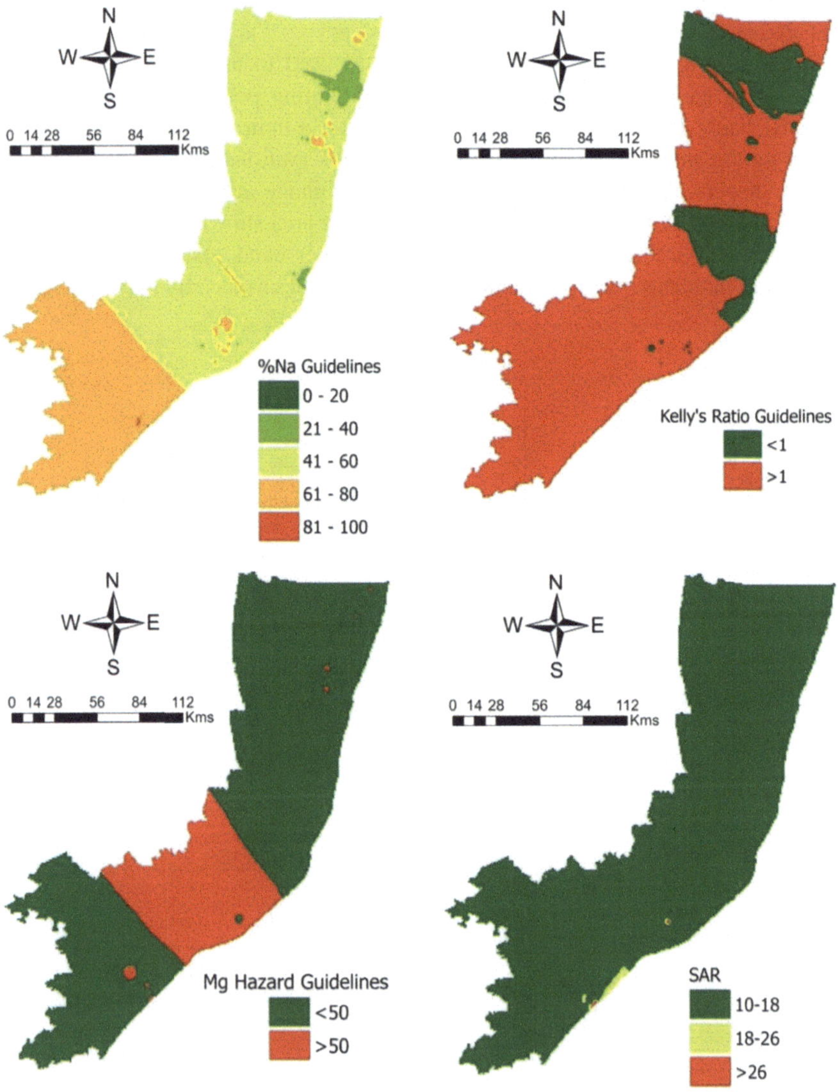

Fig. 7.4 Spatial distribution of irrigational hazard indices

The Fig. 7.5 depicts the spatial distribution of corrosivity ratio, indicating the potential of water to corrode industrial infrastructure. A significant portion of the area is classified as "unsuitable" (shown in red), suggesting that water in these regions poses a high risk of corrosion to industrial equipment and pipelines. In contrast, a smaller portion is classified as "safe-moderate" (shown in green), indicating a lower risk of corrosion with limited but unavoidable tendency to cause corrosion in long term usage (Ahmed et al. 2021). This distribution of corrosivity potential has significant implications for industrial development and infrastructure planning, as it highlights the need for corrosion-resistant materials and mitigation strategies in areas with unsuitable water quality (Kurdi et al. 2015; Rabbani et al. 2009).

The Fig. 7.5 depicts the spatial distribution of corrosivity ratio, indicating the potential of water to corrode industrial infrastructure. A significant portion of the area is classified as "unsuitable" (shown in red), suggesting that water in these regions poses a high risk of corrosion to industrial equipment and pipelines. In contrast, a smaller portion is classified as "safe-moderate" (shown in green), indicating a lower risk of corrosion with limited but unavoidable tendency to cause corrosion in long term usage (Ahmed et al. 2021). This distribution of corrosivity potential has significant implications for industrial development and infrastructure planning, as it highlights the need for corrosion-resistant materials and mitigation strategies in areas with unsuitable water quality (Kurdi et al. 2015; Rabbani et al. 2009).

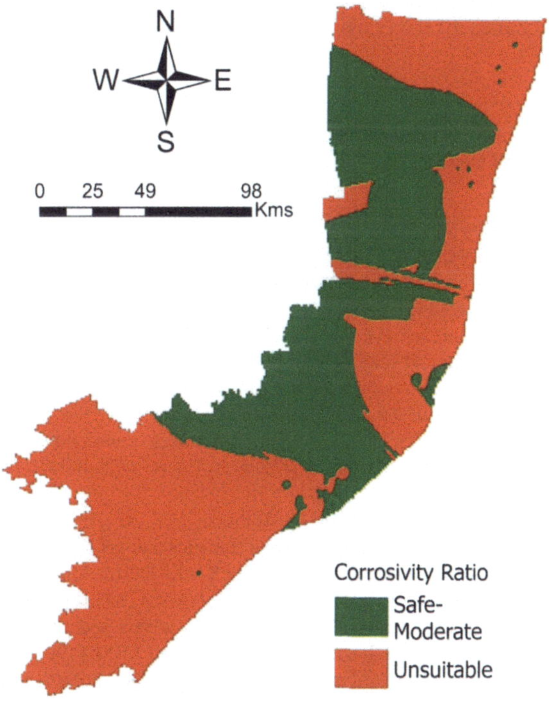

Fig. 7.5 Spatial distribution of Corrosivity ratio for industrial usage

7.6 Conclusions

Agriculture is the predominant livelihood in the study area, highlighting the critical need to ensure the availability, consistent monitoring, and preservation of groundwater quality to maintain soil fertility and safeguard human health. This study aimed to investigate the geochemical characteristics of groundwater along the eastern coastal zones of South Africa and assess its suitability for domestic consumption, agricultural irrigation, and industrial use. The findings suggest that:

- The water in the aquifers is slightly alkaline in nature with dominating ions such as Na^+ and Cl^- especially in the southern aquifers due to increased seawater intrusion and agricultural activities.
- WQI suggested higher degree of contamination of water quality in southern zones with Tugela aquifers having the most contaminated groundwater followed by uMhlathuze aquifer. However, good water quality standard is widespread in majority of the considered area.
- Irrigational unsuitability was evident from %Na, Kelly's ratio, Mg hazard in central to southern zones with southern aquifer being least permeability. The likelihood of soil sodicity and sodium hazard dominated the central and southern zones whereas likelihood of Mg hazard was found higher and limited to central zone, especially in uMhlathuze aquifer. Corrosivity ratio suggested high corrosive nature of water with moderate to extreme levels and was found mainly in costal zones all the considered aquifers.

References

Abukhadra MR, Eid MH, El-Sherbeeny AM, Abd Elgawad AEE, Shim JJ (2022) Effective desalination of brackish groundwater using zeolitized diatomite/kaolinite geopolymer as low-cost inorganic membrane; Siwa Oasis in Egypt as a realistic case study. J Contam Hydrol 244:103923. https://doi.org/10.1016/j.jconhyd.2021.103923

Ahmed S, Sultan MW, Alam M, Hussain A, Qureshi F, Khurshid S (2021) Evaluation of corrosive behaviour and scaling potential of shallow water aquifer using corrosion indices and geospatial approaches in regions of the Yamuna river basin. J King Saud Univ Sci 33(1):101237. https://doi.org/10.1016/j.jksus.2020.101237

Akinnawo SO (2023) Eutrophication: causes, consequences, physical, chemical and biological techniques for mitigation strategies. Environ Chall 12:100733. https://doi.org/10.1016/j.envc.2023.100733

AL-Areeq AM, Chowdhury S, Rahman MT, AL-Areeq NM, Quriaa H (2023) Assessment of water quality using multivariate statistics and geographical information systems (GIS) of Wadi Aldabab, Taiz, Yemen. Appl Water Sci 13(8):169. https://doi.org/10.1007/s13201-023-01973-8

Al-Mashreki MH, Eid MH, Saeed O, Székács A, Szűcs P, Gad M, Ramadan HS (2023) Integration of geochemical modeling, multivariate analysis, and irrigation indices for assessing groundwater quality in the Al-Jawf basin, Yemen. Water 15(8):1496. https://doi.org/10.3390/w15081496

APHA (2005) Standard methods for the examination of water and wastewater, 20th edn. American Public Health Association, Washington, DC

Craswell E (2021) Fertilizers and nitrate pollution of surface and ground water: an increasingly pervasive global problem. SN Appl Sci 3(4):518. https://doi.org/10.1007/s42452-021-04521-8

Doneen LD (1964) Notes on water quality in agriculture published as a water science and engineering. Department of Water Sciences and Engineering, University of California, paper, 4001

Elumalai V, Brindha K, Sithole B, Lakshmanan E (2017) Spatial interpolation methods and geostatistics for mapping groundwater contamination in a coastal area. Environ Sci Pollut Res 24:11601–11617. https://doi.org/10.1007/s11356-017-8681-6

Elumalai V, Rajmohan N, Sithole B, Li P, Uthandi S, van Tol J (2023) Geochemical evolution and the processes controlling groundwater chemistry using ionic ratios, geochemical modelling and chemometric analysis in uMhlathuze catchment, KwaZulu-Natal, South Africa. Chemosphere 312:137179. https://doi.org/10.1016/j.chemosphere.2022.137179

Gastmans D, Chang HK, Hutcheon I (2010) Groundwater geochemical evolution in the northern portion of the guarani aquifer system (Brazil) and its relationship to diagenetic features. Appl Geochem 25(1):16–33. https://doi.org/10.1016/j.apgeochem.2009.09.024

Hakami RA, Naser RS, El-Bakkali M, Othman MDM, Yahya MS, Raweh S, Belghyti D (2024) Groundwater quality deterioration evaluation for irrigation using several indices and geographic information systems: a case study. Desalin Water Treat 320:100645. https://doi.org/10.1016/j.dwt.2024.100645

Irfeey AMM, Najim MM, Alotaibi BA, Traore A (2023) Groundwater pollution impact on food security. Sustainability 15(5):4202. https://doi.org/10.3390/su15054202

Kaushal SS, Likens GE, Pace ML, Reimer JE, Maas CM, Galella JG, Woglo SA (2021) Freshwater salinization syndrome: from emerging global problem to managing risks. Biogeochemistry 154:255–292. https://doi.org/10.1007/s10533-021-00784-w

Kelly WP (1940) Permissible composition and concentration of irrigated waters. Proc ASCF 66:607. https://doi.org/10.1061/TACEAT.0005384

Kelly WP (1963) Use of saline irrigation water. Soil Sci 95(6):385–391

Kundzewicz ZW, Doell P (2009) Will groundwater ease freshwater stress under climate change? Hydrol Sci J 54(4):665–675. https://doi.org/10.1623/hysj.54.4.665

Kurdi M, Ferdows MS, Maghsoudi A (2015) Sensitivity of corrosion and scaling indices based on ions, case study Iran. Water Qual Expo Health 7(3):363–372. https://doi.org/10.1007/s12403-015-0156-8

Miller JD, Workman CL, Panchang SV, Sneegas G, Adams EA, Young SL, Thompson AL (2021) Water security and nutrition: current knowledge and research opportunities. Adv Nutr 12(6):2525–2539. https://doi.org/10.1093/advances/nmab075

Mthembu PP, Elumalai V, Brindha K, Li P (2020) Hydrogeochemical processes and trace metal contamination in groundwater: impact on human health in the Maputaland coastal aquifer, South Africa. Exposure Health 12(3):403–426. https://doi.org/10.1007/s12403-020-00369-2

Ntanganedzeni B, Elumalai V, Rajmohan N (2018) Coastal aquifer contamination and geochemical processes evaluation in Tugela catchment, South Africa—geochemical and statistical approaches. Water 10(6):687. https://doi.org/10.3390/w10060687

Panthi J, Pradhanang SM, Nolte A, Boving TB (2022) Saltwater intrusion into coastal aquifers in the contiguous United States—a systematic review of investigation approaches and monitoring networks. Sci Total Environ 836:155641. https://doi.org/10.1016/j.scitotenv.2022.155641

Pitchaimani VS, Narayanan MSS, Abishek SR, Shyamala G (2024) Assessment of coastal aquifer vulnerability to seawater intrusion: insights from Kadaladi region using the Galdit model. Case Stud Chem Environ Eng 10:100772. https://doi.org/10.1016/j.cscee.2024.100772

Ponnusamy D, Elumalai V (2022) Determination of potential recharge zones and its validation against groundwater quality parameters through the application of GIS and remote sensing techniques in uMhlathuze catchment, KwaZulu-Natal, South Africa. Chemosphere 307:136121. https://doi.org/10.1016/j.chemosphere.2022.136121

Qadir M, Sposito G, Smith CJ, Oster JD (2021) Reassessing irrigation water quality guidelines for sodicity hazard. Agric Water Manag 255:107054. https://doi.org/10.1016/j.agwat.2021.107054

Rabbani Khan D, Zade M, Avar P (2009) Studying rural water corrosion and sedimentation indices under Abfar company authority, Kashan, 2007–2009. Feiz Sci Res Q XV 4:382–388
Rajmohan N, Patel N, Singh G, Amarasinghe UA (2017) Hydrochemical evaluation and identification of geochemical processes in the shallow and deep wells in the Ramganga sub-basin, India. Environ Sci Pollut Res 24:21459–21475. https://doi.org/10.1007/s11356-017-9704-z
Ramalingam S, Panneerselvam B, Kaliappan SP (2022) Effect of high nitrate contamination of groundwater on human health and water quality index in semi-arid region, South India. Arab J Geosci 15(3):242. https://doi.org/10.1007/s12517-022-09553-x
Raman Y (1985) Impact of corrosion in the conveyance and distribution of water. J Indian Water Work Assoc 15(1):115–121
Richards LA (1954) Diagnosis and improvement of saline soils, vol 60, p 160. US Department of Agriculture Handbook. https://doi.org/10.1097/00010694-195408000-00012
Salam MA, Adlii A, Eid MH, Abukhadra MR (2021) Effective decontamination of Ca^{2+} and Mg^{2+} hardness from groundwater using innovative muscovite based sodalite in batch and fixed-bed column studies, dynamic and equilibrium studies. J Contam Hydrol 241:103817. https://doi.org/10.1016/j.jconhyd.2021.103817
Salcedo FP, Cutillas PP, Cabañero JJA, Vivaldi AG (2022) Use of remote sensing to evaluate the effects of environmental factors on soil salinity in a semi-arid area. Sci Total Environ 815:152524. https://doi.org/10.1016/j.scitotenv.2021.152524
Senarathne S, Jayawardana JMCK, Edirisinghe EANV, Chandrajith R (2021) Geochemical and isotope evidence for groundwater mineralization in a semi-arid river basin, Sri Lanka. Appl Geochem 124:104799. https://doi.org/10.1016/j.apgeochem.2020.104799
Shaki AA, Adeloye AJ (2006) Evaluation of quantity and quality of irrigation water at Gadowa irrigation project in Murzuq basin, southwest Libya. Agric Water Manag 84(1–2):193–201. https://doi.org/10.1016/j.agwat.2006.01.012
Singh BP, Setia R, Wiesmeier M, Kunhikrishnan A (2018) Agricultural management practices and soil organic carbon storage. In: Soil carbon storage, pp 207–244. Academic Press
Szabolcs I, Darab C (1964) The influence of irrigation water of high sodium carbonate content of soils. In: Proceedings of 8th international congress of ISSS, Transaction II, pp 803–810
Torres-Martínez JA, Mora A, Mahlknecht J, Kaown D, Barceló D (2021) Determining nitrate and sulfate pollution sources and transformations in a coastal aquifer impacted by seawater intrusion—a multi-isotopic approach combined with self-organizing maps and a Bayesian mixing model. J Hazard Mater 417:126103. https://doi.org/10.1016/j.jhazmat.2021.126103
Verma A, Sharma A, Kumar R, Sharma P (2023) Nitrate contamination in groundwater and associated health risk assessment for Indo-Gangetic plain, India. Groundw Sustain Dev 23:100978. https://doi.org/10.1016/j.gsd.2023.100978
WHO (2011) Guidelines for drinking-water quality, 4th edn. World Health Organization, Geneva
Wilcox LV (1955) Classification and use of irrigation waters, USDA, circular 969, Washington DC
Yin X, Feng Q, Li Y, Deo RC, Liu W, Zhu M, Liu R (2022) An interplay of soil salinization and groundwater degradation threatening coexistence of oasis-desert ecosystems. Sci Total Environ 806:150599. https://doi.org/10.1016/j.scitotenv.2021.150599

Chapter 8
Health Risk Assessment of Groundwater in KwaZulu-Natal Coastal Aquifers

Groundwater is a vital resource for drinking and agricultural needs globally, yet its quality is increasingly threatened by anthropogenic activities. This chapter investigates the health risks associated with heavy metal contamination in the coastal aquifers of KwaZulu-Natal, South Africa, a region reliant on groundwater for agriculture and domestic use. Employing a multidisciplinary approach, 128 groundwater samples were analyzed for metals such as Cd, Pb, Cr, and Zn. Source apportionment via Positive Matrix Factorization (PMF) identified agricultural fertilizers (e.g., phosphate-based inputs) and industrial activities as primary contamination sources. Probabilistic health risk assessment, integrated with Monte Carlo simulations, quantified non-carcinogenic and carcinogenic risks for adults and children. Results revealed elevated risks, particularly for children, with cadmium and chromium posing significant carcinogenic threats. Spatial mapping highlighted contamination hotspots in southern and central regions, underscoring the urgency for targeted mitigation strategies. The study bridges hydrochemical data with public health insights, offering actionable recommendations to align groundwater management with sustainable development goals.

8.1 Introduction

Groundwater serves as lifeline for billions and is a vital source for drinking water needs for more than half of the global population (Gleeson et al. 2012). It plays a significant and critical role in rural and agricultural zones of water-stressed regions around the world. The storage extensive aquifer systems allow localized extraction making it conveniently available to support domestic and agricultural suage. Nearly 40% of the world's groundwater has been utilized for agriculture, especially in arid and semi-arid regions, where water scarcity prevails as major concern (Alao et al. 2024). Its potential to act as an alternative depleting surface water resource, which are

P. Li and V. Elumalai, *Groundwater Quality Under Agricultural Activities—Cases from China and South Africa*, Professional Practice in Earth Sciences,
https://doi.org/10.1007/978-3-031-98993-3_8

comparatively volatile makes a highly dependable resource for survival of humanity and their sustainability. Groundwater use in agriculture as a foundation for food security of the world, was key element of the green revolution of the twentieth century (Karandish et al. 2025; Kopittke et al. 2019). In South Africa, the availability of groundwater played an intriguing role due to easy access to coastal regions, small farmers and commercial enterprises especially during dry seasons or draught periods with crops like sugarcane, citrus, maize, and vegetables (Ncube 2020). Such availability effectively compensated for changing threats such as changing rainfall patterns and higher dependency on surface water resources in the country while utilizing more than 20% of the available groundwater resources (Ingrao et al. 2023; Kusangaya et al. 2014).

In addition, with intensifying agriculture to meet global food demand, groundwater systems face mounting pressure due to over extraction, contamination, excess utilization of pesticides, fertilizers and animal waste loading (Kumar et al. 2021). On of such emerging threats are heavy and trace metals emanating from intense agricultural activities and rapid urbanization (Rezapour et al. 2022; Yang et al. 2022; Xu et al. 2014). They are one of the most insidious threats towards groundwater quality, originating from geogenic and anthropogenic sources (Ullah et al. 2022). The intricacies between them are very closely related as geogenic leaching of metals might be controlled or regulated by ongoing anthropogenic activities. For example, the weathering of metal containing rocks in soil could be intensified with the introduction of mining, industrial and agricultural activities, simultaneously instructing toxic metals into the environment from fertilizers and agrochemicals (Alengebawy et al. 2021; Ali et al. 2019). Modern agricultural practices heavily rely on the use of fertilizers containing trace metals such as phosphate, which contains Cd and Pb, increases the mobility in the soil eventually reaching groundwater table through leaching (Alengebawy et al. 2021; Khan et al. 2018; Gantayat and Elumalai 2024). The non-biodegradable nature of metals makes significant threats as they can persist for decades while increasing their accumulation levels gradually posing a serious threat in terms of health and safety towards humans and surrounding ecosystems (Saravanan et al. 2024). Once groundwater tainted with metal pollution, they bioaccumulate in human bodies and get absorbed by various organs causing serious diseases (Briffa et al. 2020; Aziz et al. 2023). According to WHO (2017), millions of people worldwide are exposed to toxic metals such as As, Cd, Pb and Hg through direct consumption of water, leading to various chronic diseases such as cancer, kidney dysfunction, neurological disorders and developmental impairments in adults and children (Hassan et al. 2025; Balali-Mood et al. 2021). Contamination of As has impacted over 150 million people especially in south and southeast Asia (Shaji et al. 2021). Pb has been specifically linked to cognitive decline and cardiovascular issues (Kohler et al. 2024). Several studies around South Africa have documented elevated levels of metals such as Cd, Pb, Mn, As, Se, V etc. on groundwater resources raising concerns over public health consequences (Malan et al. 2015; Utembe et al. 2015; Knüppe 2011; Conrad et al. 1999). Moreover, such issues are closely interconnected with ongoing environmental change, land use shifts and agricultural intensification globally.

Health risk assessment of water is a tool for quantitative assessment of the direct impact of environmental contamination on human health through various pathways such as ingestion and inhalation (Ayejoto et al. 2023; Robson et al. 2022; Villanuevaet al. 2014). It bridges the gap between hydrochemical research and their impact on public health concerns. It enables scientists to identify specific associated risk with existing metals in groundwater helping farmers, policymakers and stakeholders to make information decisions for future sustainability and strategize for global water management, groundwater use, and agricultural management. This satisfies the broader view of sustainable development goals (SDGs) 3 and 6 adopted by United nations in 2015, which are good health and wellbeing along with clean and water and sanitation.

8.2 Sample Collection and Methodologies

For the evaluation of health risks associated with metal concentration and distribution in KwaZulu-Natal coastal aquifers of South Africa, a total of 128 groundwater samples were collected from Maputoland, uMhlathuze and Tugela aquifer. During sampling predetermined precautions were taken to avoid contamination among the samples. All boreholes were pumped for at least 5 min to remove stock rainwater water from top and better representation of geochemical condition surrounding the borehole. Sample were collected in high density polyethylene bottles after filtration through 0.45 μm membrane filters with the help of a portable hand pump. The sample bottles were thoroughly rinsed for at least 3 times with the collected water to avoid any influence of pre-existing contamination. 1 L of sample were collected from each borehole and transported back to the International Research Centre for Water Science and Technology (IRC-WST) laboratories at the University of Zululand. The samples were acidified with HNO_3 and pH levels were brought to 2 to avoid any organic matter influence and stored at 4 °C further analysis. The analysis for metals such as Ag, Al, B, Cd, Co, Cr, Cu, Fe, Mn, Ni, Pb and Zn using Perkin Elmar Nexon 2000 ICP-MS system. The instrument was calibrated using multi-elemental standard of 1000 ppm with 5% HNO_3 solution provided by Perkin Elmar. All analytical procedures were done using ultrapure water from Millipore Milli-Q advantage A10 system (resistivity > 18.2 MΩ·cm) to keep the analytical contamination to minimum levels. Certified reference materials such as SRM 2702 from NIST were utilized to check the recovery and accuracy of the instrument.

8.2.1 Source Apportionment Using Positive Matrix Factorization (PMF)

The receptor models are a group of mathematical and statistical tools designed to identify and apportion sources of considered environmental pollutants. These models mainly rely on the chemical composition of sampled data to infer the nature of contributing source rather than depending on emission rate or source characteristics. Tools such as Positive Matrix Factorization (PMF) has emerged as one of the widely used and robust software's, that works around the receptor model concept and developed by U.S. Environmental Protection Agency for environmental analysis. The major strength of PMF lies in its ability to process and decompose environmental datasets into factor contribution and profiles. These matrices are derived under constraint of non-negativity and account for uncertainty in the dataset by incorporating error estimates for each observation, enhancing the reliability of the results. The model identifies various latent factors representing potential sources based on their unique signatures and estimates the total source contribution of pollutants at measured sites. PMF was utilized in this study to apportion the source of considered trace and heavy metals based on the following chemical balance principle with the help of Eq. (8.1) and (8.2).

$$X_{ij} = \sum_{k=1}^{p} g_{ik} f_{kj} + e_{ij} \tag{8.1}$$

$$Q = \sum_{i=1}^{n} \sum_{j=1}^{m} \left[\frac{x_{ij} - \sum_{k=1}^{p} g_{ik} f_{kj}}{u_{ij}} \right] \tag{8.2}$$

where, X_{ij} denotes the concentration of the jth metal in the ith sample, and p represents the total number of factors considered in the study. The term g_{ik} indicates the mass concentration contribution from the kth factor to the ith sample, while f_{kj} refers to the mass concentration of the jth metal in the kth factor. The residual component for the ith sample and jth metal is represented by e_{ij}. A weighted least squares method is employed in the model to minimize the objective function (Q), ensuring that no negative loadings occur in any sample during the analysis. u_{ij} is the measured uncertainty of jth metal in ith sample.

8.2.2 Health Risk Assessment (HRA) and Monte Carlo Simulation

Health risk assessment is computation of level of risk associated with a parameter or metals, due to ingestion and dermal exposure over a long period of time. HRA is computed based on regulation given by U.S. Environmental Protection Agency

(USEPA 2011), by taking various factors such as age, body weight, duration of exposure, and frequency of water consumption, which influence the extent of potential health risks posed to individuals through domestic water use. In this study, the HRA approach was employed to assess both non-carcinogenic and carcinogenic risks linked to the presence of heavy metals in water sources. The non-carcinogenic risk (NCR) evaluation is particularly relevant for chronic exposures and is capable of addressing potential adverse health effects such as developmental, genetic, and teratogenic outcomes). Given the physiological differences that affect susceptibility to toxic substances, the assessment was conducted separately for adults and children as the target populations. To identify these risks, key indices such as total hazard index (HI), hazard quotient (HQ) were utilized for non-carcinogenic risk, whereas total carcinogenic risk (TCR) and carcinogenic risk (CR) were utilized for the calculation of carcinogenic impact and are calculated as per Eqs. 8.3–8.6 (Liang et al. 2025; Xie et al. 2025). An unique approach was considered in addition to the traditional methods where, HRA model was integrated with PMF source profiles, where health risk were simulated using Monte Carlo techniques to assess the probability of the risk along with the uncertainties associated with HRA assessment. Together these indices are very effective in proving comprehensive understanding of both immediate and long-term health threats associated with groundwater in the study area.

$$\mathrm{ADD_{ing}} = \frac{X_{\mathrm{ij}}^{k} \times IR_{\mathrm{ing}} \times \mathrm{CF} \times \mathrm{EF} \times \mathrm{ED}}{\mathrm{BW} \times \mathrm{AT}} \tag{8.3}$$

$$\mathrm{ADD_{der}} = \frac{X_{\mathrm{ij}}^{k} \times \mathrm{CF} \times \mathrm{SA} \times \mathrm{AF} \times \mathrm{ABF} \times \mathrm{EF} \times \mathrm{ED}}{\mathrm{BW} \times \mathrm{AT}} \tag{8.4}$$

$$\mathrm{THI} = \sum HQ_{ij}^{k} = \sum \frac{ADD_{ij}^{k}}{RfD_{i}} \tag{8.5}$$

$$\mathrm{TCR} = \sum CR_{ij}^{k} = \sum ADD_{ij}^{k} \times SF_{i} \tag{8.6}$$

where X_{ij}^{k} represents metal concentration of *i*th metal from *k*th site/source. ADD represents the average daily dose of *i*th metal from *j*th source for ingestion and dermal pathways. HQ represents hazard quotient of the individual concentration of *i*th metal in *k*th site/source whereas CR represents carcinogenic risk of *i*th metal from *k*th site/source in water.

Monte Carlo Simulation (MCS) was applied to perform source-based HRA derived from PMF. In this approach, the mean and standard deviation values of metal concentrations were transformed using a logarithmic scale ($\log_{10}$) and modelled using a lognormal distribution, which is commonly suited for environmental contaminant data. Deterministic parameters including ingestion rate (IR), exposure duration (ED), and body weight (BW) were assigned specific probability distributions to reflect variability: IR and BW were modeled using uniform distributions, while ED followed a triangular distribution. The simulation was conducted considering 10,000 iterations

Table 8.1 Deterministic parameters utilized for calculation and simulation of HRA

Parameters		Units	Children	Adult	Distribution	Reference
X_{ij}^{k}	Concentration of heavy metals	mg/l	Measured	Measured	Log normal	Current study
BW	Average body weight	Kg	(1.96,42)	(60.7, 70)	Uniform	USEPA (2000)
IR	Ingestion rate	L/day	(1, 1.32)	(1, 3)	Uniform	USEPA (2011)
EF	Exposure frequency	Day/year	(180, 345, 365)		Triangle	USEPA (2011)
ED	Exposure duration	Year	10	70	Fixed	USEPA (2011)
ET	Exposure time	Hour/day	0.18	0.33	Fixed	CAEPM (2013)
AT	Average time	Days	ED × EF		Fixed	USEPA (2000)
SA	Skin surface area	Cm^2	13,300	17,000	Fixed	USEPA (2011)
CF	Conversion factor	L/cm^3	0.001	0.001	Fixed	USEPA (2000)
PC	Skin permeability coefficient	Cm/h	0.001(Fe, Cu, Co), 0.0001 (Pb, Mn), 0.0004 (Ni)		Fixed	USEPA (2011, 2004)

for each index assessment, where cumulative probabilities and frequency distributions were recorded to capture the range and likelihood of potential health risk outcomes across different exposure scenarios. To evaluate the uncertainty and influence of each variable on the model output, a sensitivity analysis was conducted. This analysis assessed the contribution of individual metals as well as the deterministic input parameters to the overall uncertainty in risk estimates, thereby identifying the most influential factors driving risk variability. The values and range of the parameters utilized in the simulation are presented in Tables 8.1 and 8.2.

8.2.3 Data Processing

The spatial distribution of various metals was interpolated using Inverse distance weighting (IDW) method in ArcGIS pro-3.4.5. Monte Carlo simulation (MCS) for all the models were performed using XLRisk plugin in Microsoft excel. Acquired data from MCS were further refined and processed using R programming in R Studio 2024. Packages such as "ggplot2", "tidyr" and "dplyr" (Wickham 2016; Wickham et al. 2023, 2024) were utilized, whereas acquired sensitivity data was represented using Microsoft excel.

Table 8.2 Reference dose (RfD) and Slope factor (SF) utilized for HQ and CR calculation

Metals	RfD (mg/kg/d)		SF (mg/kg/day)		References
	Ingestion	Dermal	Ingestion	Dermal	
Pb	3.6	0.52	0.0085	0.00077	USEPA (2011)
Cd	0.5	25	6.1	1.22	USEPA (2011)
Cu	5	12	–	–	USEPA (2011)
Ni	20	20.6	1.7	0.43	USEPA (2011)
Mn	140	0.8	–	–	USEPA (2011)
Zn	300	60	–	–	USEPA (2011)
Co	0.3	0.2	–	–	USEPA (2011)
Fe	300	45	–	–	USEPA (2011)
Al	1000	1260	–	–	ATSDR (2008)
Cr	3	0.075	0.5	0.20	USEPA (2011)
Ag	5	5	–	–	USEPA (2011)

8.3 Identification of Sources of Metals in Groundwater

8.3.1 Spatial Distribution and Descriptive Statistics of Metals

The descriptive statistics of metal concentrations in water samples are presented in Table 8.3. They reveal considerable variability and distribution asymmetry across the dataset. Mean concentrations indicate that Zn (309.20 μg/L), Mn (319.50 μg/L), and Al (274.00 μg/L) are among the most prevalent metals, while standard deviations especially for Zn (1787.82), Mn (1529.58), and Fe (595.53) suggest high variability in concentrations. The dominance of metals in descending order with respect to their mean values can be seen as Mn > Zn > Al > Ag > Pb > Cu > Fe > Co > Cd > B > Ni > Cr (Table 8.3). The coefficient of variation (CV%) further confirms this, with Zn (555.45%), Mn (463.33%), and B (336.49%) exhibiting extreme variability, pointing to uneven spatial distribution and possible localized anthropogenic pollution sources. Most metals display positive skewness, particularly Zn (10.68), Mn (10.14), and B (7.79), indicating a strong right-tailed distribution due to occasional high concentration events or hotspots of contamination. The negative skewness in parameters like Ag and Pb suggest a symmetric distribution in the study area. Elevated kurtosis values for these same metals Zn (117.64), Mn (108.80), and B (70.22) reflect the presence of sharp peaks and heavy tails, further supporting the influence of point-source anthropogenic influence in the dataset. The negative kurtosis values of Ag, Cd and Co suggest more flatter distribution indicating widespread distribution, which might be of geogenic or widespread anthropogenic origin. Sptial distribution of metals suggests higher concentration of metals in the southern aquifers, especially Cd, Cr, B and Co (Fig. 8.1). Groundwater in central or middle region of the study area indicates higher accumulation of Pb, Fe, Cu, Mn and Ag, especially nearer to

Table 8.3 Descriptive statistics of metal concentrations in the study area (μg/L for metals)

Parameters	Min	Max	Mean	Std. dev	WHO (2011)	Skewness		Kurtosis		CV%
						Statistic	Std. error	Statistic	Std. error	
Ag	BDL	506.50	247.35	210.87	100	−0.28	0.21	−1.85	0.42	85.02
Al	BDL	3052.49	274.00	320.52	200	5.14	0.21	43.95	0.42	116.37
B	BDL	3693.01	108.30	372.74	2400	7.79	0.21	70.22	0.42	336.49
Cd	BDL	273.52	77.53	122.76	3	0.98	0.21	−1.05	0.42	157.02
Co	BDL	284.88	85.37	119.75	40	0.93	0.21	−1.13	0.42	139.31
Cr	BDL	9.97	3.05	4.13	50	0.72	0.21	−1.38	0.42	134.39
Cu	BDL	1265.16	131.61	185.80	2000	3.08	0.21	15.15	0.42	140.18
Fe	2.74	5487.80	180.68	595.53	300	6.93	0.21	54.78	0.42	322.60
Mn	BDL	16,814.90	319.50	1529.58	500	10.14	0.21	108.80	0.42	463.33
Ni	BDL	225.44	34.81	39.69	70	1.29	0.21	2.53	0.42	113.45
Pb	BDL	581.76	194.17	168.78	10	−0.02	0.21	−1.54	0.42	86.67
Zn	BDL	19,964.50	309.20	1787.82	3000	10.68	0.21	117.64	0.42	555.45

the coast. On the other hand, metals such as the highest concentration of Al and Zn were observed in the northern aquifer (Fig. 8.1).

8.3.2 Source Apportionment

During the model execution, the minimum signal-to-noise (S/N) ratio recorded was 0.1, and all species were classified as "strong" with 0% modeling uncertainty, indicating a high degree of data reliability. The base run of the Positive Matrix Factorization (PMF) model began with three factors, which were incrementally increased by one in each subsequent run. Each run was initialized with a random start and repeated 20 times. All base model runs successfully converged, demonstrating the robustness and stability of the PMF solution (Zhou et al. 2023). A bootstrap analysis of the base model was performed with a maximum of 100 iterations, also starting from random initialization. A minimum R^2 value of 0.6 was set as a threshold, and the optimal model configuration was identified with four factors, which provided the best bootstrap mapping. The results from the base run exhibited normal residual distributions across all parameters, and detailed statistics, including signal-to-noise ratios. Similarly, the displacement (DISP) analysis was conducted to assess the model's rotational ambiguity. The DISP output revealed no errors (Error code: 0), no factor swapping at $dQ_{max} = 4$, and a decrease in the Q value of less than 1%, confirming the model's reliability and lack of rotational ambiguity. The resolved source profiles (fingerprints) for the identified factors are illustrated in Fig. 8.2.

Factor 1 was found to be associated with higher loading of Zn (88.2%), Cd (87%), B (69%), Al (55.8%), Mn (54.7%) and Cu (54.2%) (Fig. 8.2). The spatial distribution

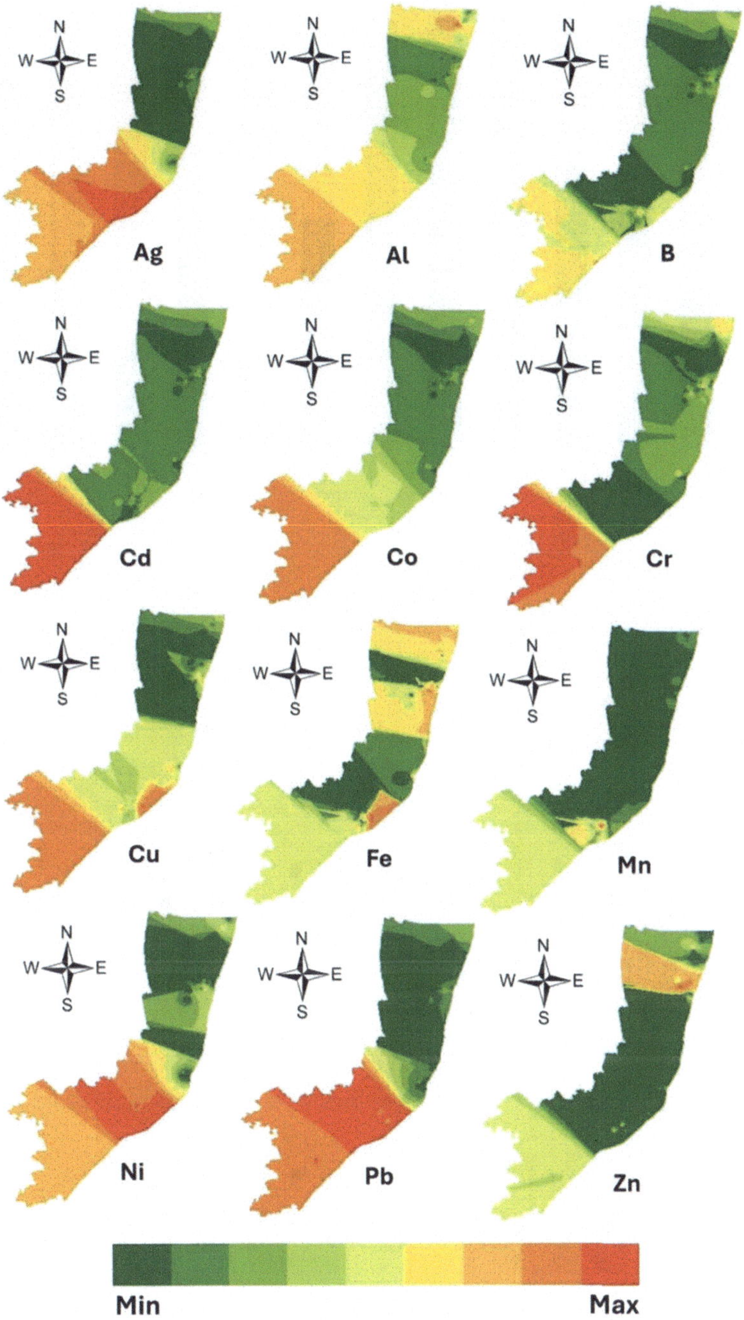

Fig. 8.1 Spatial distribution of metal concentration in coastal aquifers

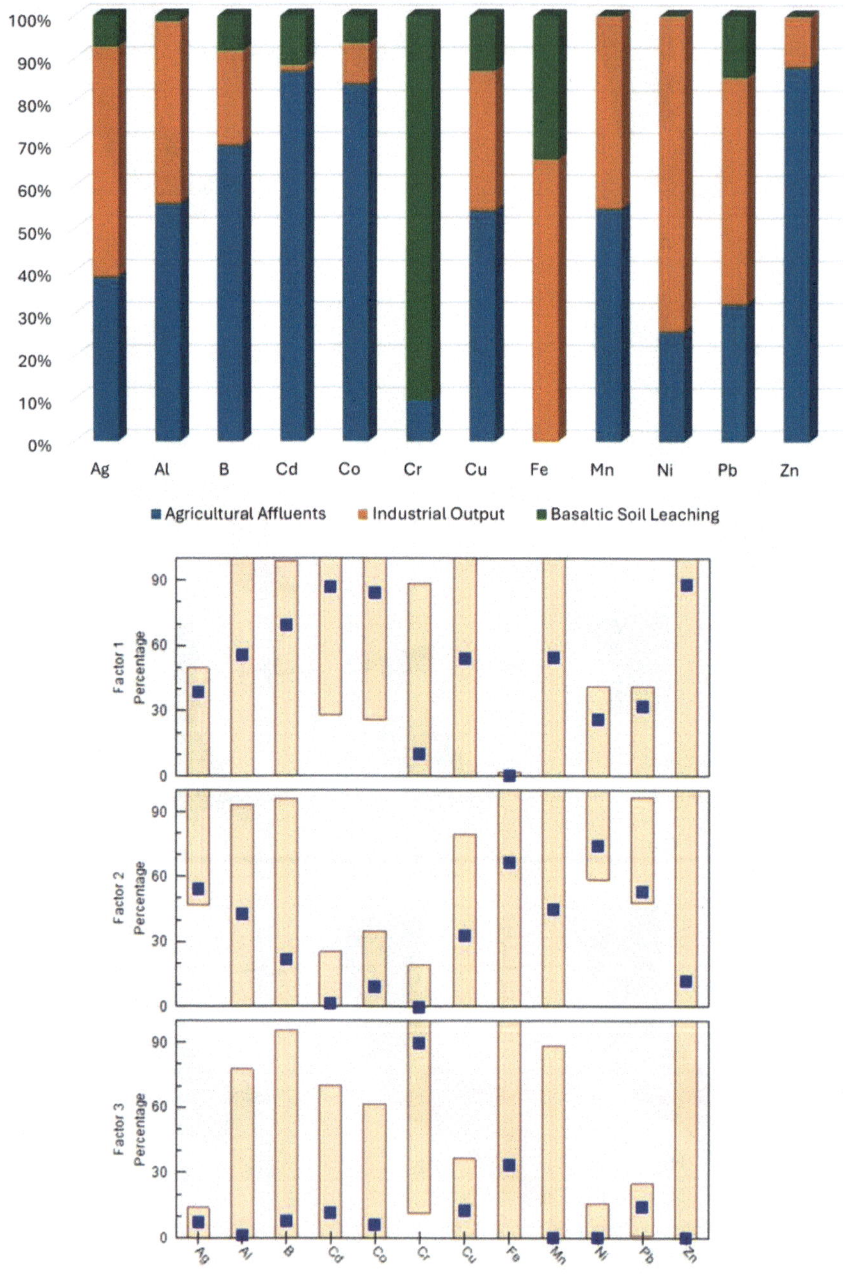

Fig. 8.2 Source contribution and profiles acquired from PMF model

suggests higher enrichment in southern catchments compared to northern catchments. In this region, commercial dry land agriculture is common and widespread (Suciu et al. 2022; Lin et al. 2019; Ntanganedzeni et al. 2018; Ogut et al. 2010) and fertilizer applications are comparatively high in this region including phosphate fertilizer derived from phosphate rocks which is higher in trace metal concentration (Gantayat et al. 2024; Khatun et al. 2022; Vetrimurugan et al. 2019). They help regulate bacterial diversity in soil, especially through application of Boron phosphate (Zhao et al. 2021). Fertilizers such as products produced include phosphate-based granular fertilizers such as Mono Ammonium Phosphate (MAP), Diammonium Phosphate (DAP) and Single Super Phosphate (SSP) are mainly used in agricultural activities in this region, which are the byproducts of Phosphoric acid produced through reaction of phosphate rock with sulphuric acid (Louw 2020; Andrunik et al. 2020). Such observation confirms the source of associated metals is agricultural effluents especially from phosphate-based fertilizer applications.

Factor 2 has significant loading of Ni (74.1%), Fe (66.3%), Ag (54.2%) and Pb (53.4%) (Fig. 8.2) and were found higher in concentration in central or middle region of the study area (Fig. 8.1). This area is highly industrialized with existing industrial towns such as Empangeni and Richards bay, which are known of industries such as smelting facilities and ort activities. Elevated concentration of Fe and Ni suggest higher influence of the Fe smelting facilities in this zone especially in Richards's bay. Ni is known to be part of deposition in non-ferrous and ferrous metals industrial area (Cempel & Nikel, 2006).

Factor 3 has a major load of metals such as Cr (90.0%) and low contribution from Fe (33.7%) along with Pb (14.4%) (Fig. 8.2). The metals found in higher concentration in the western part of the southern aquifer, especially influenced by Tugela river catchment (Fig. 8.1). This region has exposed the form of Basaltic lava of the Drakensberg (Ntanganedzeni et al. 2018), which has enriched the basaltic derived soil with chromium in trace amounts (Manson 2018). Basalts are known for containing of Cr-bearing minerals like chromite and Fe-bearing minerals in the rock's composition (Sun et al. 2022) and leaching from these soils might be increasing the concentration in the groundwater in this region.

8.4 Source Specific Health Risk Assessment

The descriptive statistics derived from the Monte Carlo Simulation provide critical insights into the non-carcinogenic and carcinogenic health risks associated with heavy metal exposure for both adults and children and is represented in Table 8.4. For non-carcinogenic risk (NCR), the mean Total Hazard Index (THI) for adults was 0.0443, while for children it was significantly higher at 0.1381. This clearly demonstrates the increased vulnerability of children to heavy metal exposure, likely due to their lower body weight and higher intake rates relative to body mass. Among individual Hazard Quotients (HQs), cadmium (Cd) exhibited the highest mean HQ for both adults (0.0112) and children (0.0294), making it the most dominant contributor to non-carcinogenic risk across age groups. In children, cobalt (Co) followed closely with a mean HQ of 0.0536, and copper (Cu) and chromium (Cr) also showed elevated mean HQs of 0.0156 and 0.0120, respectively. In adults, after Cd, cobalt (0.0204) and chromium (0.0049) were notable contributors. Metals such as manganese (Mn), zinc (Zn), aluminium (Al), and iron (Fe) had very low HQs across both groups, with values in the range of 10^{-5} to 10^{-4} (<1; USEPA guidelines), suggesting minimal concern for these elements under the current exposure conditions. The histogram of THI values and its frequencies in all 10,000 scenarios place the most optimistic (5th percentile) and most conservative scenarios (95th percentile) in negligible risk scenarios for both adult and children (Fig. 8.3). Similarly, cumulative probability of risk scenarios was found to be 0% and were well below USEPA guidelines for all metals as shown in Fig. 8.3. The sensitivity analysis for THI suggested high variability of ingestion rate (IR) are sensitive towards the risk calculation, where variable is higher in case of adults compared to children, as ingestion amount is lower in children (Fig. 8.4). On the contrary, metals such as Cd and Co dominated the risk with high variability in increasing trend, indicating priority metals to be kept in observation for future non carcinogenic risk mitigation (Fig. 8.4).

Carcinogenic risk (TCR) assessment revealed a more alarming pattern. The Total Cancer Risk (TCR) for adults averaged 0.0603 (>10^{-4}; USEPA guidelines), while for children the total cancer risk (TCR) was markedly higher at 0.1753 (>10^{-4}; USEPA guidelines). Cadmium again emerged as the most concerning metal, with a mean cancer risk of 0.0313 in adults and 0.0913 in children, exceeding or nearing the United States Environmental Protection Agency's (USEPA) acceptable risk range of 1×10^{-4} to 1×10^{-6}. Nickel (Ni) and chromium (Cr) also presented significant mean cancer risks, with Ni values at 0.0228 (adults) and 0.0663 (children), and Cr at 0.0061 (adults) and 0.0176 (children). Lead (Pb), while less dominant in terms of magnitude, still contributed to the overall TCR, with values of 7.76×10^{-5} (adults) and 2.24×10^{-4} (children). The simulated risk scenarios of TCR suggested risk in 100% of the scenarios stipulating towards extreme carcinogenic risk. The most optimistic scenarios for both adults and children were noted above the USEPA guidelines (Fig. 8.5). The higher TCR values in children again reflect their increased physiological susceptibility to environmental toxins. Sensitivity analysis suggested higher dependency of carcinogenic risks on deterministic parameters like ingestion rate (IR) especially in adults, whereas metals like Cd and Cr are sensitive towards the identified cancer risk and probably the main cause of the groundwater-based risk. On the other hand, reduced weight of children makes then more susceptible towards the emerging risk compared to adults as it suggested by its negative association with TCR (Fig. 8.4). The mitigation strategies in the region have to target Cd and Cr on priority for effective reduction of carcinogenic risk in the region.

The range of risk values further supports the conclusion that children are at substantially greater risk from both non-carcinogenic and carcinogenic effects of heavy metals. The maximum THI observed in adults was 0.885, while in children it reached 2.147, nearly 2.5 times higher. Similarly, the maximum TCR values were drastically higher in children, with cadmium reaching 2.687 and nickel 3.588. These findings highlight cadmium, nickel, and chromium as the most critical elements to

Table 8.4 Descriptive statistics of simulated source-based health risk assessment from Monte Carlo simulation

Adults					Child				
Parameter	Mean	Minimum	Maximum	Std. deviation	Parameter	Mean	Minimum	Maximum	Std. deviation
THI	0.0443	0.0078	0.8849	0.0249	THI	0.1381	0.0184	2.1473	0.1559
HQ (Pb)	0.0028	0.0006	0.0100	0.0011	HQ (Pb)	0.0189	0.0028	0.2074	0.0206
HQ (Cd)	0.0112	0.0004	0.1618	0.0104	HQ (Cd)	0.0294	0.0006	0.9272	0.0446
HQ (Cu)	0.0016	0.0003	0.0058	0.0007	HQ (Cu)	0.0156	0.0019	0.1558	0.0173
HQ (Ni)	0.0007	0.0000	0.0143	0.0008	HQ (Ni)	0.0020	0.0000	0.0923	0.0036
HQ (Mn)	0.0002	0.0000	0.0009	0.0001	HQ (Mn)	0.0005	0.0001	0.0073	0.0006
HQ (Zn)	0.0001	0.0000	0.0013	0.0001	HQ (Zn)	0.0002	0.0000	0.0046	0.0002
HQ (Co)	0.0204	0.0021	0.1677	0.0128	HQ (Co)	0.0536	0.0028	0.8873	0.0688
HQ (Fe)	0.0001	0.0000	0.0005	0.0000	HQ (Fe)	0.0002	0.0000	0.0057	0.0002
HQ (Al)	0.0000	0.0000	0.0001	0.0000	HQ (Al)	0.0000	0.0000	0.0010	0.0000
HQ (Cr)	0.0049	0.0000	0.8508	0.0149	HQ (Cr)	0.0120	0.0000	1.4913	0.0392
HQ (Ag)	0.0022	0.0004	0.0106	0.0011	HQ (Ag)	0.0057	0.0005	0.0834	0.0068
TCR	0.0603	0.0048	1.0982	0.0461	TCR	0.1753	0.0091	4.3561	0.2478
CR (Cd)	0.0313	0.0014	0.5563	0.0291	CR (Cd)	0.0913	0.0017	2.6867	0.1447
CR (Cr)	0.0061	0.0000	1.0494	0.0190	CR (Cr)	0.0176	0.0000	2.6757	0.0696
CR (Pb)	0.0001	0.0000	0.0002	0.0000	CR (Pb)	0.0002	0.0000	0.0025	0.0002
CR (Ni)	0.0228	0.0005	0.4639	0.0267	CR (Ni)	0.0663	0.0006	3.5884	0.1232

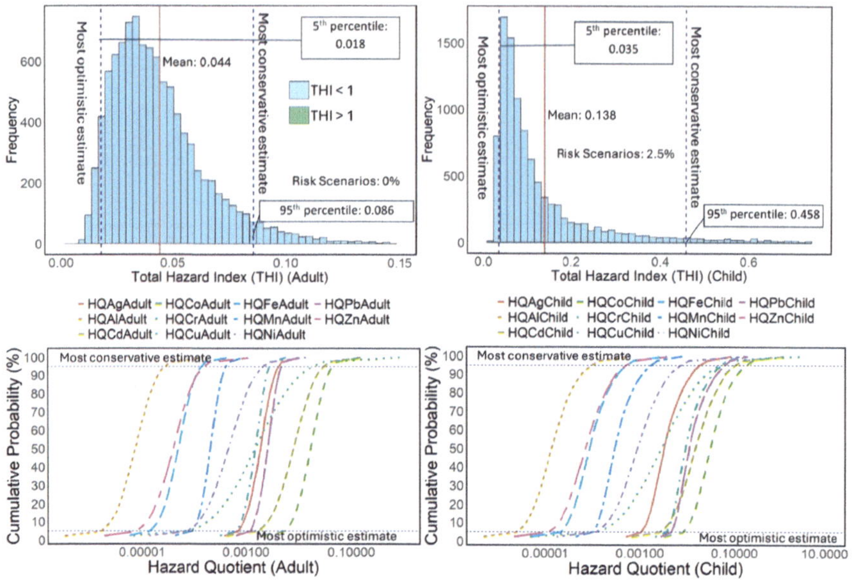

Fig. 8.3 Histogram and cumulative probabilistic representation of simulated THI and HQ values in adults and children for source based non-carcinogenic risk

monitor and control. The consistently higher standard deviations and maxima in the children's dataset underscore the greater variability and exposure sensitivity among this group. The source contribution constructed from the simulated data suggests that agricultural affluents have a bigger involvement in both non carcinogenic and carcinogenic risk, followed by industrial output. Similar patterns of contributions were found for both population groups (Fig. 8.4). This might be due to the higher toxicity potential associated with Cd and Cr in the region.

8.5 Site Specific Health Risk Assessment

The spatial site-specific non-carcinogenic health risk assessment presented across the two sets of maps reveals varying levels of health risks posed by heavy metal exposure in groundwater for both adults and children in the study area (Fig. 8.6). A consistent trend across all metals shows that children are more vulnerable, exhibiting significantly higher hazard index (HI) values compared to adults. In Fig. 8.6, Cadmium (Cd) poses the greatest risk, particularly in the central and southern zones, where children's HI values exceed 2.85 well above the safe threshold indicating a severe public health concern. Cobalt (Co) also shows high risk levels in the southern areas, again with children facing the most critical exposure. Chromium (Cr), while generally showing lower risk, still presents localized high-risk hotspots in the northern regions

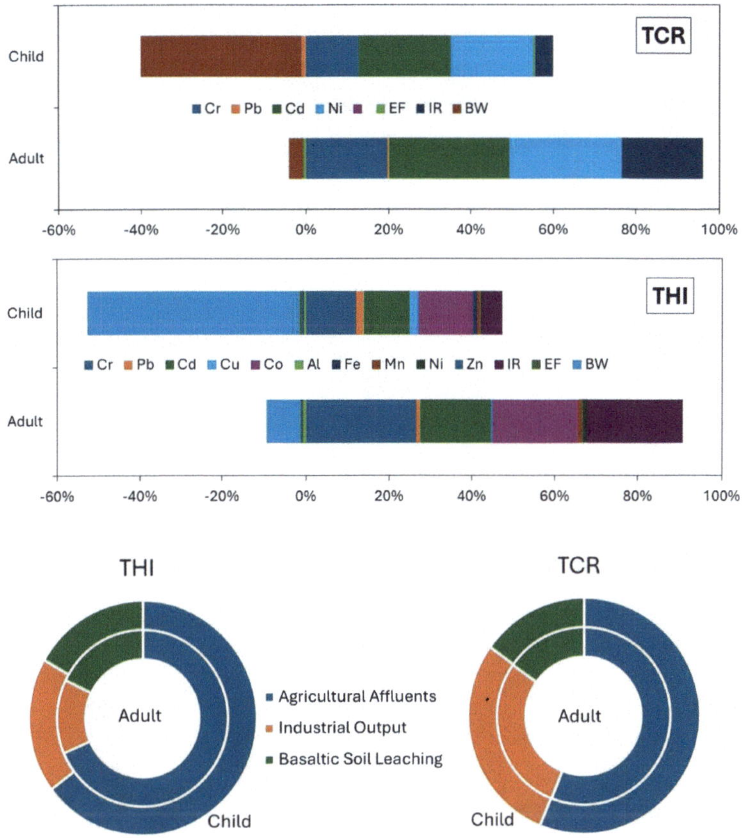

Fig. 8.4 Sensitivity analyses and source contribution of metals from Monte Carlo simulation

for children. Nickel (Ni) displays low to moderate risk, with slightly elevated values in central parts of the region. Lead (Pb), however, exhibits moderate to high risks in the central and northeastern regions, especially among adults, with even higher vulnerability for children. Metals such as Copper (Cu), Iron (Fe), and Manganese (Mn). Lead (Pb) remains a concern for children, particularly in the central-western zones, where HI values exceed 1.41, reinforcing its public health significance. Copper (Cu) presents mild to moderate risk, mainly concentrated in the central region, with children again showing higher susceptibility. Iron (Fe), on the other hand, appears to pose negligible risk across the entire area, as HI values remain extremely low or zero for both adults and children. Manganese (Mn) also shows minimal to low risk, with very limited regions crossing the higher threshold values, and only for children.

The site-specific distribution of carcinogenic risk of carcinogenic metals such as Cd, Cr, Ni and Pb suggests higher risk in groundwater in the southern region of the study area (Fig. 8.7). The carcinogenic values in children found higher compared to adults especially due to exposure of Cr followed by Cd. The CR risk values for Cr is

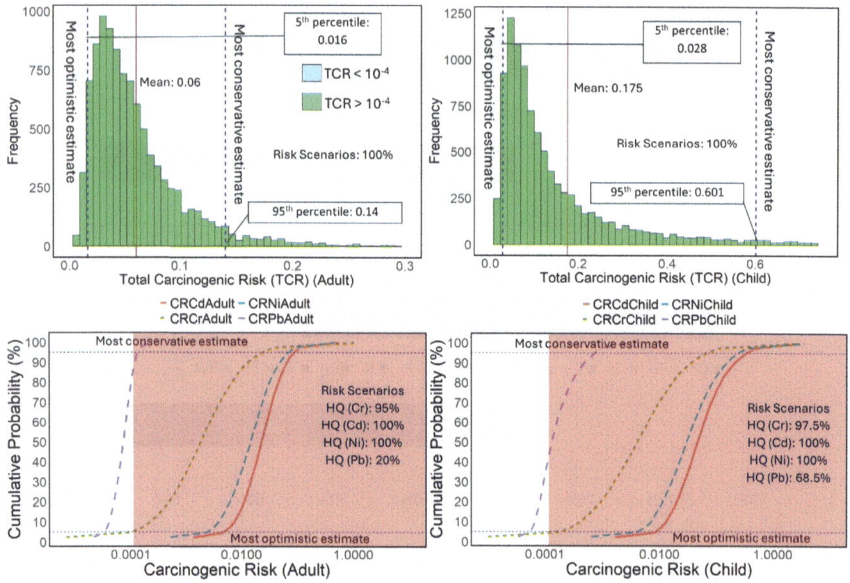

Fig. 8.5 Histogram and cumulative probabilistic representation of simulated TCR and CR values in adults and children for source based carcinogenic risk

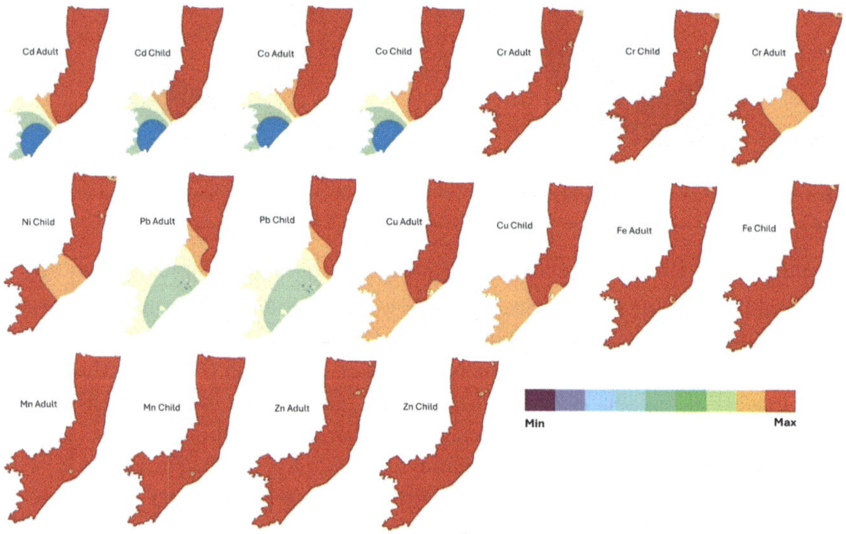

Fig. 8.6 Spatial distribution of Site-based non-carcinogenic risk in adults and children

more USEPA guidelines were observed at all sites where Cd was limited to mainly central and southern regions. The combined assessment of all mapped heavy metals underscores a distinct spatial variability in health risk, with the central and southern parts of the region showing the most severe exposure, particularly for Cd, Co, and Pb.

8.6 Conclusions

The health risk associated with metals such as Ag, Al, B, Cd, Co, Cr, Cu, Fe, Mn, Ni, Pb and Zn were studied comprehensively in 3 major coastal aquifers in the KwaZulu-Natal, South Africa. Based on the findings, the following conclusions were made.

- Metals such as Ag, Al and Cd were found higher than their guideline values at majority of the locations making the water unsuitable for drinking purposes. Metals such as Mn and Zn were found with dominant average compared to others. The dominance of metals in descending order could be seen as Mn > Zn > Al > Ag > Pb > Cu > Fe > Co > Cd > B > Ni > Cr.
- Groundwater in the southern region was observed to be highly contaminated with metals such as Zn, Cd and Mn compared to northern region of the study area. The metals in groundwater were found derived from 3 major sources, with dominating anthropogenic contribution mainly from agricultural affluents associated with phosphate fertilizer applications. Metals such as Ni, Fe and Ag were found mainly derived from industrial activities in the central region.
- The probabilistic source specific source assessment suggested negligible non-carcinogenic with increasing future threat from Cd and Co, whereas carcinogenic risk was confirmed in 100% of the simulated scenarios in Monte Carlo simulation primarily associated with Cd and Cr. The contribution of agricultural affluents was found as major source escalating both non-carcinogenic and carcinogenic risk in all aquifers and more vulnerable towards children compared to adults.
- The risk maps plotted for site specific assessment categorized increased Cd and Co heightened exposure risks, especially in the central and southern regions, with children consistently experiencing higher hazard indices than adults. Pb also poses considerable risk, notably in the central and northeastern zones, while Ni shows localized moderate risks. In contrast, Cr, Cr, Cu, and Mn generally display lower risk levels across most areas, with only minor hotspots indicating potential health threats.

The study provided a detailed overview of sources and health risk in comprehensive manner detailing information about probability of risk along with their distribution in KwaZulu-Natal coastal aquifers of South Africa. The probabilistic and zone-specific overview provides a detailed clarity of highlighted risk of Cd and Cr, which could vitally contribute towards future mitigation strategies aimed at reducing metal contamination in the region.

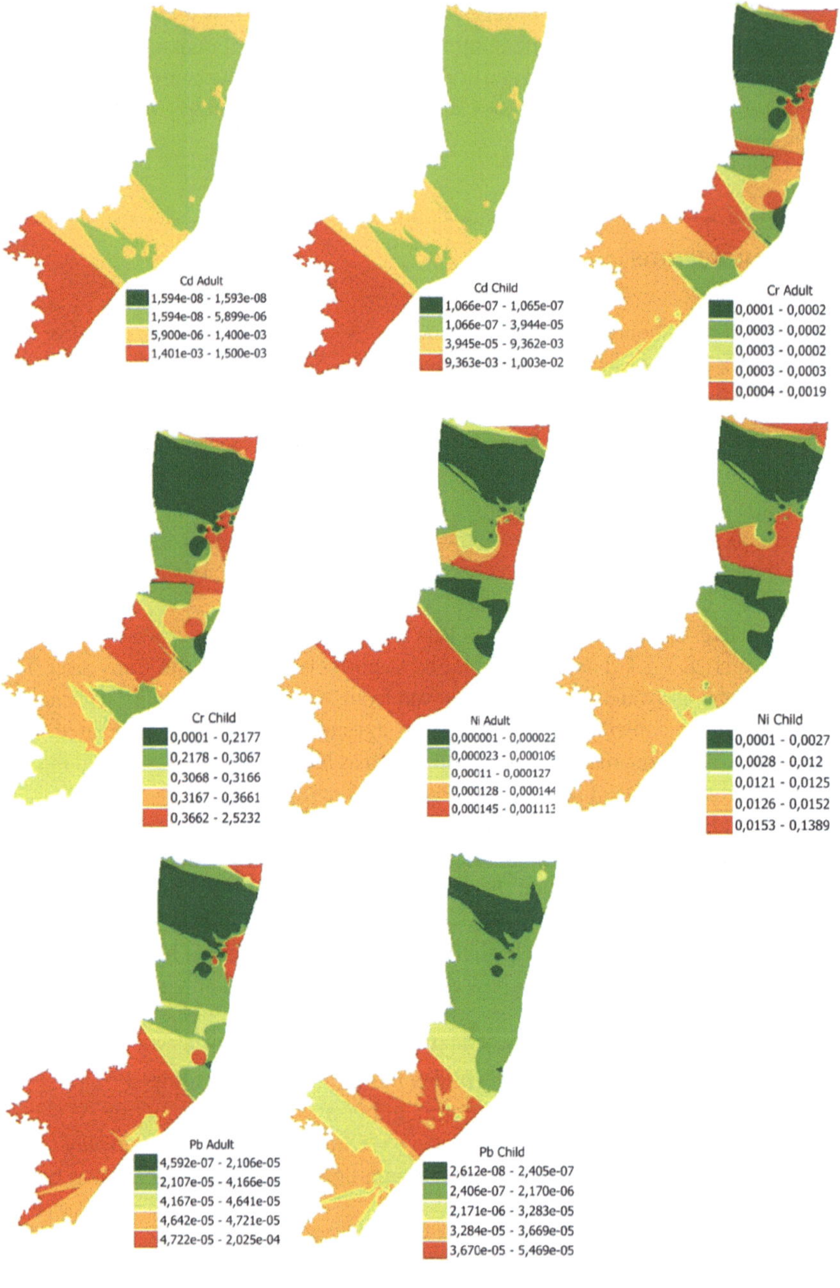

Fig. 8.7 Spatial distribution of Site-based carcinogenic risk in adults and children

References

Alao JO, Bello AY, Lawal HA, Abdullahi D (2024) Assessment of groundwater challenge and sustainable management strategies. Results Earth Sci 2:100049. https://doi.org/10.1016/j.rines.2024.100049

Alengebawy A, Abdelkhalek ST, Qureshi SR, Wang MQ (2021) Heavy metals and pesticides toxicity in agricultural soil and plants: ecological risks and human health implications. Toxics 9(3):42. https://doi.org/10.3390/toxics9030042

Ali H, Khan E, Ilahi I (2019) Environmental chemistry and ecotoxicology of hazardous heavy metals: environmental persistence, toxicity, and bioaccumulation. J Chem 2019(1):6730305. https://doi.org/10.1155/2019/6730305

Andrunik M, Wołowiec M, Wojnarski D, Zelek-Pogudz S, Bajda T (2020) Transformation of Pb, Cd, and Zn minerals using phosphates. Minerals 10(4):342. https://doi.org/10.3390/min10040342

Ayejoto DA, Agbasi JC, Egbueri JC, Abba SI (2023) Evaluation of oral and dermal health risk exposures of contaminants in groundwater resources for nine age groups in two densely populated districts, Nigeria. Heliyon 9(4):e15483. https://doi.org/10.1016/j.heliyon.2023.e15483

Aziz KHH, Mustafa FS, Omer KM, Hama S, Hamarawf RF, Rahman KO (2023) Heavy metal pollution in the aquatic environment: efficient and low-cost removal approaches to eliminate their toxicity: a review. RSC Adv 13(26):17595–17610. https://doi.org/10.1039/D3RA00723E

Balali-Mood M, Naseri K, Tahergorabi Z, Khazdair MR, Sadeghi M (2021) Toxic mechanisms of five heavy metals: mercury, lead, chromium, cadmium, and arsenic. Front Pharmacol 12:643972. https://doi.org/10.3389/fphar.2021.643972

Briffa J, Sinagra E, Blundell R (2020) Heavy metal pollution in the environment and their toxicological effects on humans. Heliyon 6(9):e04691. https://doi.org/10.1016/j.heliyon.2020.e04691

Cempel M, Nikel GJPJS (2006) Nickel: a review of its sources and environmental toxicology. Pol J Environ Stud 15(3):375–382

Conrad JE, Colvin C, Sililo O, Gorgens A, Weaver J, Reinhardt C (1999) Assessment of the impact of agricultural practices on the quality of groundwater resources in South Africa. Water Res Comm Rep 641(1):99

Gantayat RR, Elumalai V (2024) Salinity-induced changes in heavy metal behavior and mobility in semi-arid coastal aquifers: a comprehensive review. Water 16(7):1052. https://doi.org/10.3390/w16071052

Gleeson T, Wada Y, Bierkens MF, Van Beek LP (2012) Water balance of global aquifers revealed by groundwater footprint. Nature 488(7410):197–200

Hassan MN, Reza AS, Rasel MIA, Mahim MMA, Jannat MFT (2025) Pollution and health risk assessment of heavy metals in surface water of the industrial region in Gazipur, Bangladesh. Environ Chem Ecotoxicol 7:527–538. https://doi.org/10.1016/j.enceco.2025.02.014

Ingrao C, Strippoli R, Lagioia G, Huisingh D (2023) Water scarcity in agriculture: an overview of causes, impacts and approaches for reducing the risks. Heliyon 9(8):e18507. https://doi.org/10.1016/j.heliyon.2023.e18507

Karandish F, Liu S, de Graaf I (2025) Global groundwater sustainability: a critical review of strategies and future pathways. J Hydrol 657:133060. https://doi.org/10.1016/j.jhydrol.2025.133060

Khan MN, Mobin M, Abbas ZK, Alamri SA (2018) Fertilizers and their contaminants in soils, surface and groundwater. Encycl Anthr 5:225–240. https://doi.org/10.1016/B978-0-12-409548-9.09888-2

Khatun J, Intekhab A, Dhak D (2022) Effects of uncontrolled fertilization and heavy metal toxicity associated with arsenic (As), lead (Pb) and cadmium (Cd), and possible remediation. Toxicology 477:153274. https://doi.org/10.1016/j.tox.2022.153274

Knüppe K (2011) The challenges facing sustainable and adaptive groundwater management in South Africa. Water SA 37(1). https://doi.org/10.4314/wsa.v37i1.64110

Kohler K, Macheda T, Hobbs MM, Maisel MT, Rodriguez A, Farris L, Murphy MP (2024) Exposure to lead in drinking water causes cognitive impairment via an Alzheimer's disease gene-dependent mechanism in adult mice. J Alzheimer's Dis 100(s1):S291–S304. https://doi.org/10.3233/JAD-240640

Kopittke PM, Menzies NW, Wang P, McKenna BA, Lombi E (2019) Soil and the intensification of agriculture for global food security. Environ Int 132:105078. https://doi.org/10.1016/j.envint.2019.105078

Kumar N, Kumar A, Marwein BM, Verma DK, Jayabalan I, Kumar A, Kumar N (2021) Agricultural activities causing water pollution and its mitigation—a review. Int J Mod Agric 10(1):590–609

Kusangaya S, Warburton ML, Van Garderen EA, Jewitt GP (2014) Impacts of climate change on water resources in southern Africa: a review. Phys Chem Earth Parts A/B/C 67:47–54. https://doi.org/10.1016/j.pce.2013.09.014

Liang H, Li P, Elumalai V, Tian Y, Kou X (2025) Hydrochemical characteristics, groundwater nitrate sources and potential health risks in a typical alluvial plain of northwest China. Phys Chem Earth 139:103903. https://doi.org/10.1016/j.pce.2025.103903

Louw I (2020) Potential radiological impact of the phosphate industry in South Africa on the public and the environment (paper 1). J Environ Radioact 217:106214. https://doi.org/10.1016/j.jenvrad.2020.106214

Malan M, Müller F, Cyster L, Raitt L, Aalbers J (2015) Heavy metals in the irrigation water, soils and vegetables in the Philippi horticultural area in the western cape province of South Africa. Environ Monit Assess 187:4085. https://doi.org/10.1007/s10661-014-4085-y

Manson A (2018) Soil organic matter. Res Technol Bull:20017–20018. https://doi.org/10.13140/RG.2.2.22073.83047

Ncube B (2020) Smallholder farmer drought coping and adaptation strategies in Limpopo and western cape provinces. Water Res Comm 2716(1):20

Ntanganedzeni B, Elumalai V, Rajmohan N (2018) Coastal aquifer contamination and geochemical processes evaluation in Tugela catchment, South Africa—geochemical and statistical approaches. Water 10(6):687. https://doi.org/10.3390/w10060687

Ogut M, Er F, Brohi A (2010) Excessive phosphorus fertilization does not increase cadmium concentrations in soil or carrots (*Daucus carota* L.) grown in Konya (Turkey). Acta Agric Scand Sect B–Soil Plant Sci 60(5):420–426. https://doi.org/10.1080/09064710903071130

Rezapour S, Siavash Moghaddam S, Nouri A, Khosravi Aqdam K (2022) Urbanization influences the distribution, enrichment, and ecological health risk of heavy metals in croplands. Sci Rep 12(1):3868. https://doi.org/10.1038/s41598-022-07789-x

Robson MG, Toscano WA, Meng QY, Kaden DA (2022) Risk assessment for environmental health. CRC Press. https://doi.org/10.1201/9780429291722

Saravanan P, Saravanan V, Rajeshkannan R, Arnica G, Rajasimman M, Gurunathan B, Pugazhendhi A (2024) Comprehensive review on toxic heavy metals in the aquatic system: sources, identification, treatment strategies, and health risk assessment. Environ Res 258:119440. https://doi.org/10.1016/j.envres.2024.119440

Shaji E, Santosh M, Sarath KV, Prakash P, Deepchand V, Divya BV (2021) Arsenic contamination of groundwater: a global synopsis with focus on the Indian peninsula. Geosci Front 12(3):101079. https://doi.org/10.1016/j.gsf.2020.08.015

Suciu NA, De Vivo R, Rizzati N, Capri E (2022) Cd content in phosphate fertilizer: which potential risk for the environment and human health? Curr Opin Environ Sci Health 30:100392. https://doi.org/10.1016/j.coesh.2022.100392

Sun SS, Ao M, Geng KR, Chen JQ, Deng THB, Li JJ, Qiu RL (2022) Enrichment and speciation of chromium during basalt weathering: insights from variably weathered profiles in the Leizhou peninsula, South China. Sci Total Environ 822:153304. https://doi.org/10.1016/j.scitotenv.2022.153304

USEPA, 2011. Exposure Factors Handbook 2011 Edition (Final). U.S. Environmental Protection Agency, Washington, D.C.

Ullah Z, Rashid A, Ghani J, Nawab J, Zeng XC, Shah M, Iqbal J (2022) Groundwater contamination through potentially harmful metals and its implications in groundwater management. Front Environ Sci 10:1021596. https://doi.org/10.3389/fenvs.2022.1021596

Utembe W, Faustman EM, Matatiele P, Gulumian M (2015) Hazards identified and the need for health risk assessment in the South African mining industry. Hum Exp Toxicol 34(12):1212–1221. https://doi.org/10.1177/0960327115600370

Vetrimurugan E, Shruti VC, Jonathan MP, Roy PD, Sarkar SK, Rawlins BK, Villegas LEC (2019) Comprehensive study on metal contents and their ecological risks in beach sediments of KwaZulu-Natal province, South Africa. Mar Pollut Bull 149:110555. https://doi.org/10.1016/j.marpolbul.2019.110555

Villanueva CM, Kogevinas M, Cordier S, Templeton MR, Vermeulen R, Nuckols JR, Levallois P (2014) Assessing exposure and health consequences of chemicals in drinking water: current state of knowledge and research needs. Environ Health Perspect 122(3):213–221. https://doi.org/10.1289/ehp.1206229

WHO (2011) Guidelines for drinking-water quality. WHO Chron 38(4):104–108

WHO (2017) Water quality and health-review of turbidity: information for regulators and water suppliers.

Wickham H (2016) ggplot2: elegant graphics for data analysis. Springer-Verlag, New York. https://ggplot2.tidyverse.org

Wickham H, François R, Henry L, Müller K, Vaughan D (2023) dplyr: a grammar of data manipulation (R package version 1.1.4). https://CRAN.R-project.org/package=dplyr

Wickham H, Vaughan D, Girlich M (2024) TIDYR: Tidy, messy data (R package version 1.3.1). https://CRAN.R-project.org/package=tidyr

Xu X, Zhao Y, Zhao X, Wang Y, Deng W (2014) Sources of heavy metal pollution in agricultural soils of a rapidly industrializing area in the Yangtze Delta of China. Ecotoxicol Environ Saf 108:161–167. https://doi.org/10.1016/j.ecoenv.2014.07.001

Xie L, Li P, Fida M, Elumalai V (2025) Characteristics and potential health risk of inorganic nitrogen in phreatic water in the central and southern parts of Yinchuan plain (northwest China). Exposure Health 17(2):581–598. https://doi.org/10.1007/s12403-024-00679-9

Yang L, Meng F, Ma C, Hou D (2022) Elucidating the spatial determinants of heavy metals pollution in different agricultural soils using geographically weighted regression. Sci Total Environ 853:158628. https://doi.org/10.1016/j.scitotenv.2022.158628

Zhao Z, Wang Y, Shi J, Wang S, White PJ, Shi L, Xu F (2021) Effect of balanced application of boron and phosphorus fertilizers on soil bacterial community, seed yield and phosphorus use efficiency of *Brassica napus*. Sci Total Environ 751:141644. https://doi.org/10.1016/j.scitotenv.2020.141644

Zhou H, Chen Y, Yur X, Ren D, Liu Y, Yang K (2023) Identification and hazard analysis of heavy metal sources in agricultural soils in ancient mining areas: a quantitative method based on the receptor model and risk assessment. J Hazard Mater 445:130528. https://doi.org/10.1016/j.jhazmat.2022.130528

USEPA (2011) Exposure Factors Handbook 2011 Edition (Final). U.S. Environmental Protection Agency, Washington, D.C.

ATSDR (2008) Toxicological Profile for Aluminum. US Department of Health and Human Services, Agency for Toxic Substances and Disease Registry, Atlanta. Available at https://www.atsdr.cdc.gov/ToxProfiles/tp22.pdf, accessed July 10, 2025

Chapter 9
Impact of Land Use and Land Cover Changes on Groundwater Quality in the KwaZulu-Natal Coastal Region

Groundwater quality in the KwaZulu-Natal coastal region of South Africa faces escalating threats from rapid land use and land cover (LULC) changes driven by urbanization, agricultural expansion, and deforestation. This chapter investigates the spatiotemporal dynamics of LULC transformations between 2001, 2013, and 2023 using Landsat imagery and GIS techniques, and evaluates their impacts on groundwater quality. Six LULC classes, i.e., agricultural land, built-up areas, forests, bushland, barren land, and water bodies, were analyzed, revealing significant increases in agricultural (24 to 29%) and urbanized areas (0.3 to 1.2%), alongside declines in bushland (45 to 33%). Groundwater quality parameters, including physicochemical properties, major ions, and trace metals, were assessed against WHO standards. Key findings identified elevated levels of electrical conductivity (EC), total dissolved solids (TDS), chloride (Cl^-), sodium (Na^+), and lead (Pb) in agricultural and urban zones, correlating with anthropogenic activities like fertilizer use and industrial runoff. Forests, primarily commercial plantations, exhibited high sulfate and chloride concentrations, while barren lands showed minimal contamination. Statistical analyses (Kruskal–Wallis tests) confirmed significant variations in contaminants across LULC classes, emphasizing the role of land use in water quality degradation. The study underscores the urgent need for integrated land management policies, sustainable agricultural practices, and continuous monitoring to mitigate contamination risks and safeguard groundwater resources in this ecologically sensitive coastal region.

9.1 Introduction

Groundwater is the most significant source of freshwater for drinking, domestic, agricultural, and industrial purposes, especially in arid and semi-arid countries. It is the preferred and primary source of potable water for about half of the world's population (He et al. 2020; Sadeghi et al. 2021). The survival and well-being of

P. Li and V. Elumalai, *Groundwater Quality Under Agricultural Activities—Cases from China and South Africa*, Professional Practice in Earth Sciences,
https://doi.org/10.1007/978-3-031-98993-3_9

humans depend on groundwater, which is also crucial for advancing public health. Studies on groundwater quality have been carried out all around the world. Nevertheless, there isn't a comprehensive study that combines the sources and variables that endanger the quality of groundwater (Abanyie et al. 2023). Groundwater is considered less contaminated and safe compared to surface water resources due to its limited contact with the external environment and natural filtration of contaminants by the soil during percolation (Elumalai et al. 2017; Mutileni et al. 2023). However, various studies have indicated multiple cases of groundwater contamination due to various sources (Igboama et al. 2022).

Water quality serves as a vital indicator of ecosystem health and a cornerstone for public health and biodiversity (Su et al. 2025). Population, salinity, and increasing water consumption have caused high pressure on groundwater resources (Sadeghi et al. 2021). The groundwater quality is influenced by anthropogenic pollution and activities such as LULC change, agricultural return flow, disposal of domestic and industrial wastes, and hydrogeochemical processes that occur beneath the surface, and vary temporally and spatially (Mutileni et al. 2023). Urbanization and LULC patterns significantly affect groundwater quality, especially in metropolitan areas with intense anthropogenic activities (He et al. 2020; Su et al. 2025). LULC changes such as deforestation, agricultural area expansion, and urbanization are primary drivers of global environmental transformation directly impacting water resource conditions primarily through modifications to watershed hydrology processes (Sadeghi et al. 2021). Thus, exploring the nexus between land use change and water quality is of considerable scientific and practical importance (Su et al. 2025).

Several studies have been carried out to evaluate the impact of LULC change on groundwater quality (He et al. 2020; Ahmad et al. 2021; Sadeghi et al. 2021; Siddik et al. 2022). For instance, Siddik et al. (2022) indicated that several human activities, including changes in LULC, have an impact on groundwater recharge. Among these, the dynamics of groundwater flow are significantly impacted by long-term temporal and seasonal variations in LULC. LULC changes in the area have considerable control over the degradation of water quality. This may lead to an overabundance of nutrients being released into surface water sources through agricultural runoff (Zena et al. 2024). Elmahdy et al. (2020) show that the groundwater levels in the residential and agricultural regions of the study area are rapidly declining. The NO_3^- concentration in the groundwater was shown to be strongly linked with human and agricultural activities ($r = 0.91$), however, there was a negative association between NO_3 and groundwater level. LULC transformations are among the most significant global environmental challenges, as they directly affect critical natural resources and ecosystem functions. For example, changes in LULC can disturb hydrological cycles, affecting water availability and quality; intensify soil erosion, diminishing agricultural productivity; and fragment habitats, threatening biodiversity, which ultimately affects human well-being (Zadbagher et al. 2018; Yang et al. 2024). Thus, understanding the spatial and temporal dynamics of LULC change is essential for effective environmental management and sustainable development to prevent further declines in natural resources (Kindu et al. 2015; Yang et al. 2024).

LULC change is a complex issue primarily driven by various factors, including population growth, economic development, agricultural practices, urbanization, and climate variability (Assede et al. 2023; Gebeyehu et al. 2023). For example, population growth increases the demand for food, housing, and infrastructure, leading to the conversion of natural and agricultural land. Economic development often involves industrialization and resource extraction, which can degrade ecosystems. The drivers of LULC are often caused by agricultural practices, such as reduced fallow periods and intensive farming, which can deplete soil and water resources to meet growing demands for food resources. Urbanization and infrastructure develop fragments of landscapes and reduce biodiversity. Climate variability, including changing temperature and precipitation patterns, can alter land suitability and exacerbate land degradation. These factors contribute to LULC change, resulting in negative impacts such as biodiversity loss, soil degradation, water scarcity, and climate change. Understanding these drivers and implementing sustainable land-use management practices is crucial to mitigate these impacts and ensure a sustainable future. The loss of vegetation cover, soil erosion, a decrease in water availability and declined agricultural production are some of the effects of LULC change (Tang et al. 2021; Zhou et al. 2008). Understanding the factors that influence LULC changes is essential for predicting future events or developing management plans to prevent or reduce the loss of the ecosystem (Kindu et al. 2015).

The primary cause of the significant differences in LULC is the conversion of natural areas into industrial or agricultural areas (Jat et al. 2008). Anthropogenic activities including agriculture, mining, urbanization, and industrialization can affect changes in land use and water quality (Li et al. 2017; Lipczynska-Kochany 2018). It is believed that these LULC patterns significantly affect the quality of groundwater (Khan and Jhariya 2018). Many human activities, including changes in LULC, have an impact on groundwater recharge (Siddik et al. 2022). Shooshtarian et al. (2018) demonstrated the ability to forecast the likely effects of LULC change on changes in drinking water quality. The findings clearly showed that changes in urban and agriculture, such as expansion in size and conversion of other land uses to industrial, had contributed to a decline in water quality. Understanding important LULC movements and trends is made easier by the use of multi-temporal Landsat datasets (Roy and Inamdar, 2019). It is now simpler to track the developments and changes that have occurred over the previous few decades because of the advent of Landsat satellites and breakthroughs in computer technology.

The study of LUCC and water quality often involves integrating remote sensing, geographic information systems (GIS), and various statistical techniques. The integration of these techniques is crucial in studying LULC change impacts on groundwater quality, as it allows for comprehensive assessments of the changes. These technologies facilitate the analysis of spatial and temporal dynamics, enabling researchers to understand the impacts of human activities on the environment. Remote sensing provides satellite imagery that captures land cover changes over time, whereas GIS techniques enable the classification and visualization of land use changes, enhancing the understanding of spatial patterns and trends (Bill Donatien et al. 2024). The statistical approaches also quantify the impacts of LULC change on groundwater qualities, revealing significant changes in direct radiative forcing due to land cover alterations (Wang et al. 2024). Thus, this study focuses on the coastal region of KwaZulu-Natal, South Africa. The primary objective of this study is to quantify and analyse the impact of LULC changes on groundwater quality in the study area between the years 2001, 2013 and 2023. The study determined the spatial patterns of LULC changes and their impact on groundwater quality in the study area by employing Landsat satellite imagery. The findings of this research will contribute to a deeper understanding of the factors influencing LULC changes on groundwater quality and provide valuable insights into informed decision-making and future planning in the region.

9.2 Data and Methods

9.2.1 Study Area

The study area is located on the KwaZulu-Natal coast of South Africa (Fig. 9.1). Geographically, this coastal area is situated approximately 180 km northeast of Durban, the capital city of KwaZulu-Natal provinces, South Africa. The area covering a total area of approximately 25,987 km^2, it stretches between latitudes of 27.5°S and 29.5°S and longitudes of 30.5°E and 33.0°E. The coastal plain is bounded by the Indian Ocean in the east and the Lebombo Mountains in the west (Ramjeawon et al. 2020). The area is of strategic significance, with its diverse natural and human-made features. It is a subtropical climate, characterized by humid summers and mild winters. During the summer months (January to March), temperatures range from 21 to 29 °C, while in the winter season (June to August), they range from 12 to 23 °C. This climatic pattern supports various ecological and human activities. The region is among South Africa's highest rainfall zones (Schulze 1982), with annual precipitation varying significantly from 850 mm in the western parts to over 1,200 mm along the coastline (DWAF 2000). Rainfall predominantly occurs between October and March, contributing to the lush vegetation and agricultural viability of the area. Surface water bodies in the region experience seasonal evaporation rates, with summer values between 150 and 250 mm per month, reducing to around 100 mm per month during the winter (Kelbe and Germishuyse 2001).

The landscape of the region is defined by flat to gently undulating terrain interspersed with low hills and flat-bottomed drainage features. The uMhlathuze River watershed, the largest river in the area, originates in the Babanango Mountains. It traverses key towns such as Empangeni and Richards Bay before discharging into the Indian Ocean. This river system is vital for both ecological sustainability and human activities, including agriculture and industry. Industrially, the Richards Bay area is a hub of economic activity, hosting a range of industries such as aluminium smelters, fertilizer plants, and mineral ore mining. It is home to South Africa's largest coal export facility, playing a crucial role in the country's export economy. Agriculturally, the inland regions of this coastal plain are productive, with crops like sugarcane,

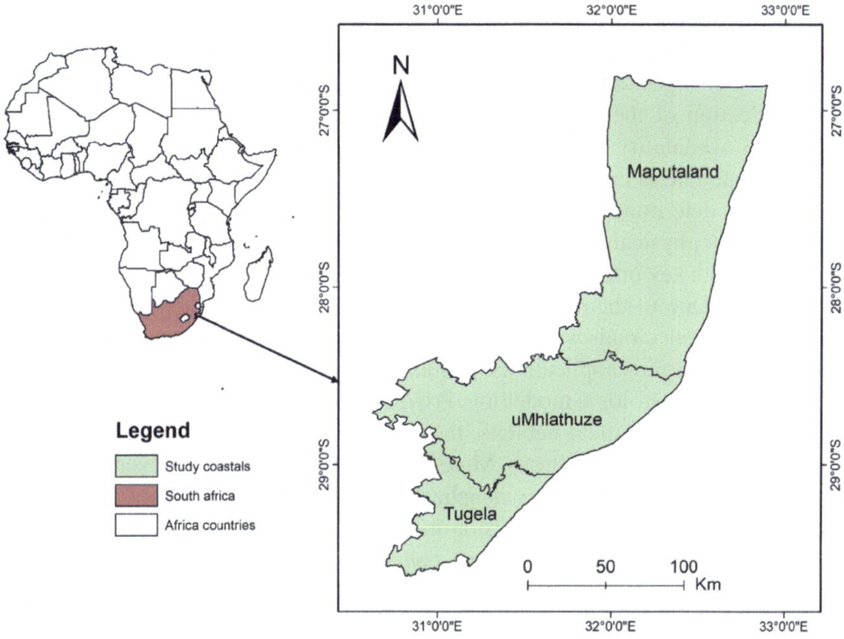

Fig. 9.1 Locations of the study coastal area

citrus, vegetables, and maize cultivated extensively. Additionally, commercial plantations of pine and eucalyptus are significant contributors to the region's economy. Environmental concerns are also evident, with a landfill site located in the central part of the study area. This site underscores the need for sustainable waste management practices, given the ecological sensitivity of the region. The collocation of natural beauty and significant human activity makes this area a critical zone for studying interactions between the environment and socio-economic development. The number of industries that function in the Richards Bay area include aluminium smelters, fertilizer plants, and the mining of mineral ores. It has the largest coal export facility in South Africa. Agriculture in the region is inland and includes sugarcane, citrus, vegetables, maize, etc. apart from commercial products such as pine and eucalyptus. There is also a landfill site located in the centre of the study area.

9.2.2 Data Sources and Collection

In this study, LULC maps from 2001, 2013, and 20,233 were used to assess its changes. The Landsat images were used to identify the changes and patterns in LULC spatiotemporal distribution and magnitude in the KwaZulu-Natal coastal area of South Africa. These Landsat image datasets were downloaded freely from the

United States Geological Survey (USGS) https://earthexplorer.usgs.gov/ (Table 9.1) (Accessed on November 20, 2024) and extracted according to the study watershed to assess the changes in LULC (Karra et al. 2021).

The selection of the Landsat images was based on the following criteria: the quality and availability of satellite images for each decade in the USGS database that were available in the study area; the period of significant LULC changes in the study area as determined by field observation and interviews with local communities; and the biophysical environment and socioeconomic activities as determined by discussions with key informants, administrative agricultural office experts, and local communities. Due to their ease of analysis, Landsat images taken during the dry season and without clouds were chosen. The LULC map offers a wealth of information crucial for various applications, including conservation planning, food security assessment, and hydrologic modelling. Providing detailed land cover data allows for the identification of critical habitats, the monitoring of agricultural land use, and the analysis of water flow patterns. Moreover, this dataset's global coverage enables researchers and policymakers to visualize and analyse LULC dynamics anywhere on Earth, facilitating informed decision-making at local and global scales.

The groundwater quality data were collected in the study area The measured parameters include physical parameters: electrical conductivity (EC), total dissolved solids (TDS), pH, and oxidation–reduction potential (ORP). Major ions (mg/L): chloride (Cl^-), bicarbonate (HCO_3^-), sulfate (SO_4^{2-}), nitrate (NO_3^-), calcium (Ca^{2+}), potassium (K^+), magnesium (Mg^{2+}), and sodium (Na^+). Trace metals (μg/l): cadmium (Cd), cobalt (Co), copper (Cu), iron (Fe), lithium (Li), nickel (Ni), lead (Pb), and zinc (Zn).

Groundwater quality data on a variety of physical characteristics, including electrical conductivity EC (μS/cm), total dissolved solids TDS (mg/L), pH, and oxidation–reduction potential ORP in (mV), were collected to evaluate the quality of the groundwater in the study area. The analysis also included trace metals like cadmium (Cd), cobalt (Co), copper (Cu), iron (Fe), lithium (Li), nickel (Ni), lead (Pb), and zinc (Zn), measured in μg/L, as well as major ions like sodium (Na^+), calcium (Ca^{2+}), potassium (K^+), magnesium (Mg^{2+}), sodium chloride (Cl^-), bicarbonate (HCO_3^-), sulphate (SO_4^{2-}), nitrate (NO_3^-), calcium (Ca^{2+}), potassium (K^+), magnesium (Mg^{2+}), and sodium (Na^+), all reported in mg/L.

Table 9.1 Data acquisition from satellite images for the study area

Source	Sensors	Path/row	Spatial resolution	Date
Earth explorer	Landsat TM	168/080	30 m × 30 m	2001/04/02
	Landsat TM	168/080	30 m × 30 m	2013/04/07
	Landsat TM	168/080	30 m × 30 m	2023/08/23

9.2.3 LULC Types

The study coastal region was divided into six main LULC classes or types based on the spectral responses of features. These are agricultural land, barren land, built-up area, bushland, forest and water bodies. The shape, texture, colour, and land cover discrimination are examples of spatial-spectral band features that are used in the LULC classification to divide land covers into various classes. Table 9.2 illustrates the name and description of LULC classes found in the study area (Kindu et al. 2013).

9.2.3.1 LULC Classification and Accuracy Assessment

The accuracy of the image classification was evaluated to ensure the reliability of subsequent analysis and applications. Within these sampled polygons, 150 random points were generated. The classification's accuracy was then determined by comparing these points against the classified image. Following this initial accuracy assessment, a post-classification refinement was performed, resulting in improved classification accuracy. To rigorously evaluate the classification accuracy, a confusion matrix was generated, from which key metrics such as the Kappa coefficient (KC), User's Accuracy (UA), Producer's Accuracy (PA), and Overall Accuracy (OA) were computed. The OA, providing a comprehensive measure of the model's ability to correctly classify across all categories, was calculated using Eq. 9.1. The KC further assesses the classification's performance by comparing its accuracy to that expected from random assignment (Eq. 9.2). Ranging from 0 to 1, K quantifies the

Table 9.2 Description of LULC classes (Karra et al. 2021)

Name	Descriptions
Water bodies	This class includes large bodies of water like oceans, seas, large lakes, and rivers
Forest	Any significant clustering of tall (~15 feet or higher) dense vegetation, typically with a closed; examples: wooded vegetation, clusters of dense tall vegetation within savannas, plantations
Agricultural land	It is a region of land used for farming, including cultivating crops and raising livestock. It encompasses various types of land, such as arable land for growing crops, pastureland for grazing animals, and land used for specialized agricultural practices like orchards or vineyards
Built up area	Human-made structures; major road and rail networks; large homogeneous impervious surfaces including parking structures, office buildings, and residential housing; examples: houses, dense villages/towns/cities, paved roads, and asphalt
Barren land	Areas of rock or soil with very sparse to no vegetation for the entire year; large areas of sand and deserts with not too little vegetation; examples: exposed rock or soil, desert, and dunes, dry salt flats/pans
Bushland	It describes land covered with shrubs, small trees, and grass. It's often found in areas with dry climates, such as savannas and arid regions

level of agreement between the classification and the ground truth, with 1 indicating perfect agreement and 0 indicating no agreement (Andualem et al. 2023). The classified Landsat imagery was successfully analyzed and features were detected using ArcGIS 10.8 and Google Earth Pro (Abuhay et al. 2023).

$$\text{OA} = \frac{\text{Total number of correctly classified pixels (Diagonal)}}{\text{Total number of reference pixels}} \times 100 \tag{9.1}$$

$$\text{KC} = \frac{(\text{Total sample} \times \text{Total corrected sample} - \Sigma(\text{Column total} \times \text{Row total}))}{\text{Total sample}^2 - \Sigma(\text{Column total} - \text{Row total})} \times 100 \tag{9.2}$$

9.2.3.2 LULC Change Detection

Two separate periods of classified images are needed to detect changes in land use and cover. After the classes are created, an area of comparison between the two classifications can be performed. The post-reclassification comparison technique was employed separately to determine the LULC changes in the research area. The Geographic Information System (GIS) environment was utilized to carry out the LULC change detection investigation. LULC changes were reclassified and analyzed using the open attribute table that appeared in the land using class names and values. The LULC variations between the initial and final periods are detailed in the class statistics. This was done to show the difference in LULC modifications using the classed images from 2001 to 2023. Analyzing the status of LULC changes and drivers of change is important for identifying vulnerable areas for change and designing sustainable ecosystem services (Abebe et al. 2022). The percentage and rate of LULC change were calculated by the following formula (Eqs. 9.3 and 9.4):

$$\text{Rate of change}\left(\text{km}^2/\text{year}\right) = \frac{\text{A2} - \text{A1}}{\text{z}} \tag{9.3}$$

$$\%\text{ of change} = \frac{\text{A1} - \text{A2}}{\text{A2}} \times 100 \tag{9.4}$$

where, A2 is the area of LULC (km^2) in period 2, A1 is the area of LULC (km^2) in period 1, and z is the period interval in years between A_2 and A_1.

9.2.3.3 Impacts of LULC Change on Water Quality Parameters

The selection of buffer zones is a critical step in studying the relationship between LULC and water quality, as it directly influences the accuracy and reliability of the overall analysis (Su et al. 2025). Thus, a majority analysis of the LULC class in the 300 m circular buffer zone was used to assess the impact of land use on water quality parameters. Simple descriptive statistics were performed such as range, mean and standard deviation to evaluate the central tendency and distribution of water quality parameters across LULC classes. The physicochemical parameters were compared with the World Health Organization (WHO) standard of drinking water quality (WHO 2011) to check the possible contamination levels and to suggest appropriate remediations. The relationship between LULC and water quality was analysed using the Kruskal–Wallis test (Eq. 9.5), a nonparametric method of testing whether samples come from the same distribution.

$$H = \frac{12}{N(N+1)} \sum_{i=1}^{K} \frac{R_i^2}{n_i} - 3(N+1) \tag{9.5}$$

where, H = Kruskal–Wallis test statistic, N = total number of observations across all groups; K = number of groups (e.g., different LULC types); R_i^2= sum of the ranks for the ith group, and n_i = number of observations in the ith group.

9.3 Results and Discussion

9.3.1 Analysis of Land Use and Land Cover Change

There are six LULC classes identified in the study area: agricultural land, barren land, built-up area, bushland, forest and water bodies between 2001, 2013 and 2023 (Fig. 9.2). According to the LULC classification maps from 2001, 2013, and 2023, the amount of agricultural land and built-up areas has consistently increased over time. At the same time from 2001 to 2023, there has been a decrease in the amount of forest and bushland, which suggests that land conversion and urbanization are continuing. During this time, water bodies have been largely constant. Agricultural land, barren land and bushland covered the largest area in the watershed than the other LULC types in all three periods (2001, 2013 and 2023), whereas built-up area, forest and water bodies covered less area. This indicated that agriculture is crucial to the socioeconomic development of the study area (Malede et al. 2022). Table 9.3 summarizes the area coverage in kilometre squares (km^2) and percentage (%) of each LULC class over the study area for the three periods (2001, 2013, and 2023). In 2001, Bushland constituted most of the study area, covering an area of 45%, whereas the built-up area occupied the smallest area nearly 0.3%.

Similarly, bushland comprised 36% of all LULC in 2013, but this percentage was lower than that of LULC classes in 2001. Agriculture land and barren land occupied an area of 26% and 25% respectively during the same period. Furthermore, 2023 LULC classification classes also indicate that agricultural land, barren land and bushland comprise the greatest share of all other LULC classes, which occupied

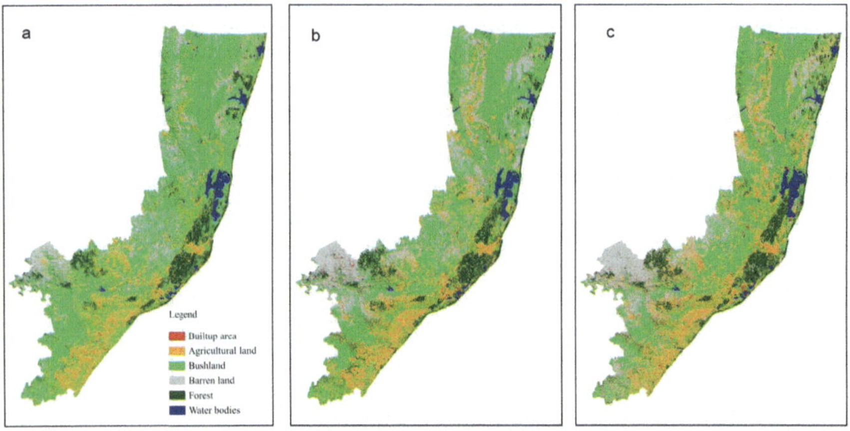

Fig. 9.2 LULC change classification **a** 2001, **b** 2013 and **c** 2023

Table 9.3 The proportional area coverage in kilometres squares (km^2) and percentage (%) of LULC classes in the study area

Land use classes	2001		2013		2023	
	km^2	%	km^2	%	km^2	%
Agricultural land	5777	24	6218	26	7126	29
Barren land	4773	20	6011	25	5569	23
Built-up area	62	0.3	154	0.6	284	1.2
Bushland	10,852	45	8837	36	8127	33
Forest area	2276	9	2582	10.6	2678	11
Water bodies	562	2.3	501	2.1	519	2.1

29%, 23% and 33%, respectively, while built-up, forest and water bodies LULC classes occupied an area of 1.2%, 11% and 2.1%, respectively (Table 9.3).

Agricultural land, built-up area and forest showed a consistent increase in LULC classes between 2001 and 2023 from 24–29%, 0.3–1.2% and 9–11%, respectively. Built-up area exhibited the most significant relative increase, growing from a small base of 62 km^2 (0.3%) in 2001 to 284 km^2 (1.2%) in 2023, whereas forest experienced a modest but steady increase in both area (from 2276 km^2 to 2678 km^2) and percentage (from 9 to 11%). Urbanization and the expansion of forests are the primary causes of land degradation (Abbas et al. 2021). On the other hand, barren land, bushland and water bodies revealed a decreasing LULC classes between 2001 and 2023. However, barren land increased between 2001 and 2013 but then decreased by 2023, yet the overall trend from 2013 to 2023 was a decreasing trend. Bushland and water bodies showed a consistent decrease from 2001 to 2023. Developing successful management strategies to reduce the depletion of natural resources requires an understanding of the factors influencing changes in LULC (Kindu et al. 2015). Feng et al. (2024) reported that rangelands, which make up about 40% of the planet's land area, are essential for local livelihoods as well as for the services provided by ecosystems worldwide. The structure and magnitude of ecosystem and non-ecosystem service supply and demand are changed by densification and population growth, which may erode direct feedback mechanisms between ecosystems and societies and cause social-ecological systems to collapse (Cumming et al. 2014). Additionally, the interaction of population expansion, urbanization, and technological development results in feedback loops that reinforce ecosystem overexploitation, which has a major influence on sustainable resource usage.

Accurate LULC classification is an essential precondition for studies that attempt to forecast LULC changes (Appiah et al. 2015). One of the most widely used techniques for estimating a model's prediction ability is the Kappa statistics (Hua 2017). The Kappa values have been divided into various groups: <0 indicates no agreement, 0–0.2 as slight, 0.0–0.41 as poor, 0.41–0.60 as moderate, 0.60–0.80 as significant, and 0.81–1.0 as almost perfect agreement (Manonmani and Suganya 2010). The accuracy of the classified LULC map from 2001, 2013 and 2023 was evaluated. The kappa coefficients were 0.88, 0.91, and 0.89, respectively, while the overall accuracies were 90.78, 93.04, and 90.54% based on the LULC maps from 2001, 2013, and 2023 (Tables 9.4, 9.5 and 9.6). These findings thus show that the ground truth LULC classes and the classified map were accepted. Additionally, the classified map set the minimal level of accuracy needed for the ensuing post-classification procedures (Munthali et al. 2019).

Land use changes between 2001, 2013, and 2023 are depicted in the (Fig. 9.3). LULC were observed over this period. Notable increases in built-up areas and agricultural land are seen throughout all periods, reflecting growing human activity. The greatest decrease was seen in bushland, particularly between 2001 and 2023, indicating a loss of vegetation on a large scale. Over the past ten years, there has been a little rise in forest areas and a slight decrease in barren terrain. During the study period, water bodies stayed mostly unaltered. All things considered, the pattern shows

Table 9.4 Accuracy assessment of LULC map in 2001

Land use classes	Agricultural land	Barren land	Built-up area	Bushland	Forest area	Water bodies	Total users
Agricultural land	26	1	1	1	2	0	31
Barren land	0	26	0	2	1	0	29
Built-up area	0	0	3	0	0	0	3
Bushland	0	1	0	31	0	0	32
Forest area	0	1	0	3	20	0	24
Water bodies	0	0	0	0	1	22	23
Total producers	26	29	4	37	24	22	141

Overall accuracy = 90.78 and Kappa coefficient = 88.83

Table 9.5 Accuracy assessment of LULC map of 2013

Land use classes	Agricultural land	Barren land	Built-up area	Bushland	Forest area	Water bodies	Total users
Agricultural land	12	0	0	2	0	0	14
Barren land	1	32		1	0	0	34
Built-up area	0	2	6	0	0	0	8
Bushland	2	1	0	51	0	0	54
Forest area	0	0	1	0	26	0	27
Water bodies	0	0	0	0	1	20	21
Total producers	15	35	7	54	27	20	158

Overall accuracy = 93.04 and Kappa coefficient = 91.06

Table 9.6 Accuracy assessment of LULC map of 2023

Land use classes	Agricultural land	Barren land	Built-up area	Bushland	Forest area	Water bodies	Total users
Agricultural land	23	1	0	0	0	0	24
Barren land	3	24	1	4	0	0	32
Built-up area	1	0	6	0	0	0	7
Bushland	0	0	0	36	0	0	36
Forest area	0	1	0	1	28		30
Water bodies	0	0	0	1	1	17	19
Total producers	27	26	7	32	29	17	148

Overall accuracy = 90.54, and Kappa coefficient = 88.51

how, throughout the last 20 years, land uses have shifted from natural vegetation to urban and agricultural areas. The most significant change is seen in the built-up area, which increased by 5% between 2001 and 2013 and by another 1% between 2013 and 2023. This suggests rapid urbanization in the region. These findings align with previous research that has also observed similar trends in land use change, particularly the dominance of rangeland, crops, and trees in many regions (Winkler et al. 2021; Annan et al. 2024). A similar study by Malede et al. (2023) indicated that settlements increased from 0.73% to 1.42% between 1986 and 2018, whereas agricultural land increased significantly from 56.39% to 70.19%. On the other hand, the research area's bushlands, forest regions, and barren land LULC classes saw declines from 26.18% to 19.30%, 4.94% to 1.92%, and 11.77% to 7.16%, respectively. Population and water consumption increased along with the growth of vegetated and built-up areas, whereas desert and coastal areas decreased (Elmahdy et al. 2020).

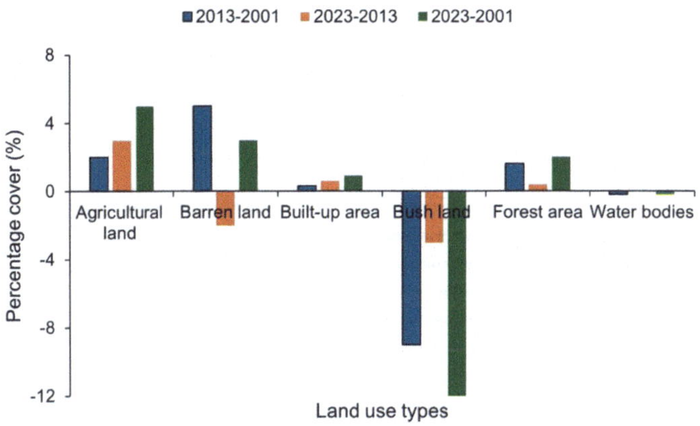

Fig. 9.3 Changes in different land use classes from the year 2001, 2013 to 2023

9.3.1.1 LULC Change Detection and Analysis

Figure 9.4 presents the LULC transitions in the study area from 2001 to 2023, showing the area in square kilometres for each type of LULC change. The most significant transition is the persistence of bushland (Bushland→ Bushland) covering approximately 5800 sq. km, indicating substantial stability in this land cover land use type. The other major stable categories include Agricultural land → Agricultural land and Barren land→ Barren land, each around 2900 and 2300 km^2, respectively. Notably, a large portion of bushland has been converted to agricultural land (~2700 km^2), highlighting a key land use change trend. Smaller transitions such as Forest → Bushland, Forest → Agricultural land, and Water bodies → Water bodies account for minimal areas, reflecting either limited conversion or stable minor land classes in the study area. The changes have potential environmental and socioeconomic impacts, requiring careful consideration for sustainable land management and policy development. The observed land use transitions of bare ground to rangeland and crop to trees align with the trends and findings from other studies (Asenso Barnieh et al. 2020; Winkler et al. 2021). Environmental issues are triggered by human activities such as land degradation, particularly in arid and semi-arid regions (Prăvălie 2016; Prăvălie et al. 2021). However, there are also growing efforts towards afforestation and reforestation, driven by various factors (Mohan et al. 2021).

9.3.1.2 LULC Map Conversion Matrix

Confusion matrices for the years 2001 and 2023 were created to have a better understanding of the development of the LULC classes in the study area. Table 9.7 presents a confusion matrix of the transition changes that the various land use categories experienced between 2001 and 2023. It provides a detailed breakdown of how each

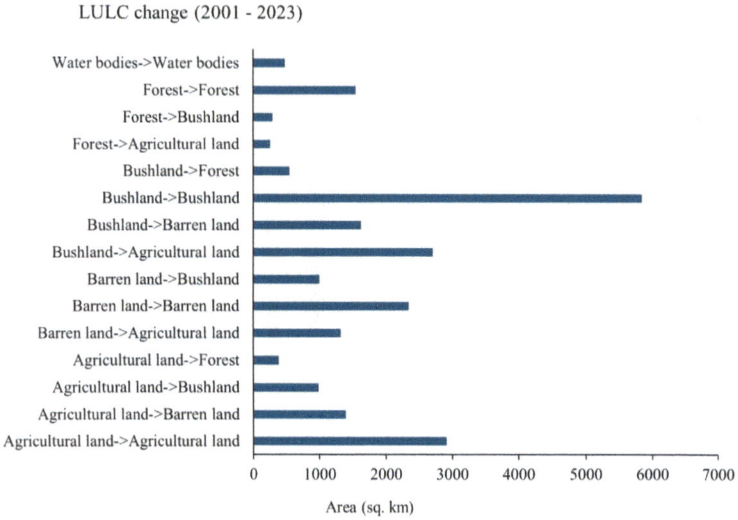

Fig. 9.4 Details of LULC transition of the study area from 2001 to 2023

land class in 2001 transitioned to different land classes in 2023. The diagonal bold elements present the area of each land class that remained unchanged between the two years (2001 and 2023). For example, 2915.10 km^2 of agriculture in 2001 remained agriculture in 2023. Off-diagonal elements are the area of one land class that is converted to another land class. For instance, 1313.97 km^2 of barren land in 2001 was converted to agricultural land in 2023. The total column and row show the total area of each land class in 2001 and 2023, respectively.

Agriculture land to barren land, buildup area to agricultural land, bushland to forest, forest to bushland, and water to built-up regions are some of the major transitions as indicated in Table 6. The built-up area, which made up 30.74 km^2 (81.73%), of its whole area in 2001, had the highest conversion matrix from 2001

Table 9.7 Conversion matrix of LULC maps from the years 2001, 2013 and 2023

LULC classes		Land class 2023						
		Agriculture	Barren	Built up	Bushland	Forest	Water	Total
Land class 2001	Agriculture	**2915.10**	1391.59	81.50	985.05	387.93	15.47	5776.64
	Barren	1313.97	**2348.07**	17.40	1003.41	89.73	0.36	4772.95
	Built up	5.57	22.90	**6.87**	0.76	1.00	0.51	37.61
	Bushland	2710.81	1623.20	132.23	**5835.13**	554.32	20.32	10,876.01
	Forest	258.79	160.30	22.05	293.89	**1539.04**	2.05	2276.11
	Water	21.46	22.68	23.55	8.72	5.37	**480.14**	561.92
	Total	7225.69	5568.73	283.60	8126.96	2577.40	518.86	**24,301.24**

to 2023. In the same time frame, agricultural land, barren land, bushland, forest, and water bodies were transformed into various classes over 2861.54 km^2 (49.54%), 2424.87 km^2 (50.80%), 5040.88 km^2 (46.35%), 737.08 km^2 (32.38%), and 60.32 km^2 (10.73%), respectively. These results revealed significant transitions, particularly in areas covered by built-up areas, which experienced the most substantial conversion 81.73%. Other LULC classes, including barren land, agricultural land, and bushland also underwent notable transformations. Rapid urbanization is changing the environment, which frequently leads to a decline in ecosystem services and the standard of living in cities (Wang et al. 2024).

9.3.1.3 Spatial Distribution of Groundwater Variation

The analysis of water quality parameters concerning the guidelines established by the World Health Organization (WHO, 2011) indicates significant deviations across water quality parameters (Table 9.8). Specifically, 21.26% of the sampled water fell outside the acceptable pH range of 6.5–8.5, which can facilitate leaching and promote microbial growth. Furthermore, 40.94% of TDS samples exceeded the permissible limit of 500 mg/L. Additionally, 22.83% of EC measurements surpassed the 1500 μS/cm, while 25.98% of Cl^- concentrations exceeded the 250 mg/L limit. Notably, 10% of Pb samples were found to exceed the limit of 10 μg/L. The distribution of these water quality parameters with the highest percentage exceeding permissible limits is presented in (Fig. 9.5). 4.72% of Fe samples were above the 300 μg/L threshold. NO_3^-, Cd, Cu, Zn and Ca^{2+}. Nitrate (NO_3^-) are within safe limits. Only 1.57% exceeded 45 mg/L, suggesting minimal agricultural runoff or sewage contamination, and 99.21% of samples had Ca^{2+} within the 200 mg/L limit. All trace metals (Cd, Cu, Ni, Zn) are within the permissible limit.

9.3.1.4 Impacts of LULC Types on Physical Groundwater Quality Parameters

The statistical comparison of physical water quality parameters across various land use types reveals significant variations attributable to anthropogenic activities and natural land cover characteristics (Table 9.9). Agricultural lands exhibit a wide range of EC values (140–26,260 μS/cm), indicating substantial variability likely due to differences in management practices. The EC over agricultural lands is marginally within the limit (1357.23 μS/cm) with the highest standard deviation (2998.88 μS/cm). The pH ranges from 4.8 to 9.5, averaging at 7.38, suggesting occasional acidic conditions possibly resulting from agrochemical runoff. TDS levels are also elevated (875.99 mg/L), reflecting increased solute concentrations from agricultural inputs (Rao et al. 2024; Rajmohan et al. 2025). ORP varied widely, indicating fluctuating redox conditions influenced by organic matter decomposition and nutrient loading. Built-up areas had EC (1,737.8 μS/cm) and TDS (1,112 mg/L) above the permissible limit due to impervious surfaces and stormwater runoff carrying various pollutants

Table 9.8 The number and percentage of samples exceeding allowable limits

Parameters	WHO (2011)	Samples within the permissible limit	Samples above the permissible limit	Samples below the permissible limit	% exceeding permissible limit	% within the permissible limit
EC (μS/cm)	1500	98	29	0	22.83	77.17
pH	6.5–8.5	100	6	21	21.26	78.74
TDS (mg/L)	500	75	52	0	40.94	59.06
OVR (mV))	Not specified	–	–	–	–	–
Cl (mg/L)	250	94	33	0	25.98	74.02
NO_3 (mg/L)	45	125	2	0	1.57	98.43
K (mg/L)	12	122	5	0	3.94	96.06
Mg (mg/L)	50	121	6	0	4.72	95.28
Na (mg/L)	200	99	28	0	22.05	77.95
Ca (mg/L)	200	126	1	0	0.79	99.21
HCO_3 (mg/L)	Not specified	–	–	–	–	–
SO_4 (mg/L)	250	125	2	0	1.57	98.43
Cd (μg/L)	3	127	0	0	0	100
Co (μg/L)	Not specified	–	–	–	–	–
Cu (μg/L)	2000	127	0	0	0	100
Fe (μg/L)	300	121	6	0	4.72	95.28
Li (μg/L)	Not specified	–	–	–	–	–
Ni (μg/L)	70	127	0	0	0	100
Pb (μg/L)	10	114	13	0	10.24	89.76
Zn (μg/L)	5000	126	1	0	0.79	99.21

(Xu et al. 2022). The pH is slightly alkaline (6.08–8.91; mean: 7.15), and the negative ORP values (−6.50 mV) suggest reducing conditions, potentially from organic waste accumulation. Forested regions exhibit the highest EC 5,583.57 μS/cm) and TDS (3,641.57 mg/L), which can be attributed to commercial forestry plantations such as pine and eucalyptus are widely practiced (Elumalai et al. 2017; Mthembu et al. 2020). The pH was in a neutral to slightly alkaline range (6.46–8.5; mean: 7.47), and ORP values (−1.35 mV) indicate mildly reducing conditions, consistent with dense vegetation cover.

Bushlands have moderate EC (1250.13 μS/cm) and TDS levels (802.50 mg/L), suggesting limited anthropogenic influence. The pH (7.13) had a wide range (5.36–8.12), and ORP values (39.79 mV) indicate stable oxidative conditions, reflecting the buffering capacity of natural vegetation. Barren land demonstrates the lowest

Fig. 9.5 The spatial distribution of groundwater quality parameters

EC (550 μS/cm) and TDS (352.77 mg/L) values, reflecting minimal anthropogenic disturbance. The pH is slightly acidic to neutral (6.82), and ORP values (mean: 28.58 mV) suggest oxidative conditions, indicative of healthy, well-aerated soils.

Kruskal–Wallis test results indicated that EC ($H = 23.66$, p-value $= 0.0001$), TDS ($H = 23.67$, p-value $= 0.0001$) and pH ($H = 10.46$ and p-value $= 0.03$) show significant differences across LULC classes. ORP (mV) had an H-statistic of 9.26, and a p-value of 0.06, indicating no significant differences across the LULC classes (Fig. 9.6).

9.3.1.5 Impact of LULC Classes on Major Ions of Groundwater Quality

Over agricultural lands concentrations of Na^+ (205.38 mg/L) and Cl^- (326.7 mg/L), surpass the WHO thresholds, suggesting potential contamination from fertilizers and irrigation practices. Nitrate levels, with a maximum of 112 mg/L on agricultural lands, indicate possible leaching from agricultural runoff, posing risks to human

Table 9.9 Impact of LULC on physical parameters of water quality

Land use	Parameters	Range	Mean	SD
Agriculture	EC	140–26,260	1357.23	2998.88
	pH	4.8–9.5	7.38	0.76
	TDS	90–16,806	875.99	1921.03
	ORP	−21.6–231	34.62	82.62
Bushland	EC	185–9844	1250.13	2370.36
	pH	5.36–8.12	7.13	0.72
	TDS	118–6300	802.50	1517.34
	ORP	−14.3–227.8	39.79	89.78
Built-up	EC	557–3607	1737.8	1378.34
	pH	6.08–8.91	7.15	1.17
	TDS	356–2308	1112	881.90
	ORP	−12.8–0	−6.50	6.40
Forest	EC	645–29,240	5583.57	10,589.36
	pH	6.46–8.5	7.47	0.64
	TDS	402–19,230	3641.57	6972.64
	ORP	−4.5–4	−1.35	3.10
Barren land	EC	115–3526	550	924.30
	pH	5.67–8.34	6.82	0.80
	TDS	74–2245	352.77	593.17
	ORP	−4.1–230.7	28.58	78.55

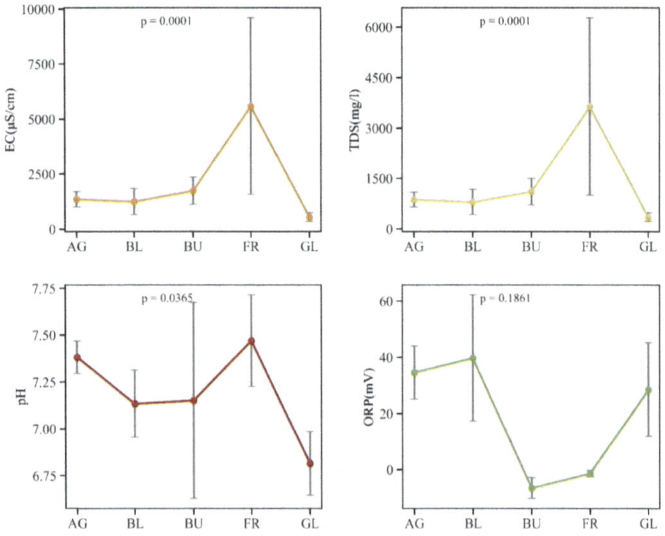

Fig. Physical parameters among different LULC classes

health. Bushlands maintain moderate water quality, with mean NO_3^- concentrations at 7.88 mg/L and sodium at 145.81 mg/l, both within WHO-acceptable limits. However, the presence of high potassium levels (47.5 mg/L) requires monitoring. Urbanized zones show increased levels of K^+ (55 mg/L) and Ca^{2+} (70.6 mg/L), likely due to domestic wastewater discharge and infrastructure-related leaching. Forest areas have the highest mean concentrations of Cl^- (1,795.14 mg/L) and Na^+ (1,186.43 mg/L), significantly exceeding WHO permissible thresholds, likely due to the presence of large-scale commercial forestry in the study area. It also has the highest concentration of SO_4^{2-}. Barrend lands maintain the most favorable water quality, with all measured parameters within WHO-recommended limits indicating minimal anthropogenic impact (Table 9.10).

Statistically significant variations were observed for chloride (Cl^-: H = 18.43, p = 0.0010), nitrate (NO_3^-: H = 14.75, $p = 0.0053$), potassium (K^+: H = 21.71, $p =$ 0.0002), magnesium (Mg^{2+}: H = 9.93, $p = 0.0417$), and sodium (Na^+: H = 20.35, $p =$ 0.0004), indicating that their concentrations differ significantly among the land cover classes (Fig. 9.7). However, no statistically significant differences were found for bicarbonate (HCO_3^-: H = 8.55, $p = 0.0734$), sulfate (SO_4^{2-}: H = 8.37, $p = 0.0789$), and calcium (Ca^{2+}: H = 3.93, $p = 0.4161$), suggesting more uniform distributions of these ions across the LULC classes. These results highlight the influence of land cover on the spatial variability of selected water quality parameters.

9.3.1.6 Trace Metals in Groundwater Quality

All the trace metals were within the WHO permissible limits (Table 9.11). However, Cd concentrations were highest in agricultural areas (0.76 μg/L) and lowest in built-up zones (0.34 μg/L). Co showed a similar pattern, showing the peak in agriculture (1.72 μg/L) and lower in built-up areas (0.57 μg/L). Cu showed a higher concentration in agriculture (26.99 μg/L) and bushland (18.05 μg/L), while built-up and forest had the lowest amount. Fe showed the highest concentration in bushland (100.22 μg/L), followed by built-up areas (57.75 μg/L), with forest showing the lowest concentration (5.74 mg/L). Li also peaked in the bushland (17.99 μg/L) and agriculture (9.97 μg/L), indicating possible natural sources. Ni and Pb were moderately distributed, with agriculture showing the highest levels (5.17 and 5.33 μg/L, respectively), while built-up areas had the lowest for both (1.58 and 1.66 μg/L). A contrast was observed in Zn, where bushland exhibited a high mean (865.11 μg/L), far exceeding all other land uses. This suggests either a significant natural source or a localized anomaly. In comparison, Zn levels in built-up (8.11 μg/L) and forest (1.76 μg/L) were much lower. Overall, agriculture and bushland tended to show high levels of trace metals, possibly due to both chemical inputs like fertilizers and pesticides and geogenic factors (Mthembu et al. 2020).

The Kruskal–Wallis test results reveal statistically significant differences in the concentrations of several trace metals across the LULC classes. Significant variations were observed for iron (Fe: H = 26.77, $p < 0.0001$), lithium (Li: H = 29.57, $p < 0.0001$), nickel (Ni: H = 25.91, $p < 0.0001$), lead (Pb: H = 24.85, $p = 0.0001$),

Table 9.10 Impact of LULC types on major ions of groundwater quality parameters

Land Use	Parameters	Range	Mean	Std
Agriculture	Cl^-	40–7657	326.7	878.64
	NO_3^-	0–112	9.42	13.48
	K^+	0–632	13.25	71.52
	Mg^{2+}	2–184	18.52	22.27
	Na^+	15–5121	205.38	585.97
	Ca^{2+}	5–217	32.62	29.58
	HCO_3^-	18–532	121.03	87.94
	SO_4^{2-}	2–374	26.69	47.48
Bushland	Cl^-	37–1531	219	368.23
	NO_3^-	0–14	7.88	4.94
	K^+	2–634	47.5	156.97
	Mg^{2+}	2–101	18.88	24.58
	Na^+	19–1144	145.81	281.76
	Ca^{2+}	4–115	29.29	28.02
	HCO_3^-	12–1903	196.69	459.73
	SO_4^{2-}	3–155	24.81	39.94
Built-up	Cl^-	50–524	247.6	225.63
	NO_3^-	2–19	10.8	6.83
	K^+	6–124	55	61.69
	Mg^{2+}	12–33	26.2	8.2
	Na^+	22–278	129.8	127.47
	Ca^{2+}	15–175	70.6	66.25
	HCO_3^-	78–459	242.4	163.58
	SO_4^{2-}	6 – 45	24.4	15.87
Forest	Cl^-	91–9926	1795.14	3645.33
	NO_3^-	0–19	11.83	6.84
	K^+	4–15	7.57	3.99
	Mg^{2+}	10–112	40	44.05
	Na^+	52–6971	1186.43	2570.16
	Ca^{2+}	12–154	68.43	66.34
	HCO_3^-	85–241	128	55.38
	SO_4^{2-}	7–1330	232.29	490.83
Barren land	Cl^-	27–931	141.86	237.6
	NO_3^-	0–62	5.6	13.25
	K^+	1–10	3	2.14
	Mg^{2+}	2–56	12.32	13.7
	Na^+	14–604	80.18	149.1

(continued)

Table 9.10 (continued)

Land Use	Parameters	Range	Mean	Std
	Ca^{2+}	4–106	27.64	28.17
	HCO_3^{-}	14–312	81.23	91.38
	SO_4^{2-}	3–176	25.5	45.15

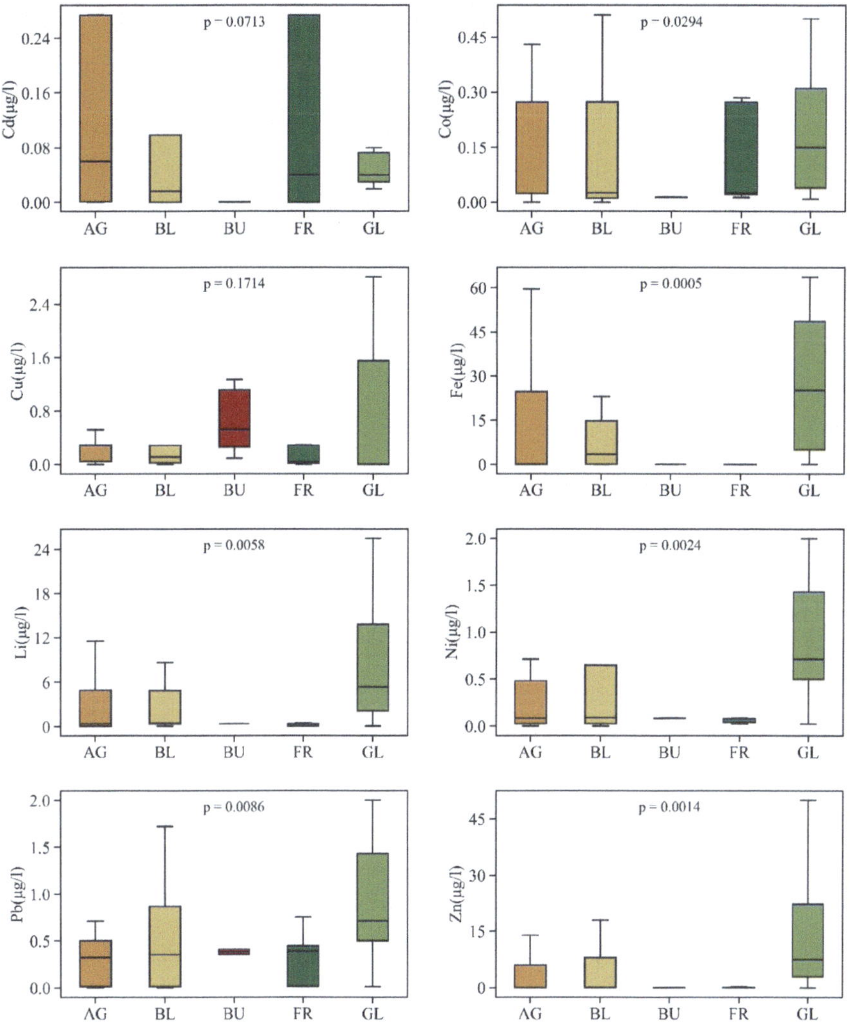

Fig. 9.7 Major distribution among different LULC classes

Table 9.11 LULC and trace metal concentrations in the study area

LULC	Parameters	Range	Mean	Std
Agriculture	Cd	0–4.59	0.52	1.2
	Co	0–8.16	0.95	2.22
	Cu	0–279.53	12.33	47.23
	Fe	0.021–1848.64	58.41	228.58
	Li	0.045–116.08	8.62	20.63
	Ni	0–25.96	2.97	7.22
	Pb	0–25.96	3.05	7.19
	Zn	0–731	28.55	102.43
Bushland	Cd	0–3.78	0.3	0.93
	Co	0–6.08	0.5	1.5
	Cu	0–30.6	2.76	7.81
	Fe	0.025–981.11	80.7	246.62
	Li	0.045–33.65	6.4	11.33
	Ni	0–22.14	2.27	5.8
	Pb	0–22.14	2.38	5.76
	Zn	0–85	12.77	26.59
Built-up	Cd	0–0.011	0	0
	Co	0.011–0.025	0.02	0.01
	Cu	0.092–1.265	0.65	0.52
	Fe	0.024–0.096	0.07	0.03
	Li	0.37–0.391	0.38	0.01
	Ni	0.077–0.1	0.09	0.01
	Pb	0.355–0.41	0.38	0.03
	Zn	0–0.185	0.04	0.08
Forest	Cd	0–0.27321	0.12	0.14
	Co	0.014–0.28488	0.13	0.14
	Cu	0–0.28727	0.13	0.14
	Fe	0.022–9.25	1.37	3.47
	Li	0.046–0.422	0.23	0.18
	Ni	0.02583–0.75	0.16	0.26
	Pb	0.011–0.75	0.29	0.29
	Zn	0–48	6.91	18.12
Barren land	Cd	0.02–4.48	0.62	1.46
	Co	0.01–8.36	1.18	2.66
	Cu	0–199.64	9.96	42.4
	Fe	0.075–592.13	74.02	147.7
	Li	0.047–32.18	9.44	9.32

(continued)

Table 9.11 (continued)

LULC	Parameters	Range	Mean	Std
	Ni	0.02483–24.05	4.01	7.86
	Pb	0.013–24.05	4.01	7.86
	Zn	0.007–19,965	921.59	4253.45

and zinc (Zn: $H = 28.97$, $p < 0.0001$), indicating that these elements exhibit spatial variability linked to land cover types. In contrast, no statistically significant differences were found for cadmium (Cd: $H = 4.55$, $p = 0.3367$), cobalt (Co: $H = 3.64$, $p = 0.4565$), and copper (Cu: $H = 8.03$, $p = 0.0906$), suggesting more uniform distributions across different LULC classes (Fig. 9.8). These findings underscore the differential influence of land cover on trace metal concentrations in the study area.

9.3.1.7 LULC Change Transitions on Water Quality Parameters

Results suggested that different LULC change patterns had different effects on water quality (Table 9.12). EC ranged from a low of 264.83 μS/cm in the areas that have transitioned from barren land to Agriculture to a high of 5536 μS/cm in areas that transitioned from forest to agriculture. TDS exhibited a similar pattern, with the lowest value of 169.67 mg/L and the highest of 3543 mg/L. The pH varied between a low of 5.9 in areas that transitioned from forest to agriculture and a high of 7.9 in forest areas that did not transition to other LULC classes. Lastly, the oxidation–reduction potential (ORP) showed the opposite trend, with the lowest value of -3.04 mV recorded on agricultural land transitioned from forest and the highest of 73.13 mV on agricultural land transitioned from barren land.

Cl concentration ranged from a minimum of 59 mg/L in areas that changed to agriculture from barren lands to a maximum of 865 mg/L in agricultural areas that transitioned from forest. HCO_3^- concentrations also peaked at the agricultural lands changed from the forest (1009.5 mg/L) and were lowest at the agricultural lands transitioned from barren lands (47.5 mg/L). In contrast, SO_4^{2-} levels were highest at agricultural lands that were on the same LULC class since 200 (37.17 mg/L) and lowest at forest areas that were forests since 2001(8.75 mg/L). NO_3^- concentrations varied from 4.17 mg/L on agricultural lands and changed from barren land to 15 mg/L on forest areas that didn't transition to other uses. For major cations, calcium (Ca^{2+}), potassium (K^+), magnesium (Mg^{2+}), and sodium (Na^+) all exhibited their highest concentrations on agricultural land changed from forest (80 mg/L, 327 mg/L, 63.5 mg/L, and 597 mg/L, respectively) and their lowest concentrations at the areas that transitioned to agriculture from barren lands (12.83 mg/L, 2.82 mg/L at GRGR, 7.33 mg/L, and 32 mg/L, respectively).

The distribution of trace metals revealed varying patterns across the sampling sites. Cadmium (Cd) concentrations ranged from undetectable levels on agricultural lands and transitioned from forest to a maximum of 1.08 μg/L in forest areas. Cobalt

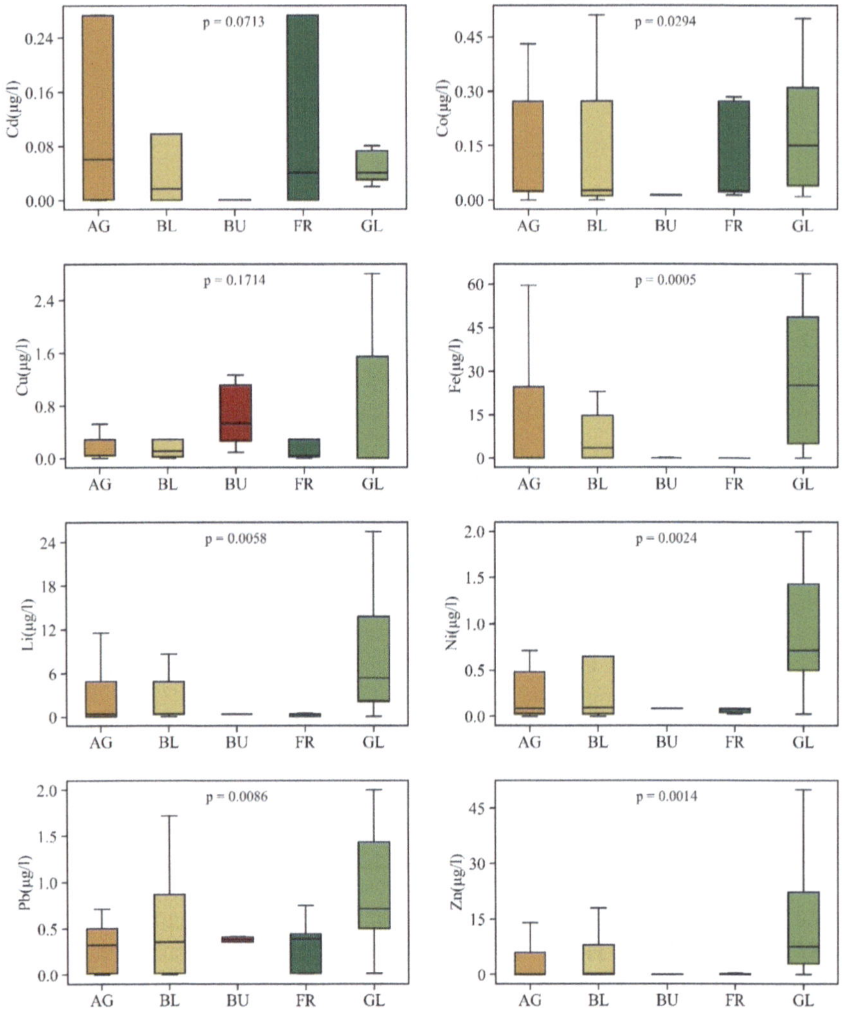

Fig. 9.8 Variations of trace metal elements among LULC classes

(Co) was found to have the lowest concentration in agricultural lands transitioned from forest (0.02 µg/L) and peaked in forest areas that were in the same LULC class in 2001(1.82 µg/L). Copper (Cu) exhibited the highest concentration in bushlands (36.91 µg/L) and the lowest in agricultural lands transitioned from forest (0.07 µg/L). Iron (Fe) concentrations were highest at agricultural lands transitioned from barren lands (333.47 µg/L) and lowest at agricultural lands changed from forest (0.71 µg/L). Lithium (Li) concentrations varied from 0.39 µg/L in agricultural lands that transitioned from forest to 16.78 µg/L in areas that transitioned to barren lands from agriculture. Nickel (Ni) and lead (Pb) both showed maximum concentrations at bushlands (6.78 µg/L and 6.91 µg/L, respectively) and minimum concentrations at agricultural

Table 9.12 LULC change and water quality

Parameters	AGAG	AGBU	AGBR	BLAG	BLBL	FRAG	FRFR	BRAG	BRBR
EC	1837.86	968.57	702.29	1313.6	879.71	5536	670	264.83	338.91
TDS	1182.33	619.86	458.93	853.05	563	3543	428.75	169.67	217.09
pH	7.38	7.52	6.94	7.24	7.16	5.9	7.9	6.89	6.83
ORP	20.12	45.16	30.36	32.54	33.97	−3.04	57.47	73.13	37.4
Cl	465.88	136.57	175.14	291.65	194.29	865	97.25	59	79.55
HCO_3	130.12	156	84	135.7	109.29	1009.5	118.5	47.5	93.73
SO_4	37.17	16	25.57	28.4	35.89	30.5	8.75	8.83	13.09
NO_3	9.18	6.29	5.04	14.15	6.49	8	15	4.17	7.55
Ca	38.05	45.14	27.21	39.61	35.86	80	20.19	12.83	29.07
K	22.95	23.43	3.79	7.8	4.43	327	7.5	3.83	2.82
Mg	22.83	17.71	12.79	16.8	24.14	63.5	13.25	7.33	13.27
Na	293.67	67	101.79	176.7	105.57	597	66.25	32	41.91
Cd	0.32	0.62	0.73	0.31	0.6	0	1.08	0.78	0.36
Co	0.54	1.11	1.29	0.56	1.46	0.02	1.82	1.02	0.7
Cu	1.55	11.96	22.16	0.43	36.91	0.07	14.68	5.19	19.17
Fe	23.12	11.33	69.14	56.35	36.01	0.71	9.99	333.47	81.73
Li	1.28	10.79	16.78	12.26	6.03	0.39	1.78	9.2	12.28
Ni	1.1	3.84	4.48	1.69	6.78	0.1	5.04	5.29	3.24
Pb	1.18	4.07	4.47	1.76	6.91	0.46	5.33	5.34	3.24
Zn	4.10	6.15	1506.72	13.71	61.29	0.12	3.07	21.5	18.45

* AGAG = agriculture unchanged, AGBU = agriculture to built-up, AGBR = agriculture to barren land, BLAG = bushland to agriculture, BLBL = bushland unchanged, FRAG = forest to agriculture, FRFR = forest unchanged, BRAG = barren land to agriculture, and BRBR = barren land unchanged

lands shifted from forest (0.1 μg/L and 0.46 μg/L, respectively). Zinc (Zn) concentrations were highest in barren land that had changed from agriculture (1506.72 μg/L) and lowest in agricultural lands that transitioned from forest (0.12 μg/L).

9.4 Conclusions

The study identified the changes in the LULC patterns in the study area of the KwaZulu-Natal Coastal Region of, South Africa from 2001 to 2023. Landsat images of 2001, 2013, and 2023 and GIS techniques were employed to analyze the spatial and temporal dynamics of LULC changes on groundwater quality parameters. It is essential to identify adverse trends from previous years to enhance landscape planning and sustainable management. Water resources may potentially be negatively impacted by shifting LULC patterns. This study successfully employed remote sensing techniques and GIS analysis to investigate LULC changes in the coastal region.

The analysis revealed significant changes in LULC patterns, particularly a notable and consistent increase in agriculture, built-up and forest areas, driven by socioeconomic development. Conversely, bushland experienced a notable decrease, while forest cover showed a modest increase, indicating dynamic vegetation changes. The LULC change detection, and analysis further detailed these shifts, with a substantial conversion of bushland to agricultural land identified as a key trend. The findings suggest a change in LULC patterns in the region and highlight the importance of continuous monitoring of its dynamics to inform sustainable land management practices.

Amongst the parameters, EC, pH, TDS, Cl^-, Na^+, and Pb had a higher number of samples above the WHO permissible limit. Thus, Prioritize remediation for TDS, EC, and Pb hotspots through filtration or source control. Amongst LULC classes agriculture, built-up areas and forests (the forest is predominantly commercial in the area) have higher concentrations compared to the bush and barren lands indicating there is an anthropogenic impact on groundwater quality in the coastal plains. Thus, continuous monitoring of water quality parameters and the use of agrochemicals and waste management is required by regulating authorities to address emerging contaminants and related health impacts.

References

Abanyie SK, Apea OB, Abagale SA, Amuah EEY, Sunkari ED (2023) Sources and factors influencing groundwater quality and associated health implications: a review. Emerg Contam 9(2):100207. https://doi.org/10.1016/j.emcon.2023.100207

Abbas Z, Yang G, Zhong Y, Zhao Y (2021) Spatiotemporal change analysis and future scenario of LULC using the CA-ANN approach: a case study of the Greater bay area, China. Land 10(6):584. https://doi.org/10.3390/land10060584

Abebe G, Getachew D, Ewunetu A (2022) Analysing land use/land cover changes and its dynamics using remote sensing and GIS in Gubalafito district, Northeastern Ethiopia. SN Appl Sci 4(1):30. https://doi.org/10.1007/s42452-021-04915-8

Abuhay W, Gashaw T, Tsegaye L (2023) Assessing impacts of land use/land cover changes on the hydrology of upper Gilgel Abbay watershed using the SWAT model. J Agric Food Res 12:100535. https://doi.org/10.1016/j.jafr.2023.100535

Ahmad W, Iqbal J, Nasir MJ, Ahmad B, Khan MT, Khan SN, Adnan S (2021) Impact of land use/land cover changes on water quality and human health in district Peshawar Pakistan. Sci Rep 11(1):16526. https://doi.org/10.1038/s41598-021-96075-3

Andualem TG, Peters S, Hewa GA, Boland J, Myers BR (2023) Spatiotemporal trends of urban-induced land use and land cover change and implications on catchment surface imperviousness. Appl Water Sci 13(12):223. https://doi.org/10.1007/s13201-023-02029-7

Annan E, Amponsah W, Adjei KA, Disse M, Hounkpè J, Biney E, Agbenorhevi AE, Agyare WA (2024) Spatio-temporal land use and land cover change assessment: insights from the Ouémé river basin. Sci Afr 25:e02262. https://doi.org/10.1016/j.sciaf.2024.e02262

Appiah D, Schröder D, Forkuo E, Bugri J (2015) Application of geo-information techniques in land use and land cover change analysis in a Peri-urban district of Ghana. ISPRS Int J Geo-Inf 4(3):1265–1289. https://doi.org/10.3390/ijgi4031265

Asenso Barnieh B, Jia L, Menenti M, Zhou J, Zeng Y (2020) Mapping land use land cover transitions at different spatiotemporal scales in west Africa. Sustainability 12(20):8565. https://doi.org/10.3390/su12208565

Assede ESP, Orou H, Biaou SSH, Geldenhuys CJ, Ahononga FC, Chirwa PW (2023) Understanding drivers of land use and land cover change in Africa: a review. Curr Landsc Ecol Rep 8(2):62–72. https://doi.org/10.1007/s40823-023-00087-w

Bill Donatien LM, Clobite BB, Meris Midel ML (2024) Land use land cover change detection using multi-temporal Landsat imagery in the north of Congo Republic: a case study in Sangha region. Geocarto Int 39(1):2425184. https://doi.org/10.1080/10106049.2024.2425184

Cumming GS, Buerkert A, Hoffmann EM, Schlecht E, Von Cramon-Taubadel S, Tscharntke T (2014) Implications of agricultural transitions and urbanization for ecosystem services. Nature 515(7525):50–57. https://doi.org/10.1038/nature13945

DWAF (2000) Strategic environmental assessment for water use Mhlathuze catchment—KZN

Elmahdy S, Mohamed M, Ali T (2020) Land use/land cover changes impact on groundwater level and quality in the northern part of the United Arab Emirates. Remote Sens 12(11):1715. https://doi.org/10.3390/rs12111715

Elumalai V, Brindha K, Sithole B, Lakshmanan E (2017) Spatial interpolation methods and geostatistics for mapping groundwater contamination in a coastal area. Environ Sci Pollut Res 24(12):11601–11617. https://doi.org/10.1007/s11356-017-8681-6

Feng L, Naz I, Quddoos A, Zafar Z, Gan M, Aslam M, Hussain ZK, Soufan W, Almutairi KF (2024) Exploring rangeland dynamics in Punjab, Pakistan: integrating LULC, LST, and remote sensing for ecosystem analysis (2000–2020). Rangel Ecol Manag 98:377–388. https://doi.org/10.1016/j.rama.2024.09.008

Gebeyehu AK, Snelder D, Sonneveld B (2023) Land use-land cover dynamics, and local perceptions of change drivers among Nyangatom agro-pastoralists, southwest Ethiopia. Land Use Policy 131:106745. https://doi.org/10.1016/j.landusepol.2023.106745

He S, Li P, Wu J, Elumalai V, Adimalla N (2020) Groundwater quality under land use/land cover changes: a temporal study from 2005 to 2015 in Xi'an, northwest China. Hum Ecol Risk Assess Int J 26(10):2771–2797. https://doi.org/10.1080/10807039.2019.1684186

Hua AK (2017) Land use land cover changes in detection of water quality: a study based on remote sensing and multivariate statistics. J Environ Public Health 2017:1–12. https://doi.org/10.1155/2017/7515130

Igboama WN, Hammed OS, Fatoba JO, Aroyehun MT, Ehiabhili JC (2022) Review article on impact of groundwater contamination due to dumpsites using geophysical and physiochemical methods. Appl Water Sci 12(6):1–14. https://doi.org/10.1007/s13201-022-01653-z

Jat MK, Garg PK, Khare D (2008) Monitoring and modelling of urban sprawl using remote sensing and GIS techniques. Int J Appl Earth Obs Geoinf 10(1):26–43. https://doi.org/10.1016/j.jag.2007.04.002

Karra K, Kontgis C, Statman-Weil Z, Mazzariello JC, Mathis M, Brumby SP (2021) Global land use/land cover with sentinel 2 and deep learning. In: IEEE International geoscience and remote sensing symposium IGARSS, Brussels, Belgium, pp 4704–4707. https://ieeexplore.ieee.org/document/9553499

Kelbe B, Germishuyse T (2001) Geohydrological studies of the primary coastal aquifer in Zululand: Richards bay region. Water Res Comm

Khan R, Jhariya DC (2018) Assessment of land-use and land-cover change and its impact on groundwater quality using remote sensing and GIS techniques in Raipur City, Chhattisgarh, India. J Geol Soc India 92(1):59–66. https://doi.org/10.1007/s12594-018-0953-3

Kindu M, Schneider T, Teketay D, Knoke T (2015) Drivers of land use/land cover changes in Munessa-Shashemene landscape of the south-central highlands of Ethiopia. Environ Monit Assess 187(7):452. https://doi.org/10.1007/s10661-015-4671-7

Kindu M, Schneider T, Teketay D, Knoke T (2013) Land use/land cover change analysis using object-based classification approach in Munessa-Shashemene landscape of the Ethiopian highlands. Remote Sens 5(5):2411–2435. https://doi.org/10.3390/rs5052411

Li P, Tian R, Xue C, Wu J (2017) Progress, opportunities, and key fields for groundwater quality research under the impacts of human activities in China with a special focus on western China. Environ Sci Pollut Res 24(15):13224–13234. https://doi.org/10.1007/s11356-017-8753-7

Lipczynska-Kochany E (2018) Effect of climate change on humic substances and associated impacts on the quality of surface water and groundwater: a review. Sci Total Environ 640–641:1548–1565. https://doi.org/10.1016/j.scitotenv.2018.05.376

Malede DA, Alamirew T, Andualem TG (2022) Integrated and individual impacts of land use land cover and climate changes on hydrological flows over birr river watershed, Abbay Basin, Ethiopia. Water 15(1):166. https://doi.org/10.3390/w15010166

Malede DA, Alamirew T, Kosgie JR, Andualem TG (2023) Analysis of land use/land cover change trends over Birr river watershed, Abbay basin, Ethiopia. Environ Sustain Indic 17:100222. https://doi.org/10.1016/j.indic.2022.100222

Manonmani R, Suganya G (2010) Remote sensing and GIS application in change detection study in urban zone using multi temporal satellite. Int J Geomat Geosci 1(1):60–65

Mohan M, Rue HA, Bajaj S, Galgamuwa GAP, Adrah E, Aghai MM, Broadbent EN, Khadamkar O, Sasmito SD, Roise J, Doaemo W, Cardil A (2021) Afforestation, reforestation and new challenges from COVID-19: thirty-three recommendations to support civil society organizations (CSOs). J Environ Manage 287:112277. https://doi.org/10.1016/j.jenvman.2021.112277

Mthembu PP, Elumalai V, Brindha K, Li P (2020) Hydrogeochemical processes and trace metal contamination in groundwater: impact on human health in the Maputaland coastal aquifer, South Africa. Expo Health 12(3):403–426. https://doi.org/10.1007/s12403-020-00369-2

Munthali MG, Botai JO, Davis N, Adeola AM (2019) Multi-temporal analysis of land use and land cover change detection for Dedza district of Malawi using geospatial techniques

Mutileni N, Mudau M, Edokpayi JN (2023) Water quality, geochemistry and human health risk of groundwater in the Vyeboom region, Limpopo province, South Africa. Sci Rep 13(1):19071. https://doi.org/10.1038/s41598-023-46386-4

Prăvălie R (2016) Drylands extent and environmental issues: a global approach. Earth-Sci Rev 161:259–278. https://doi.org/10.1016/j.earscirev.2016.08.003

Prăvălie R, Patriche C, Borrelli P, Panagos P, Roşca B, Dumitraşcu M, Nita I-A, Săvulescu I, Birsan M-V, Bandoc G (2021) Arable lands under the pressure of multiple land degradation processes: a global perspective. Environ Res 194:110697. https://doi.org/10.1016/j.envres.2020.110697

Rajmohan N, Senthilkumar M, Alqarawy AM (2025) Hydrogeochemistry and its relationship with land use pattern and monsoon in hard rock aquifer. Appl Water Sci 15(3):57. https://doi.org/10.1007/s13201-025-02398-1

Ramjeawon M, Demlie M, Toucher ML, Van Rensburg SJ (2020) Analysis of three decades of land cover changes in the Maputaland coastal plain, South Africa. KOEDOE—Afr Prot Area Conserv Sci 62(1). https://doi.org/10.4102/koedoe.v62i1.1642

Rao NS, Das R, Sahoo HK, Gugulothu S (2024) Hydrochemical characterization and water quality perspectives for groundwater management for urban development. Groundw Sustain Dev 24:101071. https://doi.org/10.1016/j.gsd.2023.101071

Roy A, Inamdar AB (2019) Multi-temporal land use land cover (LULC) change analysis of a dry semi-arid river basin in western India following a robust multi-sensor satellite image calibration strategy. Heliyon 5(4):e01478. https://doi.org/10.1016/j.heliyon.2019.e01478

Sadeghi A, Galalizadeh S, Zehtabian G, Khosravi H (2021) Assessing the change of groundwater quality compared with land-use change and precipitation rate (Zrebar lake's basin). Appl Water Sci 11(11):170. https://doi.org/10.1007/s13201-021-01508-z

Schulze RE (1982) Agrohydrology and climatology of Natal. Water Res Comm

Shooshtarian MR, Dehghani M, Margherita F, Gea OC, Mortezazadeh S (2018) Land use change and conversion effects on ground water quality trends: an integration of land change modeler in GIS and a new ground water quality index developed by fuzzy multi-criteria group decision-making models. Food Chem Toxicol 114:204–214. https://doi.org/10.1016/j.fct.2018.02.025

Siddik MS, Tulip SS, Rahman A, Islam MdN, Haghighi AT, Mustafa SMT (2022) The impact of land use and land cover change on groundwater recharge in north western Bangladesh. J Environ Manage 315:115130. https://doi.org/10.1016/j.jenvman.2022.115130

Su S, Ma K, Zhou T, Yao Y, Xin H (2025) Advancing methodologies for assessing the impact of land use changes on water quality: a comprehensive review and recommendations. Environ Geochem Health 47(4):101. https://doi.org/10.1007/s10653-025-02413-z

Tang C, Liu Y, Li Z, Guo L, Xu A, Zhao J (2021) Effectiveness of vegetation cover pattern on regulating soil erosion and runoff generation in red soil environment, southern China. Ecol Indic 129:107956. https://doi.org/10.1016/j.ecolind.2021.107956

Wang W, Song H, Min R, Wang Q, Qi M (2024) LUCC-induced dust aerosol change increase surface and reduce atmospheric direct radiative forcing in northern China. J Environ Manage 368:122185. https://doi.org/10.1016/j.jenvman.2024.122185

WHO (2011) Guidelines for drinking-water quality. World Health Organization

Winkler K, Fuchs R, Rounsevell M, Herold M (2021) Global land use changes are four times greater than previously estimated. Nat Commun 12(1):2501. https://doi.org/10.1038/s41467-021-22702-2

Xu F, Li P, Chen W, He S, Li F, Mu D, Elumalai V (2022) Impacts of land use/land cover patterns on groundwater quality in the Guanzhong basin of northwest China. Geocarto Int 37(27):16769–16785. https://doi.org/10.1080/10106049.2022.2115153

Yang L, Shi L, Li J, Kong H (2024) Spatio-temporal pattern change of LULC and its response to climate in the loess plateau, China. Sci Rep 14(1):23202. https://doi.org/10.1038/s41598-024-73945-0

Zadbagher E, Becek K, Berberoglu S (2018) Modeling land use/land cover change using remote sensing and geographic information systems: case study of the Seyhan basin, Turkey. Environ Monit Assess 190(8):494. https://doi.org/10.1007/s10661-018-6877-y

Zena K, Demissie TA, Feyessa FF (2024) Evaluating long-term impacts of land use/land cover changes on pollution loads at a catchment scale. Water Sci Technol 90(1):75–102. https://doi.org/10.2166/wst.2024.206

Zhou P, Luukkanen O, Tokola T, Nieminen J (2008) Effect of vegetation cover on soil erosion in a mountainous watershed. CATENA 75(3):319–325. https://doi.org/10.1016/j.catena.2008.07.010

Chapter 10
Groundwater Risk and Management Plan Due to Agricultural Activities in China and South Africa

Agricultural activities pose significant risks to groundwater quality in both the Weining Plain, China, and the KwaZulu-Natal coastal aquifers, South Africa. This chapter summarized the main water quality issues in the study areas of China and South Africa, such as hydrochemical alterations, seawater intrusion, and health hazards linked to agricultural practices, including excessive fertilizer use, untreated wastewater irrigation, and land use/cover (LULC) changes. Mitigation strategies such as precision agriculture (GPS-based nutrient management, smart irrigation), managed aquifer recharge (MAR), biochar application, and community-driven governance were proposed. The chapter underscores the urgency of region-specific, integrated approaches to safeguard groundwater resources, advocating for stricter regulations, sustainable land use, and continuous innovation in contamination mitigation and adaptive management.

10.1 Identified Water Quality Issues in China and South Africa

10.1.1 Hydrochemical Changes and Seawater Intrusion

Groundwater hydrogeochemistry plays an important role in understanding the hydrochemical characteristics and controlling mechanisms of the groundwater ecosystems in the regions of Weining Plain, China and the coastal aquifers of KwaZulu-Natal, South Africa. In the Weining Plain, rock weathering and evaporation have a major role in the hydrochemical formation of groundwater. Agricultural irrigation in this region, particularly with polluted effluent, modifies hydrogeochemistry, elevates groundwater levels, and facilitates pollutant movement. The application of fertilizes and inadequate waste management exacerbate high concentrations of NO_3^-, PO_4^{3-},

P. Li and V. Elumalai, *Groundwater Quality Under Agricultural Activities—Cases from China and South Africa*, Professional Practice in Earth Sciences,
https://doi.org/10.1007/978-3-031-98993-3_10

Na^+, and Cl^- in groundwater, heightening the risk of salinization and contamination. Stable isotope studies of $\delta^{18}O$ and δ^2H reveal that groundwater in the study area is predominantly supplied by water from the Yellow River.

In the coastal aquifer of KwaZulu-Natal, South Africa, three connected coastal catchments were included in this study. Commercial forest planting and rural agriculture are both common practices in these regions. The groundwater chemistry in the research area is adversely affected by agricultural practices, primarily due to the overuse of fertilizers. Coastal aquifers experience salinization from excessive extraction, resulting in elevated Cl^- and Na^+ concentrations, especially in low-lying areas. The combination of freshwater and seawater modifies the inherent HCO_3–Ca facies to Na–Cl dominant types in certain regions, where Cl-Na type and Cl-Ca·Mg types are the dominant hydrochemical facies types across the three catchments. The SMI evaluation demonstrated that around 36% of the water is impacted by saline water. The hydrogeochemical processes that controlled the chemistry of groundwater in the research locations included silicate weathering, reverse ion exchange, rock-water interaction, seawater intrusion, and anthropogenic activities.

10.1.2 Water Quality and Health Risk Issues

Groundwater in both the Weining Plain of China and South Africa's KwaZulu-Natal coastal aquifers is increasingly stressed due to agricultural intensification. In China's Weining Plain, intense irrigation resulting higher fertilizer application, and the reuse of untreated or partially treated industrial wastewater particularly from paper mills have raised the groundwater pollution levels. Key pollutants such as nitrate and fluoride were found widespread exceeding safe limits for human consumption. Irrigation returns flow coupled with recharge was found primarily disrupting the natural groundwater flow regime with intense mixing of contaminants across aquifers, while over-extractions dropping the water levels compounding higher water stress in the region.

In South Africa, the coastal aquifers particularly southern regions are impacted by excessive agrochemical use and seawater intrusion. The Tugela aquifer faces the most critical conditions due to its high salinity and poor permeability, resulting in increased soil sodicity and poor irrigation potential. The uMhlathuze aquifer is also at risk, with high magnesium hazard levels affecting long-term soil and crop health. Moreover, heavy metals Cd, Cr, Mn, Pb, and Zn linked to phosphate fertilisers and industrial effluents, are emerging threats in KwaZulu-Natal aquifers. With health risk simulation predicting 100% probability of carcinogenic risk from Cd and Cr, especially affecting children, there is an alarming need for contamination mitigation strategy amid severely growing public health concern.

The land use changes in KwaZulu-Natal coastline region in South Africa and the Weining Plain in China suggest that they are both experiencing an expansion in urbanisation and agriculture, which contributes to the deterioration of groundwater quality. The nitrate pollution is linked to rural communities and agricultural areas in the Weining Plain study. This is attributed to the primary groundwater issue with rural communities and agriculture, while industrial activities show localized high concentrations of fluoride, Mn, TDS, Na^+, Cl^-, and SO_4^{2-}. In the eastern coastal region of South Africa, groundwater quality analysis showed that EC, pH, TDS, Cl, Na^+, and Pb frequently exceeded WHO permissible limits, with higher concentrations observed in areas dominated by agriculture, built-up land, and commercial forests, suggesting anthropogenic influence.

To address these issues a management plan should prioritise including establishing a comprehensive groundwater monitoring network; addressing geogenic contaminants and strengthening industrial discharge regulations; implementing agricultural best management practices and improving sanitation to mitigate pollution; encouraging sustainable land use planning to control urban expansion and protect natural mitigation zones; and involving local communities and fostering inter-agency collaboration. The study emphasises that to guarantee sustainable land management and address new contaminants in the coastal plains, it is essential to continuously monitor LULC dynamics and water quality metrics, as well as to use agrochemicals responsibly and manage waste better.

10.2 Mitigation Strategies Associated with Major Water Quality Issues

10.2.1 Water Quality and Health Risk Mitigation

Addressing these groundwater challenges requires multi-layered and region-specific mitigation strategies that directly respond to the identified contaminants and associated health risks. Fluoride and nitrate mitigation require precision nutrient management, which can be significantly enhanced by adopting GPS-based soil analysis, remote sensing, and controlled-release fertilizers (Reza et al. 2025; Raza et al. 2023; Xing et al. 2024). These practices help reduce nitrate leaching into groundwater. Additionally, regulations should mandate the treatment of industrial wastewater before it is reused in agriculture. Constructed wetlands and vegetative buffer zones can also be established to limit the migration of contaminants into water sources (Lavrnić et al. 2017; Almuktar et al. 2018). Irrigation management and aquifer replenishment are essential to maintaining sustainable water resources. Promoting technologies such as smart irrigation, moisture sensors, and fertigation, alongside subsidy schemes and training programs, can optimize water use efficiency (e.g. Xing et al. 2024; Afzaal et al. 2020; Parra-López et al. 2025). Furthermore, the over-extraction of groundwater should be controlled through regulated pumping and the application of managed aquifer recharge (MAR) techniques, such as percolation tanks and recharge wells (Meles et al. 2024; Sufyan et al. 2024). The application of MARs has been lower in compared other countries, which could be utilized more in high contaminated regions (Gantayat et al. 2025). In high-risk fluoride zones, health protection measures should include the distribution of point-of-use filtration systems, like activated alumina filters, to reduce fluoride exposure (Moloi et al. 2024; Buckley et al. 2018; Prabhakar et al. 2008).

Heavy metal and salinity control are critical to maintaining sustainable agricultural practices and protecting water resources. Immediate priorities include enforcing agrochemical import regulations and setting limits on the heavy metal content in fertilisers. Farmers must be trained and incentivised to adopt biofertilisers, green manures, and organic composts to reduce reliance on harmful chemicals (Wasil et al. 2023; Mthembu et al. 2020; Samantaray et al. 2024; Valizadeh et al. 2023). Land-use zoning must be implemented to restrict agricultural expansion into sensitive recharge zones and wetlands, supported by legislation and local governance structures (Rohr 2019). To immobilise heavy metals, materials such as biochar and organic biosorbents can be applied, while phytostabilisation using native plants can further control metal mobility in soil and water (e.g. Ghandali et al. 2024; Singh et al. 2022; Witek-Krowiak et al. 2011; Farhan and Khadom 2015). Effective interaction and information transfer are important for scientific study to align with real-world agricultural requirements and policy frameworks. Precision irrigation and the regulated application of fertilizers and pesticides are essential for minimizing agricultural runoff and averting groundwater contamination. Nitrate can be removed from groundwater using in-situ denitrification methods such as bioremediation, which uses natural microbial processes to transform nitrate (NO_3^-) into innocuous nitrogen gas (N_2) (Shakoor et al. 2020; Moloantoa et al. 2023; Borah et al. 2025). A few adopted techniques were utilized earlier to mitigate nitrate from groundwater: permeable reactive barrier, electrochhemical remediation (Abdel Rahman et al. 2023), phyto remediation (Ingle and Patel 2022), and chemical reduction approaches (Martín-Rodríguez et al. 2020). Addressing phosphate (PO_4^{3-}) contamination in groundwater originating from agricultural sources necessitates an integration of nutrient management and land-use strategies. Many efficacious strategies are Nutrient Management Planning, Phosphate-Enriched Organic Fertilizer, Conservation Tillage Techniques, and Buffer Strips and Riparian Areas (Guo et al. 2020; Haque 2021; Abdel Rahman et al. 2023).

10.2.2 Strategies for Monitoring of Seawater Intrusion

In coastal areas, seawater intrusion and salt management are vital for safeguarding aquifers. Protection efforts should involve setting abstraction limits based on hydrogeological models and ensuring that safe drawdown thresholds are maintained. Salinity monitoring infrastructure, including satellite imaging and geophysical methods, should be established for early detection of seawater intrusion (e.g. Bou-Rabee et al. 2025; Prusty and Farooq 2020). Additionally, promoting the use of salt-tolerant crops and soil amendments can help sustain agricultural productivity in salinised zones (van den Burg et al. 2024; Roychoudhury et al. 2025). Institutionalizing groundwater user associations that include farmers, hydrologists, health officials, and local leaders can foster participatory management and improve decision-making. Regular monitoring, enforcement of local regulations, and community education campaigns should be integral components of broader governance reforms aimed at improving water resource management and ensuring sustainable use. Effective interaction and information transfer are important for scientific study to align with real-world agricultural requirements and policy frameworks. To successfully mitigate waterlogging problems, the installation of drainage systems and land leveling is advised. In specific regions where freshwater lenses lie beneath saline groundwater, the promotion of multiple well-point systems equipped with automatic salinity control for accessing high-quality water may be advocated.

10.2.3 Impact of Agriculture Expansion and LULC Change on Groundwater Quality

Land use and land cover (LULC) change, specifically the decline of natural vegetation and the expanding urbanization and agricultural area impact the quantity and quality of groundwater supplies. Agricultural activities pose significant risks to groundwater quality, necessitating well-defined management plans. The risks to groundwater from agricultural activities arise from pollution from agrochemicals, improper irrigation practices, and poor farm waste management (Rao et al. 2024). Utilizing spatial interpolation methods and geostatistics can help map contaminated areas and aid in targeted interventions. Comparing results from different methods can lead to more reliable decision-making (Elumalai et al. 2017). Employing hydrogeochemical analyses to understand the origin and evolution of groundwater chemistry and identify sources of pollution (Muduli and Chattopadhyay 2025). Using geochemical modelling to simulate and predict the impact of different management scenarios on groundwater quality (Rajmohan et al. 2025). It's important to note that the most effective remedies will depend on the specific context, including the type and source of contamination, the hydrogeological setting, and the socio-economic conditions of the area. A holistic approach that integrates various strategies and considers the interlinkages between land use, water quality, and human health is recommended.

Agriculture is considered the primary source of nitrate pollution in groundwater globally (Akhtar et al. 2021; Ghoto et al. 2025). The extensive and improper management of nitrogen fertilisers leads to nitrate accumulation and high leachate rates in groundwater. Increased nitrate levels have been observed in areas with intensive agricultural activities (Ahmad et al. 2021). Irrigation return flow, along with agrochemicals, is a major pollution source in agricultural regions (Nethononda et al. 2019; Rajmohan et al. 2025). Increased groundwater recharge from irrigation return flow, coupled with agricultural development, can lead to soil and groundwater salinization, especially in shallow aquifers. The use of irrigation water with enriched salts can reduce soil fertility and hinder crop yield. Additionally, the sharp depletion of groundwater levels due to irrigated agricultural area expansion can lead to seawater intrusion in coastal areas (Elmahdy et al. 2020). Agricultural expansion affects groundwater by modifying water balance components, including recharge and runoff, leading to changes in water quality. Conversion of natural vegetation for agriculture can also impact infiltration rates and the pathways of pollutants to groundwater.

Implementing integrated strategies can help mitigate the detrimental impacts of agricultural activities on groundwater quality and contribute to sustainable water resource management. Several preventive and management measures are suggested in the sources to mitigate the risks posed by agricultural activities to groundwater resources (Elmahdy et al. 2020; Rao et al. 2024). Implementing proper farm waste management practices is necessary to prevent contamination from sources like manure and crop residues. Development of rainwater harvesting recharging structures and restoring these areas to their natural state, such as forest or grassland can help replenish groundwater aquifers with cleaner water and potentially dilute contaminants quality (Liu et al. 2024). Implementing comprehensive monitoring and stricter regulation of pollutants from industrial discharges (Muduli and Chattopadhyay, 2025). Regular monitoring of groundwater quality, especially in agricultural areas, is essential to detect contamination early. Stricter regulation of pollutants from agricultural activities is needed. Educating local people on adverse environmental conditions related to agricultural practices can promote responsible behaviour. Understanding the hydrogeochemical processes is important for developing remediation strategies (Mthembu et al. 2021). Competent authorities should consider the impacts of agricultural activities on groundwater and formulate appropriate water usage policies and regulations.

10.2.4 Cross-Cutting Strategies

Community engagement is essential for fostering sustainable water practices, and this can be achieved through groundwater literacy programs, integrating water education into school curricula, and enhancing agricultural extension services (Merei 2024). By educating communities, especially farmers, about the importance of responsible irrigation, pesticide use, and hygiene, behavioral changes can be promoted, leading to better water management and reduced contamination risks. Tools like Remote Sensing, AI, IoT, Big Data, Robotics, Smart Sensors, and Blockchain can enhance water use efficiency and sustainability. Among these, Remote Sensing and AI are the most versatile and widely used, especially for efficient irrigation. These technologies help reduce water waste, detect pollution hotspots, and optimize resource use in real time (Xing and Wang 2024). For instance, UAV-mounted multispectral cameras can monitor soil moisture for better irrigation, while AI models like random forests or neural networks can predict groundwater recharge and rainfall (Parra-López et al. 2025). On the policy and financing front, governments must take proactive steps by establishing central groundwater databases, incorporating health risk thresholds into national water standards, and investing in long-term surveillance and treatment infrastructure (Cardwell 2023). Sustainable funding models, such as public–private partnerships and conservation incentives, are crucial for ensuring the continued implementation and success of groundwater management strategies. Given the increasing climate variability affecting both regions, climate-responsive planning is vital for groundwater sustainability. This includes the development of climate-resilient groundwater strategies that integrate risk modelling, scenario planning, and adaptive management tools.

References

Abdel Rahman RO, El-Kamash AM, Hung YT (2023) Permeable concrete barriers to control water pollution: a review. Water 15(21):3867

Afzaal H, Farooque AA, Abbas F, Acharya B, Esau T (2020) Precision irrigation strategies for sustainable water budgeting of potato crop in Prince Edward Island. Sustainability 12(6):2419

Ahmad W, Iqbal J, Nasir MJ, Ahmad B, Khan MT, Khan SN, Adnan S (2021) Impact of land use/land cover changes on water quality and human health in district Peshawar Pakistan. Sci Rep 11(1):16526. https://doi.org/10.1038/s41598-021-96075-3

Almuktar SA, Abed SN, Scholz M (2018) Wetlands for wastewater treatment and subsequent recycling of treated effluent: a review. Environ Sci Pollut Res 25:23595–23623

Borah P, Borah G, Malakar A (2025) Mitigating nitrate contamination in groundwater: a comprehensive review of in-situ approaches. Groundw Sustain Dev 28:101406

Bou-Rabee F, Yogeshwar P, Burberg S, Tezkan B, Duane M, Ibraheem IM (2025) Imaging of groundwater salinity and seawater intrusion in Subiya peninsula, northern Kuwait using transient electromagnetics. Water 17(5):652

Buckley HL, Molla NJ, Cherukumilli K, Boden KS, Gadgil AJ (2018) Addressing technical barriers for reliable, safe removal of fluoride from drinking water using minimally processed bauxite ores. Dev Eng 3:175–187

Cardwell M (2023) Results-based agri-environmental scheme design: legal implications. Environ Law Rev 25(4):260–288

Elmahdy S, Mohamed M, Ali T (2020) Land use/land cover changes impact on groundwater level and quality in the northern part of the United Arab Emirates. Remote Sens 12(11):1715. https://doi.org/10.3390/rs12111715

Elumalai V, Brindha K, Sithole B, Lakshmanan E (2017) Spatial interpolation methods and geostatistics for mapping groundwater contamination in a coastal area. Environ Sci Pollut Res 24(12):11601–11617. https://doi.org/10.1007/s11356-017-8681-6

Farhan SN, Khadom AA (2015) Biosorption of heavy metals from aqueous solutions by Saccharomyces Cerevisiae. Int J Ind Chem 6:119–130

Ghandali MV, Safarzadeh S, Ghasemi-Fasaei R, Zeinali S (2024) Heavy metals immobilisation and bioavailability in multi-metal contaminated soil under ryegrass cultivation as affected by ZnO and MnO_2 nanoparticle-modified biochar. Sci Rep 14(1):10684

Ghoto SM, Abbasi H, Memon SA, Brohi KM, Chhachhar R, Ghanghlo AA (2025) Mapping and assessing impacts of land use land cover and climate conditions on groundwater quality using RS & GIS. Appl Water Sci 15(3):53. https://doi.org/10.1007/s13201-024-02351-8

Guo T, Johnson LT, LaBarge GA, Penn CJ, Stumpf RP, Baker DB, Shao G (2020) Less agricultural phosphorus applied in 2019 led to less dissolved phosphorus transported to Lake Erie. Environ Sci Technol 55(1):283–291

Haque SE (2021) How effective are existing phosphorus management strategies in mitigating surface water quality problems in the US? Sustainability 13(12):6565

Ingle J, Patel UD (2022) Electrochemical reduction of nitrate in the presence of silver-coated polyvinyl alcohol beads as a spatially suspended catalyst. J Water Proc Eng 49:103082. https://doi.org/10.1016/j.jwpe.2022.103082

Lavrnić S, Zapater-Pereyra M, Mancini ML (2017) Water scarcity and wastewater reuse standards in Southern Europe: focus on agriculture. Water Air Soil Pollut 228:1–12

Liu W, Wu J, Xu F, Mu D, Zhang P (2024) Modeling the effects of land use/land cover changes on river runoff using SWAT models: a case study of the Danjiang River source area. China. Environ Res 242:117810. https://doi.org/10.1016/j.envres.2023.117810

Martín-Rodríguez AJ, Rhen M, Melican K, Richter-Dahlfors A (2020) Nitrate metabolism modulates biosynthesis of biofilm components in uropathogenic Escherichia coli and acts as a fitness factor during experimental urinary tract infection. Front Microbiol 11:26

Meles MB, Bradford S, Casillas-Trasvina A, Chen L, Osterman G, Hatch T, Kisekka I (2024) Uncovering the gaps in managed aquifer recharge for sustainable groundwater management: a focus on hillslopes and mountains. J Hydrol 639:131615

Moloantoa KM, Khetsha ZP, Kana GE, Maleke MM, Van Heerden E, Castillo JC, Cason ED (2023) Metagenomic assessment of nitrate-contaminated mine wastewaters and optimization of complete denitrification by indigenous enriched bacteria. Front Environ Sci 11:1148872

Moloi SR, Okonkwo JO, Jansen R (2024) Defluoridation of water through the application of carbonised bone as a green adsorbent: a review. S Afr J Sci 120(1–2):1–6

Mthembu PP, Elumalai V, Brindha K, Li P (2020) Hydrogeochemical processes and trace metal contamination in groundwater: impact on human health in the Maputaland coastal aquifer, South Africa. Expo Health 12(3):403–426. https://doi.org/10.1007/s12403-020-00369-2

Mthembu PP, Elumalai V, Senthilkumar M, Wu J (2021) Investigation of geochemical characterization and groundwater quality with special emphasis on health risk assessment in alluvial aquifers, South Africa. Int J Environ Sci Technol 18(12):3711–3730. https://doi.org/10.1007/s13762-021-03129-0

Muduli A, Chattopadhyay PB (2025) Implications of multi-decadal land use changes on groundwater regime in tropical coastal regions. Groundw Sustain Dev 29:101419. https://doi.org/10.1016/j.gsd.2025.101419

Nethononda VG, Elumalai V, Rajmohan N (2019) Irrigation return flow induced mineral weathering and ion exchange reactions in the aquifer, Luvuvhu catchment, South Africa. J Afr Earth Sci 149:517–528. https://doi.org/10.1016/j.jafrearsci.2018.09.001

Parra-López C, Abdallah SB, Garcia-Garcia G, Hassoun A, Trollman H, Jagtap S, Carmona-Torres C (2025) Digital technologies for water use and management in agriculture: recent applications and future outlook. Agric Water Manag 309:109347

Prabhakar AR, Raju OS, Kurthukoti AJ, Vishwas TD (2008) The effect of water purification systems on fluoride content of drinking water. J Indian Soc Pedod Prev Dent 26(1):6–11

Prusty P, Farooq SH (2020) Seawater intrusion in the coastal aquifers of India—a review. HydroResearch 3:61–74

Rajmohan N, Senthilkumar M, Alqarawy AM (2025) Hydrogeochemistry and its relationship with land use pattern and monsoon in hard rock aquifer. Appl Water Sci 15(3):57. https://doi.org/10.1007/s13201-025-02398-1

Rao NS, Das R, Sahoo HK, Gugulothu S (2024) Hydrochemical characterisation and water quality perspectives for groundwater management for urban development. Groundw Sustain Dev 24:101071. https://doi.org/10.1016/j.gsd.2023.101071

Raza I, Zubair M, Zaib M, Khalil MH, Haidar A, Sikandar A, Ashfaq M (2023) Precision nutrient application techniques to improve soil fertility and crop yield: a review with future prospect. Int Res J Educ Technol 5(8):109–123

Reza MN, Lee KH, Karim MR, Haque MA, Bicamumakuba E, Dey PK, Chung SO (2025) Trends of soil and solution nutrient sensing for open field and hydroponic cultivation in facilitated smart agriculture. Sensors 25(2):453

Samantaray A, Chattaraj S, Mitra D, Ganguly A, Kumar R, Gaur A, Thatoi H (2024) Advances in microbial based bio-inoculum for amelioration of soil health and sustainable crop production. Curr Res Microbial Sci 7:100251. https://doi.org/10.1016/j.crmicr.2024.100251

Shakoor A, Muhammad Shahzad S, Farooq TH, Ashraf F (2020) Future of ammonium nitrate after Beirut (Lebanon) explosion. Environ Pollut 267:115615. https://doi.org/10.1016/j.envpol.2020.115615

Singh BM, Singh D, Dhal NK (2022) Enhanced phytoremediation strategy for sustainable management of heavy metals and radionuclides. Case Stud Chem Environ Eng 5:100176

Sufyan M, Martelli G, Teatini P, Cherubini C, Goi D (2024) Managed aquifer recharge for sustainable groundwater management: new developments, challenges, and future prospects. Water 16(22):1–28

van den Burg S, Deolu-Ajayi AO, Nauta R, Cervi WR, van der Werf A, Poelman M, van der Meer IM (2024) Knowledge gaps on how to adapt crop production under changing saline circumstances in the Netherlands. Sci Total Environ 915:170118

Wasil AH, Shah JA, Kakar SM, Ragashtai AR, Yusuf MSA, Sadat A (2023) The influential factors of organic fertilizer adoption among farmers: a review. Int J Acad Res Bus Soc Sci 13(5):3020–2036

Witek-Krowiak A, Szafran RG, Modelski S (2011) Biosorption of heavy metals from aqueous solutions onto peanut shell as a low-cost biosorbent. Desalination 265(1–3):126–134

Xing Y, Wang X (2024) Precise application of water and fertiliser to crops: challenges and opportunities. Front Plant Sci 15:1444560

Xing Y, Zhang X, Wang X (2024) Enhancing soil health and crop yields through water-fertiliser coupling technology. Front Sustain Food Syst 8:1494819

Akhtar N, SyakirIshak MI, Bhawani SA, Umar K (2021) Various natural and anthropogenic factors responsible for water quality degradation: a review. Water 13(19):2660. https://doi.org/10.3390/w13192660

Gantayat RR, Elumalai V, Li P (2025) Future research imperatives in hydrogeology. In: Li P, Elumalai V (eds) Sustainable groundwater and environment: challenges and solutions. Springer Nature Switzerland, Cham. https://doi.org/10.1007/978-3-031-34783-2_21

Merei M (2024) Empowering communities for sustainable water management: insights from just diggit and the Paani foundation. E3S Web of Conf 550:01042. https://doi.org/10.1051/e3sconf/202455001042

Rohr HE (2019) Evidence based water sensitive planning: securing water sustainability through innovative spatial planning tools. Ph.D. dissertation of North-West University (South Africa), Potchefstroom Campus

Roychoudhury A, Mukherjee S, Banerjee R (2025) Navigating salinity challenges: enhancing resilience in underground vegetable crops. In: Lal MK, Tiwari RK, Kumar A, Kumar R, Singh B (eds) Abiotic stress in underground vegetables. Academic Press. https://doi.org/10.1016/B978-0-443-23961-8.00005-X

Valizadeh N, Jalilian S, Hallaj Z, Esfandyari Bayat S, Hayati D, Bazrafkan K, Akbari M (2023) Encouraging adoption of green manure technology to produce clean rice product. Sci Rep 13(1):8690

Index

P. Li and V. Elumalai, *Groundwater Quality Under Agricultural Activities—Cases from China and South Africa*, Professional Practice in Earth Sciences,
https://doi.org/10.1007/978-3-031-98993-3

The manufacturer's authorised representative in the EU is Springer Nature Customer Service Centre GmbH, Europaplatz 3, 69115 Heidelberg, Germany. If you have any concerns regarding our products, please contact ProductSafety@springernature.com

Printed and bound by CPI Group (UK) Ltd, Croydon, CR0 4YY
01/10/2026
02225864-0008